Green Energy and Technology

Climate change, environmental impact and the limited natural resources urge scientific research and novel technical solutions. The monograph series Green Energy and Technology serves as a publishing platform for scientific and technological approaches to "green"—i.e. environmentally friendly and sustainable—technologies. While a focus lies on energy and power supply, it also covers "green" solutions in industrial engineering and engineering design. Green Energy and Technology addresses researchers, advanced students, technical consultants as well as decision makers in industries and politics. Hence, the level of presentation spans from instructional to highly technical.

Indexed in Scopus.

More information about this series at http://www.springer.com/series/8059

Shaharin A. Sulaiman
Editor

Clean Energy Opportunities in Tropical Countries

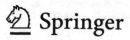 Springer

Editor
Shaharin A. Sulaiman
Department of Mechanical Engineering
Universiti Teknologi PETRONAS
Seri Iskandar, Perak, Malaysia

ISSN 1865-3529   ISSN 1865-3537   (electronic)
Green Energy and Technology
ISBN 978-981-15-9142-6   ISBN 978-981-15-9140-2   (eBook)
https://doi.org/10.1007/978-981-15-9140-2

This Springer imprint is published by the registered company Springer Nature Singapore Pte Ltd.
The registered company address is: 152 Beach Road, #21-01/04 Gateway East, Singapore 189721,
Singapore

Contents

Overview of Geography, Socio-economy, Energy and Environment of the Tropical Countries

Shaharin A. Sulaiman

Abstract Clean energy is an important issue nowadays since the global community has become more aware of the problem of emission and global warming. Despite strong efforts by many developing countries, which are mostly located in temperate zones, it has been quite unclear how countries of the tropical zone are responding. This chapter delves into the characteristics of countries within the tropical zone, looking into the aspects of demography, economy, energy, and environment. The differences in geographical locations, culture, socio-economic backgrounds, politics and many other factors make it complex for simple comparison and discussion. On the technology aspect, there have been tremendous innovations and improvements in the last 20 years. The information and communication systems have grown so fast that everybody in the world can be connected together and share valuable information. Technologies related to power generation has become more efficient and more affordable. Business models have also changed to become more competitive. In other words, things of the 1990s or earlier can no longer be perceived the same today. The simplest example is the affordability for people to fly around the world. It is hoped that the information provided in this chapter would be an eye-opener on the opportunities available in tropical countries.

1 Tropical Zone

The tropical zone occupies about 39.8% of the Earth's surface area [1]. Bearing in mind that the Earth is also occupied by the sea, the tropical zone takes 33.4% of the Earth's land surface [2]. Such a large proportion implies the importance of the tropical zones to various aspects of life on the Earth. The term tropical zone can be referring to areas that are located within the two tropic lines, but at the same time, it can be narrowed down to areas that have tropical climate. Let's analyze the

S. A. Sulaiman (✉)
Department of Mechanical Engineering, Universiti Teknologi Petronas, Block 17, 32610 Seri Iskandar, Perak, Malaysia
e-mail: shaharin@utp.edu.my

© The Author(s), under exclusive license to Springer Nature Singapore Pte Ltd. 2021
S. A. Sulaiman (ed.), *Clean Energy Opportunities in Tropical Countries*,
Green Energy and Technology, https://doi.org/10.1007/978-981-15-9140-2_1

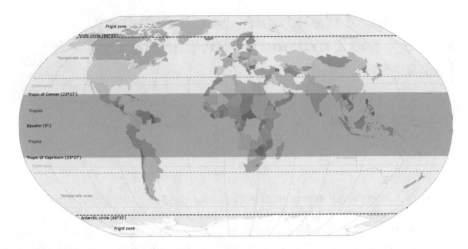

Fig. 1 The tropical zone on world map. Figure courtesy of KVDP [3]

differences here. Tropical zone is the region of the Earth surrounding the Equator. To be specific, the tropical zone lies between latitude 23° 27′ N (Tropic of Cancer) and latitude 23° 27′ S (Tropic of Capricorn), as shown in Fig. 1 [3]. One may wonder what is so special about these precise angles of latitudes. They are actually the axial tilt of the Earth, which is the angle between an object's rotational axis and its orbital axis, as depicted in Fig. 2 [4]. It is interesting to note that the tropic lines constantly change due to a slight wobble in the Earth's longitudinal alignment relative to its orbit around the sun [5]. Presently, the Tropic of Cancer drifts southward at a rate of approximately 14.5 m per year [6].

Aristotle denoted the tropical zone as the torrid zone for being the one closest to the Equator [7]. He also made an assumption that the torrid zone was too hot for human habitation because it received the sun's rays directly above. Although that was an exaggerated assumption, it has indeed been well-know today that the tropical zone receives sunlight that is more direct than the rest of Earth. Nevertheless, it must be noted that the climate within the tropical zone is somehow varied. To the general public, the term "tropical countries" nowadays would trigger an idea of countries that have hot and humid features, all year round, which result in luxuriantly growing vegetation. Figure 3 shows a forest in Malaysia at the edge of the author's workplace, which is typical scenery in many tropical countries. However, it must be noted that within the tropical zone there are also deserts and in mountainous places where there can be snows too.

Many areas within the tropical zone have distinct dry and wet seasons, though the degree may vary in certain areas. Generally, during the wet season, regions that are relatively closer to the Equatorexperience average precipitation (rainfall)of at least 60 mm per month, as defined under the Köppen climate classification [8]. The precipitation in these regions is roughly equally distributed throughout the year [9], and the natural vegetation is known as the rainforest. These regions are denoted in

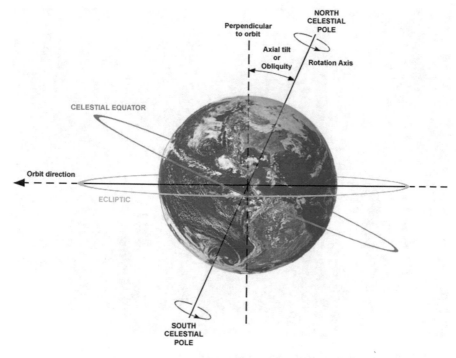

Fig. 2 Depiction of Earth's axial tilt in relation to the rotation axis, orbit path, celestial Equator, and ecliptic. Figure courtesy of Dennis Nilsson [4]

Fig. 4 by the dark blue color shades with Köppen climate classification category "Af," which stands for tropical rainforest. As denoted in the figure, the tropical rainforest covers the Amazon Basin in South America, the Congo Basin in Africa and some parts of Southeast Asia.

The tropical monsoon climate, of classification category "Am" experiences plentiful rainfall similar to that of the tropical rainforest climate, but intense during the high-sun season. The tropical monsoon climate experiences warm temperatures throughout the year. In Fig. 4 these regions are denoted by the medium blue color shades. The tropical monsoon climates cover the areas in South and Central America, and sections of South Asia, Southeast Asia, West and Central Africa, the Caribbean, North America, and North Australia.

Apart from tropical rainforest and tropical monsoon climates, there is also another class of climate within the tropical zone. It is known as tropical savanna, which experiences a very dry season. Under a more severe condition, drought would prevail. The typical annual precipitation for tropical savanna is 0.25 cm. With such a climate, the featured vegetation appears as tree-studded grasslands. The regions that experience this climate are represented by the light blue shades in Fig. 4, which include South America, West Africa, Southern Africa, India, Indochina, and the north coast of Australia.

Fig. 3 Rainforest at the edge of the author's workplace

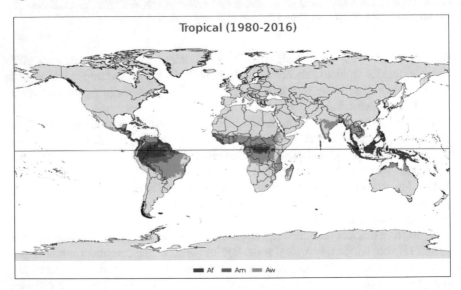

Fig. 4 Areas of the world with tropical climates: Af—rainforest, Am—monsoon, Aw—savanna. Figure courtesy of Beck et al. [10]

A careful comparison between Figs. 1 and 4 would reveal that there are regions within the tropical zone that do not have a tropical climate. There are many regions that, under the Köppen climate classification [8], are classed as dry climates. They include part of the Deserts of Australia, the Arabian Desert, the Sahara Desert, and the Atacama Desert. In contrast, there are also regions with polar and alpine climates, such as Mount Kilimanjaro, the Andes and Mauna Kea in Hawaii.

Last but not least, this book intends to also cover sub-tropic zones or the subtropics. Geographically, in the Northern hemisphere, the subtropics is located approximately between latitudes 23° 27′ N and 35° N; and vice versa for the Southern hemisphere. This is depicted in Fig. 1. The climates of the subtropics are usually featured by hot summers and mild winters. The subtropics climates have a few varieties, which depend on rainfall and temperature. The sub-tropic zone is relatively smaller as compared to the tropical zone. Furthermore, there are quite many countries that have both tropical and subtropics climates. In view of this, it has been decided that this book covers subtropics as well.

2 Demography and Socio-economy

The estimated population of tropical countries in 2020 is about 44.6% of the world's population [11]. The figure is anticipated to reach 50% by 2050 [12]. India who has the largest population within the tropical zone is also the second most populated country in the world; it is projected to be number one by 2030. Generally, the population in the tropical zone is not a straightforward number since some countries that are not quite within the zone are excluded, for example, China. If these partial portions of excluded countries were to be included, then the number of population would be higher. Nevertheless, the exact number of the population is not of prime importance in this book; it is sufficient to recognize that the tropical zone is home to the largest population in the world.

There are at least 91 countries [11] that can be regarded as having a tropical climate. Of course, geographically, there are many more countries that are situated within the tropical zone. In the Continent of Africa, all the countries in it are covered by the scope defined in this book. The countries that are within the subtropics, among others include Morocco, Tunisia, Libya, Algeria, and Egypt and South Africa. None of the European countries are located within the tropical or sub-tropic zones. Asia, which is the largest continent can be divided into a few sub-regions. Not all of the regions are located within the tropical zone. In the context of this book, Asia include West Asia, South Asia, South East Asia and China. West Asia, which can be loosely referred to as the Middle East, comprises countries like Saudi Arabia, Oman, and United Arab Emirates, Kuwait and Bahrain. Of all the West Asia countries, Yemen is the only one that is fully located in the tropical zone, while others are subtropics. Some other West Asia countries are entirely located in temperate zones, such as Azerbaijan, Turkey and Turkmenistan. In South Asia, Sri Lanka and the Maldives are entirely located in the tropics, although the latter is basically a chain of atolls. Most part of India too is located in the tropical zone, with the northern area in the subtropics passed through by the Himalayan range. Apart from India, the heavily populated Bangladesh and Pakistan are also important countries in South Asia. Towards the east is South East Asia region, comprising 10 countries that are unified under a regional intergovernmental organization that is known as the Association of Southeast Asian Nations (ASEAN). The member countries are Malaysia, Indonesia, Thailand, Singapore, the Philippines, Vietnam, Brunei, Cambodia, Myanmar, and Laos. Further east is the Oceania nations, comprising Australia, Fiji, Samoa and many other islands.

In the American continents (north, central and south), countries that are located in the tropical and sub-tropic zones include Mexico, entire Central America, the whole of Caribbean islands, and the northern half of South America(Chile and Argentina excluded).

It is interesting to note that nearly all the countries in the tropics were previously colonized by the Europeans, with the exception of Thailand and Liberia. Most of the colonized countries in the tropical zone gained independence only after World War II, and hence they are relatively young governments. In a way, the economy and wellbeing of these governments, at the time they gained independence, depended on who colonized them. For example, countries that were previously ruled by the British were left with quite good governance and infrastructure, and therefore they could be regarded as having a good start. Other countries, especially those who gained independence through war, would have to put extra effort in order to catch up with others. Due to the reason for colonization, most of these countries are regarded as either developing countries or least-developing countries. They are usually, characterized by the following undesirable features, though not necessarily all:

- low access to potable water and hygienic sewerage systems
- insufficient energy poverty
- lack of pollution prevention and control
- high number of infectious tropical diseases
- inferior infrastructure
- high-level and widespread poverty
- low education levels
- corruption.

These features have actually been the challenges faced by the countries, for which they would have to struggle in order to compete with the well-doers elsewhere.

To make up for the gaps, many developing countries welcome foreign investors to aid the growth of their economies. Foreign direct investments can be helpful in many ways, primarily in the aspect of employment. Apart from job creation, foreign investment could help to enhance the skill of the locals, and also enable the transfer of knowledge and technology. From a different perspective, the presence of sophisticated foreign companies could boost competitiveness of domestic firms. Tax exemption is one of the popular incentives offered by governments of tropical countries in securing valuable foreign direct investments. Other winning factors may include levels of salaries, education, infrastructure and the Gross Domestic Product (GDP) per capita of the countries.

Despite the pulling factors for foreign investment, there are also risks such as corruption, which is one of the biggest problems in tropical countries. Corruption is the misuse of public power for private benefit. The public power could be in the form of money or authority in illegal, dishonest, or unfair manners. Shown in Fig. 5 is the world map that indicates the Corruption Perception Index (CPI) for 2018. The figure clearly depicts that the tropical zone is filled with nations that have low CPI (high rate of corruption), and this could be one of the reasons for their slow rates of growth. Countries that have high corruption would experience situations whereby the natural laws of the economy are prevented from functioning freely. Consequently, the entire society in these countries would suffer. Emerging market economies tend to have much higher corruption levels compared to developed countries; most likely due to the greater opportunities for corruption. The other damaging effect of corruption is that investors would become hesitant to bring in their money into the countries.Even though it is known that many investors come to certain corrupt countries, they know very well that they are taking risks and would try their best to play safe.

Since their independence, the countries in the tropical zone prosper differently, depending on various aspects. Apart from the listed features and challenges, there can be other factors that affect the growth of tropical countries. Regions that are sea navigable are generally richer than nations that are land-locked. Laos, in South East Asia is one example, when compared to its ASEAN counterparts. In Africa, the situation is similar for Burkina Faso, Burundi, Central African Republic, Chad, Lesotho, Mali, Niger, Rwanda, Uganda, Zambiaand Zimbabwe, who are all not having access to the sea and at the same time are among the poorest in the world. In some other countries, communism, which fell in 1989–1992, had affected them economically and socially. With the free economy that these countries are now practicing, they are developing better and faster, as experienced by Vietnam and China, for example.

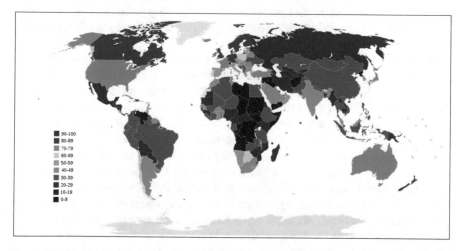

Fig. 5 World map for Corruption Perceptions Index in 2018. Figure courtesy of Young and Transparency International [13]

The Information Technology (IT) has somewhat been helpful in reducing corruption and other corporate crimes due to enhanced transparency. What used to be easily hidden from society in the past, can no longer be obscured nowadays. Amazingly, IT is not only helpful in reducing corruption. It also creates a wide range of opportunities in doing businesses. New and competitive businesses come in. Many of them were previously beyond the imagination of ordinary people. The use of e-haling, for example, has benefitted a lot of people in many countries. In South East Asia, for example, the Grab company has been widely accepted in the region, giving benefits to people at various levels. In the airline industry, the emergence of low-cost airline companies such as AirAsia has changed the way people travel in many Asian countries, making travels to other countries easy and highly affordable. All in all, the Internet has been instrumental for many sectors of the economy, resulting in faster growth in countries, and also heighten the living standards.

It has always been perceived by the Westerners that tropical areas are underdeveloped. It was suggested that warmer climates were related to primitive indigenous populations who were regarded as not having control over nature, as compared to populations in temperate zones [14]. The technologies for production, especially in agriculture and health, in the tropics are far inferior when compared with those of the temperate zones. This, along with a few other factors, contributes to a significant income gaps between the two zones. In 1995 it was estimated that the productivity of grain production was about 50% higher in temperate-zone countries [15], for instance.

More recently, in 2017 the tropics have been reported to outperform the rest of the world in economic growth for the last three decades, representing 18.7% of global economic activity [16]. The development of economy in the tropical zone would actually require support from the international community, particularly the developed nations. Ironically, the investment in research and development (R&D) by the tropics is far smaller than the global rate. According to the State of the Tropics [16], the rest of the world (other than the tropics) invested about four times in R&D than the tropics.

Only three of the tropical countries have been recognized as developed countries. Since 1997, the International Monetary Fund (IMF) regarded Hong Kong and Singapore as Advanced Economies or Developed Countries, and far earlier there was Australia on the list. In 1970 Hong Kong and Singapore were classified as among the Four Asian Tigers, who were also regarded the Newly Industrialized Countries (NIC). The NIC are those whose economies have not yet reached a developed country's status but have macroeconomically outperformed other developing countries. Nowadays, the NIC of the tropical zone is China, India, Indonesia, Malaysia, Mexico, The Philippines, South Africa and Thailand [17]. Furthermore, other tropical countries are also picking up in terms of economy due to the recent advance in technologies and information, improvement in health, improved regional networking and cooperation among countries, and other important contributing factors. Overall, many tropical countries are facing rapid growth in the economy at present time, for which 80% of the top 20 countries that are having the highest real GDP growth rate are from tropical zones [18]. Interestingly the bottom 20 countries are also dominated by tropical countries; many with negative rates of growth. All in all, it can be said that a high rate of growth in the economy is happening in most tropical countries.

3 Energy Demand

The recent changes in the socio-economy of many tropical countries have affected their economic growths, though at different rates. As a result, the need for energy is also increasing. As mentioned in the previous section, the growth in population is high too in the tropical zone, for which half of the world population is expected to reside in this zone by 2050. These fast rates of growth in population and economy would result in an increase in industrial and business activities. Consequently, many tropical countries would require a significant increase in energy, particularly electricity, in order to sustain business operations and economic growth. Furthermore, growth is also usually associated with enhanced buying power of the people.

Shown in Fig. 6 is the anticipated change in energy consumption for different economies by 2050, as demonstrated by De Cian and Sue Wing [19] in their work involving complex combination of econometric analysis of the response of energy demand to temperature and humidity exposure with future scenarios of climate change and socio-economic development. There is umpteen valuable information in Fig. 6, but the most relevant one is the fact that the projected increase in energy consumption for tropical countries is the highest. It is also shown that while most of the developed economies are expecting a reduction in their total energy consumption, countries of the tropical zones are expecting the opposite. Interestingly, in a more refined work by De Cian and Sue Wing [20], it was demonstrated that the effect of global warming on future energy consumption cannot be ignored, for which the effect is expected to be the largest in tropical regions.

In the transportation sector, more cars would be demanded since the public transportation systems in most tropical countries are yet to become efficient. The 1970 car sales in Malaysia, for example, were only 27,177 units [21] but then the number rose tremendously to 605,156 in 2010 [22]. This was the opposite in the US [23], for which the trend was generally decreasing for the same period of time (1970–2010). This trend agrees well with the earlier mentioned predictive work by De Cian and Sue Wing [20], in which the energy consumption in the transportation sector is expected to increase a lot faster in tropical countries than that in temperate countries.

At home, air-conditioners become a necessity nowadays in tropical countries, partly due to the increasing affordability resulted from an increase in income and the competitive market of air-conditioners. Everyday life, which used to be basic and simple, now becomes more sophisticated, leading to the need for more energy to satisfy new wants. Apart from homes, other types of buildings would also need air-conditioners. Cooling in buildings is essential for thermal comfort, mainly in public buildings such as offices, shopping complexes, transportation hubs, etc. A person who stays outdoor on a hot and humid day in Malaysia would usually feel more miserable than another person who is located at a place of the same temperature but of drier air.

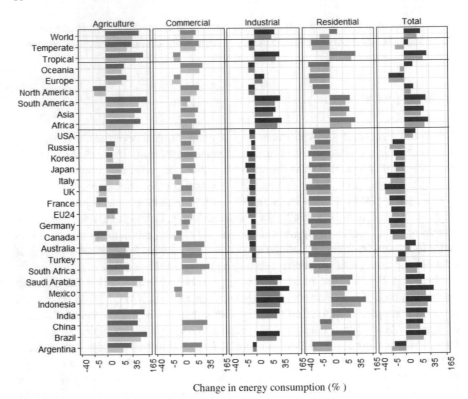

Fig. 6 Sectoral and aggregate energy demand responses of G20 nations and world regions to different warming scenario circa 2050, for different representative concentration pathways(RCP 4.5—light colors, RCP 8.5—dark colors). Figure courtesy of Enrica De Cian and Ian Sue Wing [19]

The high humidity makes the body's perspiration rate slower. When a person cannot sweat, his body cannot regulate its temperature. It is almost unimaginable today how people in the tropics could work without air-conditioners. The first Prime Minister of Singapore, Mr. Lee Kuan Yew, once quoted:

> Air conditioning was a most important invention for us, perhaps one of the signal inventions of history. It changed the nature of civilization by making development possible in the tropics. Without air conditioning you can work only in the cool early-morning hours or at dusk. The first thing I did upon becoming prime minister was to install air conditioners in buildings where the civil service worked. This was key to public efficiency.

Buildings contribute to about 40% of the total energy consumption, and the air-conditioning systems HVAC systems have the largest contributions [24]. With the fast growth in economies in the tropics, the number of newly constructed buildings have also been increasing, leading to the need for more air-conditioners. On the other hand, it has also been proven that sudden extreme hot weather, which occasionally occurred due to the El Niño effect, which would result in an increase in the installation of home

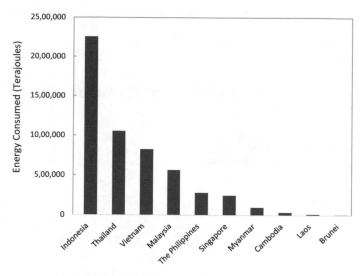

Fig. 7 Variation of energy consumption by industries (manufacturing, construction and mining) for ASEAN countries in 2015

air-conditioners tremendously. In 2016, Malaysia saw an abrupt 50% increase in sale of air-conditioners and another 30% increase in the sale of ventilation equipment [25]. This happened at the time of sudden unexpected extreme hot and dry weather with some new records of high temperature.

The vibrant local and foreign investments in many tropical countries have resulted in growth in industries, which sees the progression of facilities such as factories, fabrication yards, and other related infrastructures, in which most of them operate 24 h per day. These facilities require a huge amount of energy in order to be in operation. The energy required usually comes in the form of electricity, and there are other industries that require thermal energy too. The requirements and types of industries may vary by country. Shown in Fig. 7 is the variation of energy consumption by industries (manufacturing, construction and mining) for all the ten ASEAN countries in 2015. The data for Figs. 7 and 8 is extracted from the United Nations' 2015 Energy Balance [26]. The energy consumption here covers all forms of energy, which include electricity, oil and gas, and also renewable energy.Obviously, Indonesia had the highest energy consumption for the industry at 2,255,538 terajoules. This amount tallies with the size of the country which is large in terms of both land area and number of population. Of course, this information can also be viewed in the form of energy consumption per capita. It may be interesting to note that in 2015 Singapore had the highest industrial energy consumption at 44,677 MJ per capita, leaving far Malaysia at18,147 MJ per capita and third-place Thailand at 15,694 MJ per capita. The last in the rank went to Laos at 2,247 MJ per capita. However, these are just one of the few indicators of how the economy is performing, and the wealth is distributed. The more important aspect to note in this chapter is pertaining to where the industrial energy consumption goes to in the region, due to various reasons as such the source of CO_2

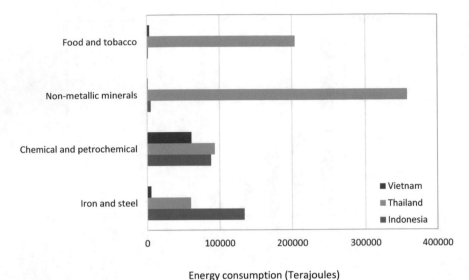

Fig. 8 Distribution of top industrial activities in relation to energy consumption for Indonesia, Thailand and Vietnam in 2015

emission and other related issues. In short, Fig. 7 implies that Indonesia, Thailand and Vietnam were the top energy consumers in industrial activities in 2015.

Shown in Fig. 8 is a histogram that represents four main industries that consume the largest amount of energy in Indonesia, Thailand and Vietnam in 2015. It is obvious that food and tobacco as well as non-metallic minerals are the groups of industries that used the most energy and they are strongly dominated by Thailand. On the other hand, the chemical and petrochemical industry is the one that shows almost fair involvement by the three nations. This is most probably caused by the fact that these countries have oil and gas resources. The other important industry is iron and steel, which sees strong domination by Indonesia. More importantly, Fig. 8 demonstrates that the nature of industries in different countries are unique due to diverse factors, and therefore, no significant generalization can be made among these countries in term of the type of industries.

While industries are known to be a major energy consumer in many countries, it is not necessarily true for all countries, especially those in the tropical zone which have a wide range of economies. Figure 9 shows the pie-charts for the breakdown of energy consumption for selected tropical countries in 2015. The countries, which are randomly selected, are Brazil, Ethiopia, Indonesia, Malaysia, Nigeria, Pakistan, Singapore and the United Arab Emirates (UAE). The consumption categories are industries, commercial, transport, residence, others and non-energy use. Detailed information for these categories can be found in the United Nations' 2015 Energy Balance [26].

It should be noted that the comparisons in Fig. 9 involve diverse economies. It can be argued that the comparison may not be a good one; for example, Singapore and

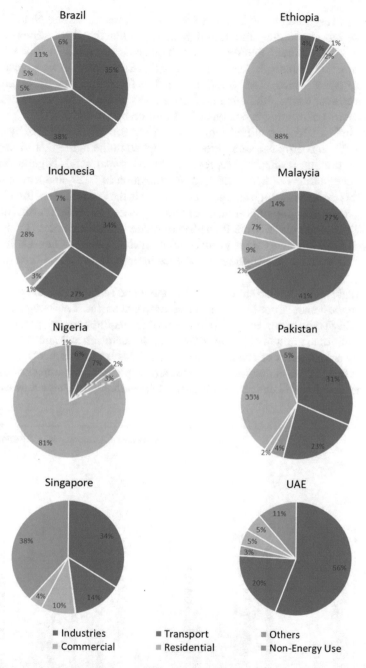

Fig. 9 Energy consumption of selected tropical countries for 2015

the UAE are very small to be compared with Brazil, Indonesia and Nigeria. Never the less, the most important point highlighted here is that the distributions of energy consumption are different, and depend much on socio-economic factors, which are not common. Interestingly, it is shown in Fig. 9 that Singapore has the largest portion of non-energy use, which refers to activities such as fuel refinery for export purposes. To the surprise of many, although not an oil and gas producer, Singapore ranks fifth in Asia in terms of refining capacity, at 1.51 million bpd in 2018 [27].

Shown in Fig. 10 is a simplified comparison of Fig. 9, which is prepared by removal of other and non-energy use categories. It is implied in the figure that Ethiopia and Nigeria are high on residential (represented by the green bars). Nigeria ranks the seventh place in terms of the world largest population in 2020, and Ethiopia ranks twelfth. This explains the high energy consumption in the residential sector. However, Indonesia has higher population than the two countries and ranks fourth. Similarly, Brazil ranks in sixth place. More than 98% of Indonesians have access to electricity but only 28% of its energy goes to residential. Such disparity in household energy consumption would require systematic studies in order to overcome the challenges involved.

Remarkably, Fig. 10 also shows that Singapore and Malaysia are the two nations with the highest energy consumption for commercial sector. This suggests that the two have a high ratio of commercial buildings, which requires energy, particularly for the air-conditioning systems. In the transportation sector, Brazil and Indonesia are on the high side in terms of energy consumption. This is most probably contributed by the fact that the two nations are vast in the size of population (number of counts of travel) and area (travel distances). Peculiarly, Malaysia has quite a high ratio of energy

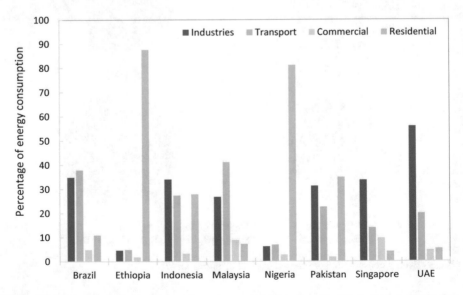

Fig. 10 Comparison of percentage of energy consumption for selected tropical countries by industries, transport, commercial and residential in 2015

consumed in the transportation sector. This could be explained by its government's promotion on the national cars(Proton and Perodua), along with the supports of good highway networks in the country and availability of cheap prices of fuels. Apart from road transportation, the aviation sector is also important for Malaysia since part of the country also lies on the island of Borneo also which has many important cities. To give an idea, there are more than twenty daily flights from Kuala Lumpur to Kuching alone, with the involvement of at least three carrier companies; something that was beyond imagination twenty years ago. The key point highlighted here is that the patterns of energy use by tropical countries are diverse and they change with time. At the time when this book is written, China is aggressively collaborating with many countries with its Belt and Road Initiatives. The pattern of economies and energy consumption presented here will definitely change!

4 Environmental Problems

Inevitably, the present state of growth in energy use leads to environmental problems. They are the side effects of growth in energy consumption activities on the biophysical environment. In 2012, the United Nations Environment Programme (UNEP) published a foresight report [28] describing 21 issues for the twenty-first century enveloping the major themes of the global environment, which include food, land, freshwater, marine, biodiversity, climate change, energy, waste, and technology as well as important cross-cutting issues such as environmental governance and human behavioural change towards the environment. In a simpler perspective, the present global environmental problems include pollution, climate change, ozone depletion, soil degradation, deforestation and loss of biodiversity. Unfortunately, nearly all of these problems are related to the energy use to sustain humans' life. This includes the use of natural resources in an unsustainable manner, leading to a negative impact on the environment.

The most prominent impact is the discharge of greenhouse gases in the Earth's atmosphere due to various human activities, especially the burning of fossil fuels. Greenhouse gases include carbon dioxide, methane, nitrous oxide, ozone, and water vapor. The problem with greenhouse gases in the atmosphere is that they absorb and emit radiant energy within the thermal infrared range, resulting in the greenhouse effect. The ultimate result increase in the Earth's average temperature leading to various problems such as a change in climate, flood, drought and many others, of which many of them are inter-related.

Certain methods of power generation and other human activities, which involve deforestations and damage to water bodies, are connected with environmental issues globally. With an increase in demand for energy due to rapid growth in human population, over-exploitation of such natural resources happens, in tandem with uncontrolled urbanization and industrialization. The other impact is on the aspect of health, which sees water becoming unsafe to drink, poor sanitation and hygiene conditions, and pollution of air and water.

Table 1 Top countries within the tropical zone that emitted the most carbon dioxide from fuel combustion in 2017

Country	World rank	CO_2 emission (MT)	CO_2 emission per capita (T)
China	1	9258	6.68
India	3	2162	1.61
Iran	8	567	6.99
Saudi Arabia	10	532	16.16
Indonesia	11	496	1.88
Mexico	12	446	3.62
Brazil	13	428	2.04
South Africa	14	422	7.44
Australia	15	385	15.63

At present, the Earth is facing a serious environmental crisis at an alarming rate, which leads to disasters and many uncertainties in the future. Urgent attention from all responsible authorities is required to overcome these issues, simply to ensure that the Earth is still livable in the future! Countries have started to realize this, though late and unhurried. This can be seen through the signing of the Paris Agreement in 2016 [29]. By the time this book was written, the agreement was signed by 197 countries. It deals with greenhouse-gas-emissions mitigation, adaptation, and finance. In principle, the Paris Agreement's enduring goal is to keep the increase in global average temperature to below 2 °C than that of the pre-industrial levels. At the same, the signatories are pursuing efforts to bring down the temperature increase further to 1.5 °C in order to minimize the risks and impacts of climate change. One of the action plans drawn in this agreement is to reduce emissions as soon as possible, in order to "achieve a balance between anthropogenic emissions by sources and removals by sinks of greenhouse gases" in the second half of the twenty-first century.

Carbon dioxide can be regarded as the primary global emission. The main emitters are generally countries that consume a high amount of energy due to fast growth in industries. Shown in Table 1 is the list of top tropical countries that emitted the most carbon dioxide in 2017, which is derived based on the data compiled by the International Energy Agency [30]. The presented numbers represent emissions by fuel combustion only. The list is presented in two forms, which are the total emissions (in metric million tonnes, MT) and the emission per capita (in metric tonnes, T). The shown rank is for the total emission. Both forms are equally important, although the former is clearly the one that would ultimately contribute to the true global damage. It must be reminded again that in this book, sub-tropic countries are regarded as tropical countries. This is important in view of the prominent influence of China (a country of sub-tropic and partially temperate climates) in the global emission. It is no secret that China, being a fast booming economy, is the highest emitter of carbon dioxide. Even the USA, who is second in the rank (not shown in Table 1), emit about half of that of China.

It is interesting to note that Saudi Arabia was the highest carbon dioxide emitter per capita in the world in 2017. Australia ranked second. Both countries are located in the tropical zone. A strong reason for this is due to the fact that both countries have large desert areas, and this results in high usage of air conditioners. Certain structures within the built-up area in the desert will retain stored heat until dawn due to the intense heat; and thus the air-conditioners would operate for 24 h. In the aspect of transportation, both Saudi Arabia and Australia are large countries, whereby the use of automobiles and aircrafts is high. As for Australia, there is another factor, which is the high consumption of coal (60%) for the generation of electricity.

The country ranks that are not shown in Table 1 are filled by nations that are already developed. At large, the table indicates that the total carbon dioxide emissions are mainly contributed by emerging economies from the tropics, and also the developed countries, of which nearly all are located in temperate climate areas. Nevertheless, in actual fact, developed nations have higher carbon dioxide emissions per capita, although the figures are not displayed in the table. This disparity becomes one of the challenges that the global community is facing in search of effective and unbiased solutions to global warming. For example, the developed nations always raise their concerns that deforestation is high in developing countries. This is a valid argument but it is also a dilemma since the former did the same long time ago, while it is undeniable also that the latter need to cut down trees for development.

It is important also to recognize the pattern of changes in emission throughout the last few decades, recognizing the increase in population, improvement in health and technologies and many other progressive developments, including the fall of communism in the 1990s. Depicted in Fig. 11 is the temporal variation of carbon dioxide emission due to fuel combustion for China, India and ASEAN countries from 1971 to 2017. The graphs are constructed based on the data compiled by the International Energy Agency [30]. These countries are selected in view of their fast rate of economic growth within tropical countries, although it is noted that many other tropical countries have been performing well too.

It is shown in Fig. 1 that within nearly 50 years, the amount of carbon dioxide emitted by China has increased by more than nine times. The global increase in the same period of time is only about two times [30]. The figure also shows that in 1999 China experiences a sudden increase in emission of carbon dioxide. This is in tandem with its economic reformation at the time of the global fall of communism, although there are reports stating that the Chinese economic reformation started in 1978. The rate of emission of China is exceptionally high as compared to other nations. It is simple to generally deduce that emission increases with economic growth. Despite the slowing down in 2013, as depicted in Fig. 1, China remains the highest emitter of carbon dioxide. On another note, the emission by other tropical countries is escalating also. It is probable that because everyone is looking at the colossal emission by China, the "crime" by other nations seem to be obscured, and this should be worrying. To add further, big economies, namely the USA, European Union, and Russia have been seeing a gradual reduction in emission of carbon dioxide. This is evidence that while these countries have lately been taking remedial actions to reduce emission, countries in the tropics have been generally doing the opposite.

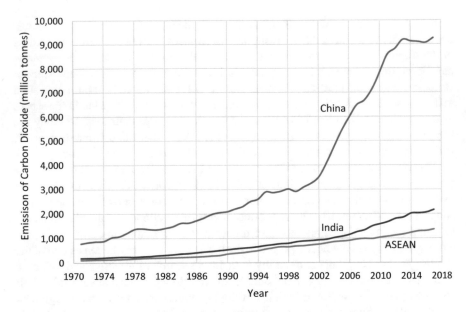

Fig. 11 Temporal variation of carbon dioxide emission by fuel combustion for China, India and ASEAN countries from 1971 to 2017

The emission of greenhouse gases like carbon dioxide is just one of the many environmental issues that are related to energy use. Another mind-boggling environmental problem is suspended particulate matter (PM), which is publicly referred to as air pollution or air quality. In comparison to the greenhouse, particulate matter affects humans in a more immediate and direct manner, especially to the lungs. Particulate matter is the combination of liquid and solid particles that are suspended in air, of which they could be natural or anthropogenic (due to human activities). They could be in the form of organic particles such as pollen and soil. Inorganic particles include soot, smoke, and liquid droplets, which are byproducts of fuel combustion. Liquid droplets are regarded as inorganic although naturally available because it has no carbon atom. The range of suspended particles can be divided into a few categories as listed in Table 2. Coarse particles larger than 10 microns are generally not regulated. The smaller one, which is known as PM10, can cause irritation to the eyes, nose, and throat. PM2.5 is more hazardous since it can penetrate deep into

Table 2 Categories of suspended particulate matter

Category	Label	Size (micron)[a]	Notes
Coarse particles	–	>10	
Coarse particles	PM10	2.5 < Size < 10	Regulated
Fine particles	PM2.5	0.1 < Size < 2.5	Regulated
Ultrafine particles	PM0.1	10	Not regulated

[a] One micron is one micrometer (μm)

the lungs and possibly into the bloodstream. The ultimate health implication can be posed by ultrafine particles (UFP), also known as PM0.1, as it can penetrate into the bloodstream.

The ability of particulates to penetrate deep into the lungs and bloodstreams may lead to respiratory diseases, heart attacks, and premature deaths. The Air Quality Guideline [31] issued by the World Health Organization (WHO) stipulates that PM2.5 should not exceed 10 $\mu g/m^3$ (annual mean) or 25 $\mu g/m^3$ (24-h mean). As for PM10, it should not exceed 20 $\mu g/m^3$ (annual mean) or 50 $\mu g/m^3$ (24-h mean). Despite the availability of guidelines from the WHO, it was reported that there has actually been no safe level of particulates; for every increase of 10 $\mu g/m^3$ in particulate matter, the rate of chance for lung cancer would rise to 22% for PM10, and 36% for PM2.5 [32]. According to the Institute for Health Metrics and Evaluation (IHME) air pollution worldwide contributed to 4.9 million deaths in 2017 [33].

There are a few versions of air quality rankings available for reference since different organizations have different ways of valuing the information. The World Air Quality Index [34], for example, provides hourly updated information for the argument that 24-h average data is misleading since the effect of wind can change measurements significantly within hours. Consequently, information such as country ranking differs. It may be misleading to relate the numbers from the site measurements as polluters, because of the wind. For example, forest fires in Sumatera have been very well known to cause heavy haze conditions almost annually from around June to September in Singapore, a small country that is generally clean in terms of internal sources of air pollution. On the other hand, in the absence of wind, the numbers would be taken as true particularly for cities that use coal as a source of energy for power generation.

In the list of top 500 cities by PM2.5 annual mean concentration measurement that was documented by the World Health Organization (WHO) [35] for the period from 2008 to 2017, China and India were the significant dominators. The first 20 countries on the list were all from the tropical zone. These cities are well known for having coal-fueled power generation plants. This is different in the US, where a high number of proposals for the development of new coal-fired power plants have been rejected, and many existing plants are scheduled for retirement. The trend is similar in Western Europe, due to movements by activists and good support by governments. Consequently, the demand for coal is presently getting higher in developing countries, especially in the Asia Pacific [36]. This is one of the latest challenges faced by tropical countries. Not only is coal a culprit for poor air quality, it is also nowadays the primary source for carbon dioxide emission (over 40%).

To delve further into the topic of the environment would make this chapter far lengthy since there are so many issues that need to be highlighted. Thus, only selected protuberant issues are discussed. Another environmental issue related to the use of energy is the release of toxic wastes. They can be in the form of heavy metals, which are discharged by coal-fired power generation plants such as mercury, arsenic, chromium, cobalt, lead, manganese, nickel, zinc, and silver. These heavy metals, which are usually discharged in the forms of coal ash and wastewater are hazardous to the environment and human health. Apart from heavy metals, other byproducts from coal-fired power generation plants like sulfur and nitrogen oxides contribute to problems such as smog and acid rain. The same threats have also resulted from the combustion of natural gas. The activity of coal mining itself is damaging to the environment as it contaminates nearby river.

Other types of power generation systems are relatively cleaner. However, some of them have considerable environmental issues and would require excellent management. Nuclear power, for example, has massive environmental risks, as evident by the renown Chernobyl and Fukushima disasters. Another clean power generation system is hydropower. The ugly side of hydropower in relation to the environment is it causes a drop in the downstream water level and damage the fragile eco-system, apart from a few other related issues. At the downstream of the hydro-station dam, water is highly needed for domestic use and irrigation, and therefore a drop in the water level could be disastrous to human activities. Even worse, there have been quite a number of cases that lead to conflict among countries that share the same river, namely the Nile, Euphrates-Tigris, Indus, and Mekong. Lastly, there are numerous forms of renewable energy which are used nowadays for power generation. They too are actually causing various environmental issues. However, the implications are far smaller compared to the damaged caused by the use of fossil fuels.

5 Energy Resources

This section is dedicated to explore the available source of energy in the tropics. Firstly, it must be made clear that the resources may be classified as primary and secondary resources. Table 3 shows a description of these resources. Secondary resources would normally involve an addition process, time, and cost before they can be used. However, with optimization, secondary resources may sometimes worth the investment. As indicated in the third column of Table 3, the energy can be further classified into renewable or non-renewable energy. A resource is generally considered as renewable if its capacity can be recovered within a period of time significant by human needs. On the other hand, non-renewable resources deplete by human usage and their potential will not be recovered significantly for a very long period of time. Basically, all fossil fuels are non-renewable. As for the secondary resources, it is

Table 3 Basic classification of energy resources

Resource	Feature	Examples
Primary	Ready for end-use	*Renewable*: solar, wind, biomass
		Non-renewable: coal, oil, natural gas
Secondary	Must be converted into another form	Battery, hydrogen, biofuels

intended here not to further categorize them mainly because they could be converted from primary sources that are renewable or non-renewable.

Globally, the supply of energy is dominated by fossil fuel, which was reported to be over 81% in 2017 [37]. Oil still dominates the share at 32%, followed by coal at 27% and natural gas at 22%. The largest contributor to renewable energy is biomass, at nearly 10%; however, a large portion of it is contributed by traditional method of burning wood for cooking in certain developing countries. Coal and gas are the dominant fuel sources for energy production in the tropics. While the supply of energy has to do with consumption, an interesting aspect to pay attention to is the available energy in tropical countries. Due to the vast area of the tropics, it has all types of energy resources and technologies known to date. In most cases, fossil fuels can be transported to different countries, known for natural gas, coal, oil, and nuclear fuels. This is different for most renewable energy, with the exception of secondary resources like biofuels and briquette. Nevertheless, the transportation of fuels will only make the fuel more expensive as opposed to local consumption.

The reserve of fossil fuel in the tropics is the largest in the world, as reported in 2018 [38]. Its proven oil reserve is about 77%, which is led by Saudi Arabia. The Kingdom holds 16.7% of the world's oil reserves. For natural gas, the tropics hold about 60% of the world's reserve. The prominent natural gas reserve is kept in the Middle East. As for coal, the United States has the biggest reserve, followed by Russia, China, Australia and India. The tropics still hold a large amount of coal at about 46% of the world's reserve. As highlighted earlier, fossil fuels cannot be renewed or recovered over a short span of time. Globally, it is estimated that presently the reserve for coal can last for approximately 115 years, and about 50 years each for oil and natural gas [39].

Different than fossil fuels, there is no such thing as reserve for renewable energy. Nevertheless, there have been some estimates on the potentially available energy from this source. Such estimates are subject to variations. Wind and ocean can sometimes be affected by changes in the global weather. Solar energy may be affected by the presence of natural clouds, latest uncertainties in weather, and also unnatural haze. Biomass energy may also be affected by changes in various issues such as seasons, change in land use, competition with other consumers such as food for humans or animals, as well as new demand from certain industries that have an interest in recyclable materials. Even hydropower may be affected by drought and various human factors.

The highest amount of renewable energy available in solar energy. The estimated amount of annual potential energy that can be converted from the sun is 23,000 TWy (Tera-Watt-years); the second-highest is the wind at only up to about 130 TWy [40]. The energy can be used to directly obtain electricity by mean of solar photovoltaic (PV) panels, which use semiconductors to convert solar radiation into direct current electricity. Shown in Fig. 12 is a world map for photovoltaic power potential, which was published by the World Bank and provided by Solargis [41]. The map displays the estimated amount of electricity (in kWh) that can be generated from a 1-kWp (peak power) free-standing crystalline silicon(c-Si) modules, which are optimally inclined towards the Equator. The displayed quantities in the map are a result of

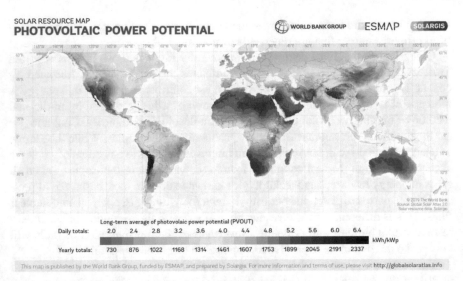

Fig. 12 World map of photovoltaic power potential [41]. Map obtained from the Global Solar Atlas 2.0, a free, web-based application is developed and operated by the company Solargiss.r.o. on behalf of the World Bank Group, utilizing Solargis data, with funding provided by the Energy Sector Management Assistance Program (ESMAP). For additional information: https://globalsol aratlas.info

long-term average that was calculated based on weather data of the last 10 years. It is clearly implied from the figure that the solar energy potential for photovoltaic (PV) is the highest within the tropics, especially in the deserts. Geographically, the closer an area is located to the Equator, the higher would be the amount of solar radiation.

Another source of renewable energy that is picking up is wind energy, with availability of up to 130 TWy [40]. Power is generated by the use of wind to turn wind turbines, which turn electric generators. Wind energy is generally stronger in temperate countries. Shown in Fig. 13 is the world map of the estimated speed of wind at 100 m above surface level [42], based on a set of models. In general, it is indicated in the figure that the equatorial zone has relatively the weakest wind energy.

The third most available renewable energy is biomass energy, which accounts for up to 6 TWy [40]. This type of energy uses materials fromplant or animal for production of electrical or thermal power, mainly by means of thermal and biochemical conversions. The source of materials can be in the form of agriculture residues, as well as plants that are purposely grown for energy such as jatropha and switchgrass. In addition, municipal solid waste and waste from sewage treatment plants are also regarded as biomass. The ecosystems in the tropics are highly diverse, comprising among others the tropical rainforests, seasonal tropical forests, dry forests, spiny forests, deserts and others. This diversity determines the types of plants that are suitable to be planted. Shown in Fig. 14 is the world map for above-ground biomass density, which is measured in tonnes of forest biomass per hectare [43]. It is implied

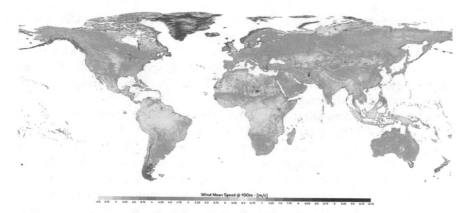

Fig. 13 Global map of wind speed at 100 m above surface level. The map is obtained from Global Wind Atlas 3.0, a free, web-based application developed, owned and operated by the Technical University of Denmark (DTU). The Global Wind Atlas 3.0 is released in partnership with the World Bank Group, utilizing data provided by Vortex, using funding provided by the Energy Sector Management Assistance Program (ESMAP). For additional information: https://globalwindatlas.info

Above-ground biomass in forest per hectare, 2015
Above ground biomass density, measured in tonnes of forest biomass per hectare.

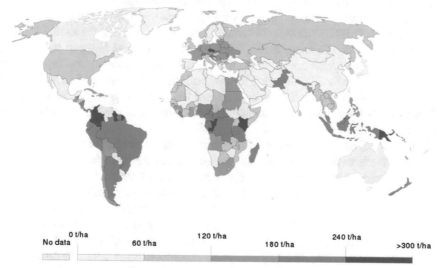

Source: UN Food and Agriculture Organization (FAO)

Fig. 14 Above-ground biomass density in 2015. Figure courtesy of Our World in Data [43]

from the figure that biomass energy is highly available in the tropics. Although not all the trees in the forest are available for use as a source of biomass energy, it is an indication that the climate and lands in the tropics are suitable for agriculture. As a result, the availability of biomass energy in the tropics can be regarded as high. Furthermore, the tropical zone does not experience winter and therefore the plants are available for use all year round.

Other forms of renewable energy available in the tropics are marine (13 TWy/y), hydropower (4 TWy/y), geothermal (3 TWy/y). Because of their small scales or low in terms of technology-readiness, they are not elaborated further in this section. Last but not least, there is also nuclear power, which is actually a type of fossil fuel but is usually treated differently due to its clean feature. Nuclear energy is theoretically available for use at any place on Earth. But there are many security, safety, social and political issues that must be resolved prior to development. Due to this, no discussion on nuclear power is presented in this chapter.

6 Present Scenarios and Challenges

The 2020s began with the coronavirus pandemic, which sees various worldwide disruptions in human activities particularly due to the urge for lockdowns in most countries [44]. Interestingly, despite the worries mentioned in the earlier section on environmental problems, the pandemic has resulted in positive impacts on the environment. Vehicles' movements came close to zero, causing the lowest emission and noise pollution in many areas. Demand for fossil fuels dropped to a level so low that it was beyond imagination. During the same period, it was reported that China experienced a 25% decrease in CO_2 emissions [45] and another 50% in NO_x emissions [46]. Although the positive impact was only temporary, it implied that it is not impossible to attain desirable healthy environmental conditions with the right efforts and supports.

In view of the threatening environmental issues, it would ideally be good for tropical countries to increase the use of renewable energy. The installed capacity of the total renewable energy in the tropics is about 50% of the world's capacity, as estimated for 2018 [47]. This capacity includes hydropower, which is not recognized as renewable by some schools of thought due to the negative consequences on the environment, particularly those involving the massive size of reservoirs and dams. Despite the issues, some governments choose to build hydropower plants due to other important reasons. In many places, reservoirs are constructed for flood control and irrigation purposes. In Malaysia, one large reservoir was uniquely developed in the 1970s to fight against communist guerillas. The pie chart in Fig. 15 shows the distribution of renewable energy capacity in the tropics for 2018. Obviously, the largest source is hydropower, a long-proven technology that has been used for over a hundred years in various parts of the world. The wind and solar have almost the same share at about 20%. Biomass energy has only a 4% share and ocean energy is almost negligible.

Fig. 15 Distribution of renewable energy capacity in the tropics for 2018

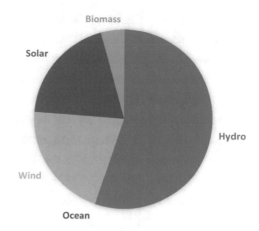

As mentioned in the earlier section, solar energy could be the best renewable energy for the tropics considering its undisputed availability, though available only during daytime. Hydropower, on the other hand, may probably better be avoided due to the disadvantages it has in terms of land use and ecosystems. Nevertheless, the selection of the type of renewable energy sources for power generation in any place or country depends on the availability of the resources. In flat islandic countries like Maldives, for example, hydropower would not be an option. On the other hand, in some areas where agricultural residues are abundant and are posing problems if unutilized, the use of biomass power generation system could be an irresistible mitigation option. External factors such as government incentives in the form of feed-in-tariff could also be an important consideration. In other words, energy sources that are perceived as good may not be favorable under certain conditions, and vice versa.

For investors, the most important factor is the cost (both capital and operational). Shown in Fig. 16 is the comparison of power generation costs for different energy sources, based on data from the US Department of Energy's Annual Energy Outlook 2019 [48]. The indicated cost in the figure is the levelized energy costs (LEC), also known as the levelized cost of electricity (LCOE) [49]. This cost reflects the total investment involved throughout the life of the power generation system over the expected period of electricity generation throughout its lifetime. The cost is comprehensive covering the complete capital and operational costs. The light green bars represent costs for power generation using renewable energy; the red bars represent those using fossil fuels. The figures could come in various versions, for example with the inclusion of government taxes or incentives for certain technologies. It must also be highlighted that the data represents the US scenario and the trend may probably vary slightly for different countries.

It is shown in Fig. 16 that even with the latest technologies the costs of electricity generation by fossil fuels are presently expensive, with the exception of natural gas-fueled power plants that are using combined cycle technology. Despite its poor reputation decades ago in term of cost, solar photovoltaic (PV) is presently the cheapest power generation technology. With proper planning, especially in handling

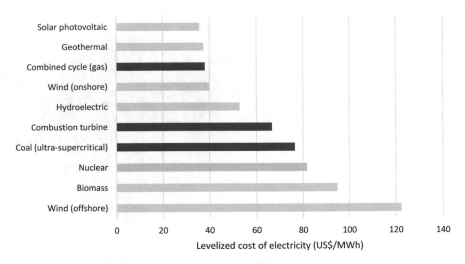

Fig. 16 Comparison of power generation costs for different sources entering service in 2025 (based on 2019 US$/MWh)

the daily operation limit, solar PV would be an advantage for tropical countries due to the abundance of sunlight in the area. Geothermal is also low cost but the availability is very limited, as implied in the previous section. Onshore wind energy is also cheap (lower than US$40/MWh) although its availability in the equatorial zone can be regarded as poor.

Hydropower, although a matured technology, has a moderate cost of generation. This is mainly due to the huge investment it takes. Nuclear energy was once regarded as a low-cost electricity generation. By now it is one of the most expensive technologies. Biomass energy is presently the second most expensive technology due to the complication involved in the conversion processes. Furthermore, there have been claims that the greenhouse gas savings from certain biofuels are less than originally anticipated. The offshore wind energy is shown in Fig. 16 to be the most expensive power generation. The cost is about three times the cost for electricity generation by each solar PV, geothermal, combined cycle and even onshore wind. Offshore wind technology is popular in Europe due to the high availability and strength of wind in the region. Despite the present energy trends, researchers, governments and investors are constantly working to improve each of the systems, especially renewable energy. Consequently, the electricity generation costs depicted in Fig. 16 may change in the future, as happened in the last few decades. In between 2010 and 2019 the electricity generation cost for solar PV, for example, decreased tremendously by 88% [48]. Onshore wind power also saw a huge reduction of 71%.

The electricity generation can be enhanced through the use of solar tracking mechanism to ensure that the photovoltaics panels are continuously tilted to positions that give the maximum irradiation. A major issue with solar power is the disappearance of sunlight due to night time and also obscuration by the clouds. This limitation is often overcome by the use of a battery or other sources of energy. The other issue with solar energy is the need for large area of land that is free from shadows. Land can be very expensive in some places and consequently it could be uneconomical to purchase such lands just for installation of solar PV. While desert areas would be ideal for installation of solar PV, it would be a bad idea to cut down forests just to give way for solar PV installations. Plants are well known to be good absorbers of carbon dioxide, which is essential in alleviating the present global greenhouse problem. An innovative idea to increase generation of electricity by solar PV is to install solar panels on the roofs.

Driven by growing demand for clean environment, various government policies and initiatives have been introduced to support utilization of renewable energy. One of the prominent initiatives is the Feed-in Tariff (FiT), which started in the USA in 1978. The FiT, which is globally accepted, is intended to support the growth of electricity generation using renewable energy sources, which has usually not been economically feasible. With the FiTmechanism, some forms of long-term contracts are offered to participating producers. The producers are not limited to corporate bodies. Anyone who has even small capital sufficient for generation of electricity, for example, homeowners, can participate. Extensive references on various frameworks and tariffs practiced worldwide can be found on the Internet. More recently some countries are shifting away from the FiT because of the significant drop in the costs of electricity generations by renewable energy sources, as implied in Fig. 16.

Despite the proven and appealing potentials of solar PV, geothermal and wind as depicted in Fig. 16, fossil fuel is still preferred in some tropical countries. For natural gas, through the use of a combined cycle, the cost of electricity generation is almost similar to that of solar PV. Furthermore, many tropical countries have natural gas resources, or at least their neighbors do. However, for coal, which is implied in Fig. 16 as no longer competitive, it is gaining interest in certain regions, especially in Asia. In Malaysia, for example, coal supplied only 7.4% of its energy for electricity generation in 1997, but it grew tremendously to 50.6% by 2017 [50] Malaysia Energy Statistics Handbook 2019. Generally, the demand for coal in Asia grows due to its availability and affordability in order to provide energy security and underpin economic development [51]. Many of these countries have large resources of coal and with low cost of labor and other resources, the use of coal may be seen as still attractive. Presently, India is having the largest growth in coal consumption, although China is the largest world consumer. Also picking up are Indonesia, Vietnam, Philippines, Malaysia and Pakistan.

A concern on energy in the tropics is on subsidized energy in certain countries that can afford to do so due to their rich fuel resources. The subsidies, which are usually patronized by the governments, are mechanisms that retain prices for consumers below the market levels for various reasons in the forms. In this situation, fuel is usually sold to the public at prices that are far lower than the world's average price. The heat can be felt at international borders where the price gap is significant, resulting in attempts to smuggle or to buy in large quantity in the country where the price is cheaper. The primary intention of the subsidy is to provide benefits to the citizen for economic and sometimes political reasons. A negative effect of the subsidy is that people take it for granted by utilizing the energy in an inappropriate or wasteful manner. One simple example is to leave a car's engine idling for hours just to benefit the cool air from the car's air-conditioner despite the availability of other options for thermal comfort. Another example is the habit of leaving electric appliances such as televisions constantly turned on even though not being watched by anyone. As a result, energy is wasted unnecessarily and emission is released. This is why elimination of fossil fuel subsidies can greatly reduce global carbon emission and help lessen the health risks of air pollution. Probably it would be good for many tropical countries to emulate certain developed countries that, in the contrary, impose high fuel taxes and at the same time improve public transportation so that fuels are consumed responsibly and consequently reduce emissions. A usual challenge for countries that are practicing subsidies is the negative response of people who have long been pampered with such luxury.

While providing sufficient and clean power is one important task, ensuring that power is efficiently utilized is another thing. Power can be wasted simply because the equipment used has low efficiency, whereby it consumes high energy in order to deliver work. In addition, human factors can also contribute to wastage in power consumption. A relatively new concept that looks into the aspect of reducing the consumption of power is known as negawatt [52]. This concept involves efforts to reduce electricity consumption instead of investing to increase supply capacity. The outcomes of this approach can be measured in many ways; for example, through avoidance of building new power stations, or in terms of cost savings and reduction in emissions. Examples of negawatt initiatives are:

1. They use light-emitting diode (LED) lamps to obtain low energy lighting.
2. Installation of thermal storage tanks in district cooling systems in order to shave peak-load.
3. The implementation of a rating system for household appliances so that consumers could make wise decisions when buying them.
4. Improved thermal insulation and airtightness in buildings in order to minimize energy loss in air-conditioning systems.

Through these initiatives, energy can be consumed wisely and therefore the power generation systems can be optimized. Eventually, this could help to protect the environment.

7 Summary

Tropical countries occupy the largest area on Earth, also known as the tropics. The climate of the tropics is generally warm and humid. Historically, most tropical countries were previously occupied by Western colonialists, and therefore many of these countries gained their independence less than 100 years ago. Economically, most tropical countries have been inferior relative to countries in temperate zones. Nevertheless, in the last few decades, there have been many changes in economy in line with global technology innovation and improved transparency. As a result, the demand for energy in tropical countries is increasing. Concurrently, the concern for environment resulted from emission and global warming is escalating. The dissimilarities in geographical locations, culture, socio-economic backgrounds, politics and many other factors make it complex for simple comparison and discussion. It is revealed that the potential for growth in tropical countries is high due to the rich resources that they have. Access to the best power generation technology in these countries is also high. Probably the greater concern today is on the choice of energy source by these countries that would not harm the public and environment. Constant awareness of environmental issues is imperative in order to attain a livable world in the future.

References

1. Lugo M (2007) How much land is in the tropics? In: God plays dice: a random walk through mathematics—mostly through the random part, 4 Dec 2007. https://godplaysdice.blogspot.com/2007/12/how-much-land-is-in-tropics.html. Accessed 2 May 2020.
2. Feeley K, Stroud JT (2018) Where on Earth are the "tropics"? Front Biogeogr 10:e38649
3. KVDP (2013) World map with the intertropical zone highlighted in crimson, https://en.wikipedia.org/wiki/Tropics#/media/File:World_map_indicating_tropics_and_subtropics.png. Accessed 2 May 2020. License: https://creativecommons.org/licenses/by-sa/3.0/deed.en
4. Nilsson D (2007) https://commons.wikimedia.org/wiki/File:AxialTiltObliquity.png. License: https://creativecommons.org/licenses/by/3.0/deed.en
5. Wahr JM (1982) The effects of the atmosphere and oceans on the Earth's wobble—I. Theory Geophys J Int 70:349–372. https://doi.org/10.1111/j.1365-246X.1982.tb04972.x
6. Main M (2013) The moving tropic of capricorn. Botswana Notes Records 45:195–200. Retrieved 30 April 2020 from www.jstor.org/stable/90024385
7. Martin C (2006) Experience of the new world and Aristotelian revisions of the earth's climates during the renaissance. History Meteorol 3:1–15
8. Peel MC, Finlayson BL, McMahon TA (2007) Updated world map of the Koppen-Geiger climate classification. Hydrol Earth Syst Sci 11:1633–1644
9. Wong C-L, Ismail ZYT (2018) Trend of daily rainfall and temperature in Peninsular Malaysia based on gridded data set. Int J GEOMATE 14:65–72
10. Beck HE, Zimmermann NE, McVicar TR, Vergopolan N, Berg A, Wood EF (2018) Present and future Köppen-Geiger climate classification maps at 1-km resolution. https://commons.wikimedia.org/wiki/File:Koppen-Geiger_Map_A_present.svg; Original source: Nature Sci Data 5, 180214. https://doi.org/10.1038/sdata.2018.214
11. World Population Review (2020)Tropical countries 2020, 7 April 2020, from https://worldpopulationreview.com/countries/tropical-countries/. Accessed 4 May 2020
12. Wilkinson A (2014) Expanding tropics will play greater global role, report predicts. Science Magazine, American Association for the Advancement of Science, 29 June 2014. Accessed 4 May 2020
13. Young SS (2019) Corruption perceptions index 2018. Adopted from Transparency International. https://commons.wikimedia.org/wiki/File:Corruption_Perception_index_2018.svg#filehistory. Accessed 20 May 2020
14. Arnold D (2000) Illusory riches: representations of the tropical world, 1840–1950. Singapore J Tropical Geogr 21(1):6–18
15. Sachs JD (2001) Tropical underdevelopment. National Bureau of Economic Research Working Paper 8119
16. State of the Tropics (2017) Sustainable infrastructure for the tropics. James Cook University, Townsville, Australia

17. Bożyk P (2006) Newly industrialized countries, globalization and the transformation of foreign economic policy. Ashgate Publishing, Ltd., Farnham
18. World Economic Outlook Database, October 2019. IMF.org. International Monetary Fund. Retrieved 14 May 2020
19. De Cian E, Sue Wing I (2016) Global energy demand in a warming climate. Centro Euro-Mediterraneo sui CambiamentiClimatici (CMCC) Research Paper No. 266, 8 Mar 2016
20. De Cian E, Sue Wing I (2019) Global energy consumption in a warming climate. Environ Resource Econ 72:365–410
21. Odaka K (1983) The motor vehicle industry in Asia: a study of ancillary firm development. Council for Asian Manpower Studies, Singapore University Press, Singapore
22. Malaysian Automotive Association (2020) Sales and production statistics. https://www.maa.org.my/statistics.html. Accessed 19 May 2020
23. Statista (2020) U.S. Car Sales from 1951 to 2019. www.statista.com/statistics/199974/us-car-sales-since-1951/. Accessed 29 May 2020
24. Pérez-Lombard L, Ortiz J, Pout C (2008) A review on buildings energy consumption information. Energy Build 40:394–398
25. Jong TX, Shan CT (2016) Electric shops running out of air-cooling devices. The Star, 6 April 2016. https://www.thestar.com.my/news/nation/2016/04/06/electric-shops-running-out-of-aircooling-devices/. Accessed 31 May 2020
26. United Nations (2015) 2015 energy balance, energy statistics. https://unstats.un.org/unsd/energystats/pubs/balance/. Accessed on 8 June 2020
27. Top five countries in Asia-Pacific region for oil refining capacities. NS Energy, 1 Jan 2020. https://www.nsenergybusiness.com/features/countries-oil-refining-asia-pacific/. Accessed on 8 June 2020
28. UNEP (2012) 21 issues for the 21st century: result of the UNEP foresight process on emerging environmental issues. United Nations Environment Programme (UNEP), Nairobi, Kenya, p 56
29. United Nations (2015) Paris agreement. https://unfccc.int/process-and-meetings/the-paris-agreement/the-paris-agreement
30. IEA (2019) CO_2 emissions from fuel combustion 2019 highlights. The International Energy Agency, 165p
31. WHO (2016) Ambient (outdoor) air quality and health: Fact sheet No. 313, World Health Organization, https://web.archive.org/web/20160104165807/http://www.who.int/mediacentre/factsheets/fs313/en/. Accessed on 14 June 2020
32. Raaschou-Nielsen O, Andersen ZJ, Beelen R, EvangeliaSamoli MS, Weinmayr G et al (2013) Air pollution and lung cancer incidence in 17 European cohorts: prospective analyses from the European Study of Cohorts for Air Pollution Effects (ESCAPE). Lancet Oncol 14:813–822
33. GBD 2017 Risk Factor Collaborators (2018) Global, regional, and national comparative risk assessment of 84 behavioural, environmental and occupational, and metabolic risks or clusters of risks for 195 countries and territories, 1990–2017: a systematic analysis for the Global Burden of Disease Study 2017. The Lancet 392:1923–1994. https://doi.org/10.1016/S0140-6736(18)32225-6
34. AQI (2020) A beginner's guide to air quality instant-cast and now-cast. The World AirQuality Project, 12 Jan 2020. https://aqicn.org/faq/2015-03-15/air-quality-nowcast-a-beginners-guide/. Accessed on 14 June 2020
35. WHO (2020) Global ambient air quality database (update 2018). World Health Organization. https://www.who.int/airpollution/data/cities/en/. Accessed 14 June 2020
36. Reuters (2019) Asia's coal addiction puts chokehold on its air-polluted cities. Thomson Reuters, 20 Mar 2019. https://www.reuters.com/article/us-asia-pollution-coal/asias-coal-addiction-puts-chokehold-on-its-air-polluted-cities-idUSKCN1R103U. Accessed on 14 June 2020
37. IEA (2019) Key world energy statistics 2019. International Energy Agency, 26 Sept 2019, pp 6, 36. Accessed 19 June 2020
38. Hannah Ritchie (2017) Fossil fuels. OurWorldin Data, https://ourworldindata.org/fossil-fuels. Accessed on 20 June 2020

39. BP (2016) Statistical review of world energy 2016. https://www.bp.com/en/global/corporate/energy-economics/statistical-review-of-world-energy/downloads.html. Accessed on 20 June 2020

40. IEA SHC (2015) A fundamental look at supply side energy reserves for the planet. Solar Update 62

41. The World Bank (2019) Photovoltaic power potential. Global Solar Atlas. https://globalsolaratlas.info. Accessed on 20 June 2020

42. Global Wind Atlas (2020) Technical University of Denmark (DTU). Accessed on 24 June 2020

43. Our World in Data (2019) Above-ground biomass in forest per hectare. Published online at OurWorldInData.org. Retrieved from: https://ourworldindata.org/grapher/above-ground-biomass-in-forest-per-hectare. Accessed on 24 June 2020

44. WHO (2020) WHO announces COVID-19 outbreak a pandemic. World Health Organization, 12 Mar 2020, https://www.euro.who.int/en/health-topics/health-emergencies/coronavirus-covid-19/news/news/2020/3/who-announces-covid-19-outbreak-a-pandemic. Accessed on 27 June 2020

45. Myllyvirta L (2020) Analysis: coronavirus has temporarily reduced China's CO_2 emissions by a quarter. CarbonBrief, 19 Feb 2020. Accessed on 27 June 2020

46. Zhang R, Zhang Y, Lin H, Feng X, Fu T-M, Wang Y (2020) (2020) NO_x emission reduction and recovery during COVID-19 in East China. Atmosphere 11:433. https://doi.org/10.3390/atmos11040433.Accessedon27June

47. IRENA (2020) Renewable capacity statistics 2020. The International Renewable Energy Agency, Abu Dhabi. https://www.irena.org/publications/2020/Mar/Renewable-Capacity-Statistics-2020. Accessed on 20 June 2020

48. EIA (2020) Levelized cost and levelized avoided cost of new generation resources. In: The Annual Energy Outlook 2020, U.S. Energy Information Administration

49. Short W, Packey DJ, Holt T (1995) A manual for the economic evaluation of energy efficiency and renewable energy technologies. National Renewable Energy Laboratory, NREL/TP-462-5173

50. Energy Commission (2020) Malaysia energy statistics handbook 2019

51. IEA (2018) Global coal demand set to remain stable through 2023, despite headwinds. News, The International Energy Agency. https://www.iea.org/news/global-coal-demand-set-to-remain-stable-through-2023-despite-headwinds. Accessed 30 June 2020

52. Lovins AB (1990) The negawatt revolution. Across the Board: Conf erence Board Mag XXVII:18–23

Hydropower Generation in Tropical Countries

Shazia Shukrullah and Muhammad Yasin Naz

Abstract The resources of renewable energy and their reserves are sources of survival in the modern era. Hydropower refers to energy being converted from running water into electricity. It is a form of renewable energy because the water cycle is constantly renewed without degrading and impacting the environment. Hydropower projects provide an opportunity to boost the social and economic development locally and globally. There is an increased understanding of the impact of dams on the ecosystem. The ecosystems enable us to make decisions about operational characteristics, size, and location of future dams. It is quite difficult to find information on hydropower dams in various tropical countries. The information presented here is collected from the reference literature and electricity planning sector. This study reports some strategies and policy decisions, which once considered, would contribute to accomplish the renewable energy goal of mitigating climate changes and providing a clean environment and energy for everyone in various tropical regions.

1 Introduction

Our present civilization, marked by a high degree of consumption, is an environment with incredibly low entropic levels that needs a tremendous amount of energy to sustain itself. Since the prices and environmental effects of oil as an energy source are growing, the route towards renewable energy sources is becoming increasingly important. The major renewable energy sources are solar energy, wind energy, hydel energy, and energy obtained from tides. These renewable energy sources are further extended to energy from biomass, energy from biofuels that are used for transportation. The conventional sources of energy are being replaced by some clean energy sources that is fuel cells and turbines at smaller scales [1–3] and wind power,

S. Shukrullah (✉) · M. Y. Naz
Department of Physics, University of Agriculture, Faisalabad 38040, Pakistan
e-mail: zshukrullah@gmail.com

M. Y. Naz
e-mail: yasin603@yahoo.com

S. A. Sulaiman (ed.), *Clean Energy Opportunities in Tropical Countries*,
Green Energy and Technology, https://doi.org/10.1007/978-981-15-9140-2_2

photovoltaics, and hydropower at larger scales. In remote areas where the generation of electricity is done by petrol or diesel-based engines, the power generation systems are being replaced by solar panels and biomass power plants. Hydropower is one of the most reliable and biggest renewable energy sources. The electrical power is produced from the water, stored in dams or running in rivers and streams, without degrading the ecosystems. There is a growing understanding of the impact of dams on the ecosystem, which enable countries to make decisions about operational characteristics, size, design, and location of future dams.

Hydropower is widely known to be the best-developed and by far the most important form of renewable energy, also playing a major role in achieving the goals set in the Paris Agreement on climate change. To minimize the emission of greenhouse gases and to limit the global temperature to well below 2 °C in the current century, more than 190 countries signed the Paris Agreement in December 2015 [4]. This agreement motivates the international community to take effective steps to combat climate change. It would not be possible to meet the objectives without adapting the renewable energy sources on a larger scale in the coming years. Hydropower is used to meet 20% of global electricity demand. Iceland is producing 83% of total electricity from hydropower. Norway uses hydropower to meet its total energy demand, and Austria produces more than half of its electricity from hydropower. Canada is the biggest hydropower producer in the world who generates more than 70% of its energy from hydroelectric sources. Hydropower stations can be graded into small, medium and large, depending on the height of the effective head. The energy extracted from the water depends on the volume and height difference between the source and the outflow of the water [5]. Generally, big hydropower plants produce electricity from the water stored in large dams [6]. Figure 1 shows schematically an arrangement of a hydropower station. The major components of any hydropower station are reservoir, penstock, valve, draft tube, tailrace water, turbine, powerhouse, transmission lines, transformers, insulators, transmission tower, and trash rack. This maximum

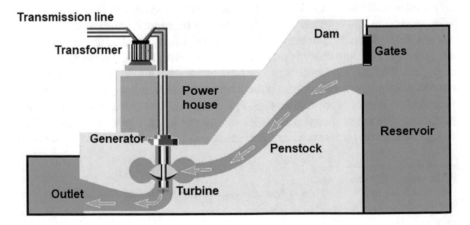

Fig. 1 Design of simple hydropower plants and its essential components

difference between source and outflow is called a head. The water potential energy linearly increases with the size of the head. To get a really high head, water is run through a wide pipe called a penstock for a hydraulic turbine.

The countries where thermal power plants supply the major part of electricity use water electricity for load following and regulation. This permits the thermal power plants to run close to the thermodynamically optimal levels rather than at continuously varying levels. The varying operation of thermal power plants lowers efficiency and raises environmental concerns. Hydropower currently represents a significant portion of the total energy supply in many countries, including Austria, Brazil, Paraguay, Canada, and Norway. Brazil has played a leading role for over ten years in the development of hydropower and hydroelectric capacity [7]. Although it is well known that fossil fuels contributor to global warming in the form of greenhouse gases, the focus on hydroelectric dams is given just in the most recent decades. According to an International Panel on Climate Change (IPCC), hydroelectric projects use large land. Countries with large land areas will have a more protected environment from changes in water-filled land. The pollutants are especially strong in big tropical reservoirs constructed over heavily forested lands, where organic matter is a high intake, the air temperature is high and there are large areas with anoxic environments at the bottom of the water column [8]. The most important climate changes impacting the energy sources are annual and seasonal changes in air temperature, precipitation and run-off [9, 10]. Some energy sources have greater up-front costs while others have greater or lesser continuing costs and impacts [11].

This chapter provides an overview of existing studies on the generation of hydropower in the world and tropical regions in particular. The information presented here is collected from reference literature and the electricity planning sector. This study reports some strategies and policy decisions, which once considered, would contribute to accomplish the renewable energy goal of mitigating climate changes and providing a clean environment and energy for everyone in various tropical regions.

2 World Hydropower Capacity

Hydropower plants have been used for decades to generate electricity. According to different studies, there are 60,000 large dams from around the world along many tropical shorelines. These dams are used for electricity generation, flood control, irrigation of farmlands, and controlling the flow of rivers. There are however many concerns about hydropower, especially the development of large dam facilities. There are more than 300 major dams worldwide, which meet one of many criteria on height and volume. The definition of size and scale of hydropower plants varies from country to country, for example, the capacity of small-scale hydropower plants in different countries is shown in Table 1. All hydropower plants except pumped store hydropower plants are controlled by the hydrological cycle, which is again governed by the sun-oriented vitality, and completely sustainable [12].

Table 1 Small scale hydropower plant capacity for various countries

Countries	Small scale hydro capacity (MW)
South Africa	≤ 10
Brazil	≤ 30
India	≤ 25
Norway	≤ 10
United Kingdom	≤ 20
Canada	< 50
China	≤ 50
USA	$5 - 100$

China has over 23,000 large dams and the United States is a second most dammed country with some 9200 large dams followed by India Japan and Brazil. During the early 1990s, the rate at which large dams are built decreased from about 1000 a year from the 1950s to the mid-1970s to around 260 a year. As of 2014, over 3700 hydropower projects are proposed or under construction on rivers around the world. If constructed, they could block more than 20% of free-flowing rivers. The construction of larger dams is not encouraged on many environmental, social, economic, and safety grounds. The key reasons for the worldwide resistance again the larger dams are the huge number of people who have been evicted from their lands and homes to make space for the reservoirs. The livelihoods of millions of locals have also suffered because of downstream effects, such as the loss of fisheries, water contamination, low water availability, and reduced fertility in farmland and forests due to the loss of natural fertilizers. Dams also spread waterborne diseases such as leishmaniasis, schistosomiasis, and malaria.

Opponents of dams often argue that the benefits of dams have been intentionally overlooked, and that other more effective and affordable means might provide the services they provide. The electricity produced through hydropower is still small as compared to some other power sources based on solar photovoltaic panels and wind turbines. However, the major concern is the need for vast land area to accommodate the power installations [13, 14]. Many solar and wind power companies are offering electricity to the consumers at a rate of USD 0.05 per kWh since 2017, which is much cheaper than hydropower [15]. The price of hydropower remains between USD 0.02 and 0.27 per kWh. Integrated cost of sustainable geothermal, wind energy, solar power, and storage energy in power systems could fall below USD 0.10 per kWh [15], which will continue to decline in the future. This data is based on a range of data sources, which can be viewed as a component of broader storage strategy. However, both solar and wind-generated electricity fluctuates and is not available throughout the day at the same rate. Solar and wind power plants are relatively expensive and need high maintenance. On the other hand, the large dams are built for irrigation and almost all of the major dams are built for hydropower. They also protect the local population from floods. Other than flood protection, dams supply water to towns and cities and help managing the river navigation. Many dams are multipurpose, offering

Table 2 An overview of the global gross production of electricity from different sources in 2018

Source	Contribution (%)
Coal	38.0
Natural gas	23.0
Hydro	16.2
Nuclear	10.1
Wind	4.8
Oil	2.9
Biofuels and waste	2.4
Solar	2.1
Geothermal, tidal, others	0.5

two or more of the advantages set out above. The dams produce approximately one-fifth of the world's electricity. An overview of the world gross electricity production in 2018 from different sources is provided in Table 2. The major part of electricity is produced through coal power plants followed by gas power plants, hydropower plants and nuclear power plants. The other sources are minor contributors towards the global electricity production.

In the coming years, renewable energy sources will be the major source of power production worldwide. As shown in Fig. 2, the situation in 2018 will totally change by 2025. The renewable energy sources will overtake the coal-fired power plants. In 2045, only 0.17 trillion kWh out of 44.25 trillion kWh will be produced from liquid

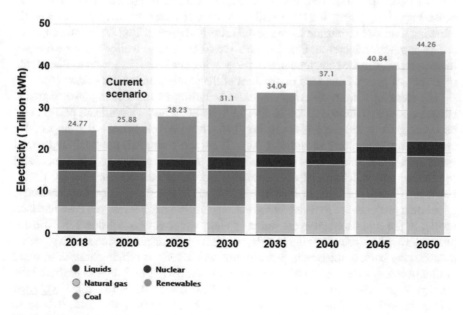

Fig. 2 Projected electricity production in the world from different sources during 2018–2050

Table 3 Projected global gross electricity production (2018–2050) in trillion kWh

Source	2018	2020	2025	2030	2035	2040	2045	2050
Liquids	0.74	063	0.44	0.32	0.24	0.19	0.17	0.17
Natural gas	5.75	5.91	6.25	6.65	7.43	8.11	8.8	9.25
Coal	8.65	8.50	8.57	8.57	8.57	8.60	8.90	9.60
Nuclear	2.64	2.71	2.78	3.04	3.14	3.28	3.41	3.58
Renewables	6.99	8.13	10.19	12.52	14.66	16.19	19.56	21.66

fuels. Renewable energy is expected to rise by as much as 21.66 trillion kWh in 2050, from nearly 7 trillion kWh in 2018. As shown in Table 3, the conventional sources of electricity are likely to be replaced by renewable sources, natural gas and nuclear energy.

Major hydropower projects can supply industry and the public with continuous electricity. These dams can ramp up their outputs on demand more rapidly than any other electricity generation sources. A massive hydropower system uses a dam to store water and hence electricity in a reservoir is the most typical form of hydroelectric power plant. Most dams are used for flood control, water storage, and irrigation purposes. The sustainable development goals of the United Nations are zero poverty, ending hunger and food security, clean water and irrigation for water savings, clean water and sanitation, urban resilience in coastal plains and deltas, and healthy river ecosystems. Dams provide benefit to river basin or if the water is diverted by trapping substantial part of the sediments, vegetation changes, the effects of bank erosion, millpond sedimentation, channel incision, and effects of stream bank erosion. In coastal aquifers, rising sea-level will enhance salt-water intrusion and eventually allowing seawater to intrude from coastal areas and diminishing freshwater supplies due to population increase. One can be affected by human wellbeing, for example, incidences of parasite schistosomiasis are growing across Africa, as dams disrupt the migration of prawns that feed on the host of the parasite and water snails [16].

The electricity from renewable sources including hydroelectric power plants is expected to reach around 21.66 trillion kWh by 2050 and a significant portion will be coming from developing countries [17]. These developments also impact the environment in a direct or indirect way. There are technical, political, and social goals associated with these developments, such as an increase of economy and local production and reduction of environmental effects of conventional energy forms. However, Gleick [18] did not agree on the role of large and small scale hydroelectric generation projects due to water losses via evaporation, sedimentation, flow, and land needs. In the hydrological cycle point of view, water constantly flows in different phases through a cycle; forming clouds, evaporating from lakes, oceans, and glaciers, then flowing back to the ocean, precipitation which falls as rain in dams, seas, other bodies of water and rivers. The hydrologic cycle shown in Fig. 3 depicts the hydraulic energy. Solar is the primary source of energy that powers the hydrological cycle and it is estimated that about 50% of all solar radiation entering the earth is used to

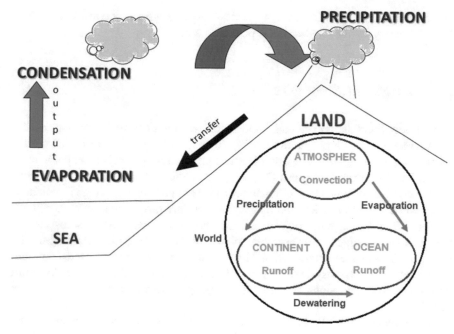

Fig. 3 Explanation of hydrological cycle and its basic components for depletion and replenishment of water resources

evaporate water in the process [4]. The hydrological and water cycle is a continuous process.

Hydropower is considered a renewable energy resource and only freshwater supplies are used to produce hydropower because of engineering considerations about the quality of structural properties for building hydroelectric power plants. The main characteristics of the hydropower potential are flow and head, as provided by the power relation:

$$P_{\text{hydro}} = C_p H Q \tag{1}$$

where C_p is the coefficient of hydropower, Q is the flow and H is the head. It can be seen in hydropower Eq. (1) that in the practical sense, flow is the only variable to produce electricity. The low-head hydropower systems use tidal flows or rivers with a head of 20 m. The flow may be controlled to produce and regulate electricity generation. For example, under certain cases, head can be modified by pumping water from other sources into a reservoir or by regulating the entry and exit from the reservoir. Hydropower plants which use flow and head as stated in Eq. (1) are called conventional hydropower plants. Modern dam development techniques generally rely on hydroelectric discharge. The flow of water can drive a turbine generator at times when demand is high, and a heavy load is placed on the system. These systems are typically assembled in waterways along with irrigation facilities, drainage systems, and high-speed rivers. Modern hydroelectric systems are also known as zero-head hydropower systems or hydrokinetic power systems.

2.1 Benefits of Reservoirs

Certain benefits of dams to humans include flood regulation to avoid periodic flooding with severe impact on people in the affected areas, storage of water during periods of drought, and long-term low-cost electrical production, which is essential in many tropical and subtropical regions [19]. Further benefits of reservoirs include recreational boating, fishing, aquaculture in net cages or tanks, water storage for intensive water-consuming industries, and irrigated agriculture production. Such advantages need to be looked at more objectively. On one side, water quality is not assured, despite different criteria, for example, the use of water for agricultural irrigation is countered by sewage contamination. Human activities, on the other hand, contradict the preservation of ecosystems, for example, development of villages, aquaculture and point-source emissions, and migration of people into the reservoir sub-watershed [20, 21].

Small dams, especially in rural areas, can be a sustainable and economic source of electricity. Most water from larger dams goes to farms and only a very small volume goes to cities. Hence, increasing the productivity of irrigated agriculture is the cheapest and most efficient way of supplying more water to cities. It is also essential to reduce leakage and wastage in urban water supply systems.

3   Tropical Rivers and Dams

Tropical rainforest regions have huge potential to produce hydropower, which is prominent in the energy development strategies of many nations. The high rainfall that enabled the growth of tropical ecosystems is also associated with large volumes of river water flow and a high potential for electricity generation through hydropower dams. As a result of this combination of rainforests and hydropower potential, several nations with large tropical rainforest areas, including Brazil, Democratic Republic of Congo, Peru, Colombia, Malaysia, and Vietnam intend to increase their hydropower capacity in coming years. On tropical streams, the number of dams is increasing over time [19, 22, 23]. In the neotropics, the development of new dams is important to meet the spurred electricity demand in the region [24, 25]. The use of power per capita in tropical nations is anticipated to increase twofold during 2005–2025 since population in tropical regions is increasing and the economy is growing at a good pace [26].

Many dam designers and hydropower advocates view tropical areas as the next place for building new dams. However, the quality of tropical freshwater may change with developing dams and hydropower plants since dams transform tropical waterways into multiple frameworks. Currently, the Brazilian government plans to construct hundreds of new dams in some of the most biodiverse tropical forest regions. But the high costs of mega dams in terms of biodiversity should be carefully balanced against any benefits of hydropower development. The Balbina Dam in the central Amazon of Brazil is one of the largest hydroelectric dams in the world in terms of the total flooded area. The construction of this dam saw a previously unbroken landscape of continuous undisturbed forest transformed into an artificial archipelago of 3,546 islands. Fires on these small islands have an animal-life knock-on effect, with extinction levels increased by the loss of habitable trees [27]. In Brazil, the large dams are termed as those with installed capacity of more than 30 MW. Tropical dams, particularly those in the wet tropics, produce more ozone harming substances than those in other climatic zones [28]. Hot environments have the highest emissions.

4 Emission of Greenhouse Gases from Reservoirs

Hydropower plants in tropics with large reservoirs can have a much greater impact on global warming than fossil fuel plants, which generate equivalent electricity. A comparison of emission of greenhouse gases from hydro reservoirs and fossil fuels is provided in Fig. 4. Steinhurst et al. [29] reported that tropical dams radiate around 1300–3000 g CO_2e per kWh, which is much higher than 160–250 g CO_2e per kWh produced by boreal dams. At the same time, a thermoelectric plant transmits 400–500 g, 790–900 g, and 900–1200 g CO_2e per kWh when utilizing gas, oil, and coal, respectively. As shown schematically in Fig. 5, greenhouse gases come from the surface of the reservoirs, spillways, turbines, and tens of kilometers downstream. The calculation of reservoir emissions is difficult since so many factors are involved in emission process. The key factors are air temperature, water temperature, time of

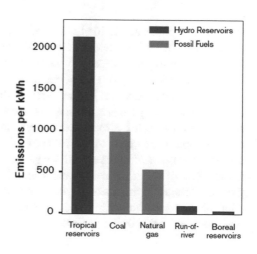

Fig. 4 Comparison of emission of greenhouse gases from hydro reservoirs and fossil fuels

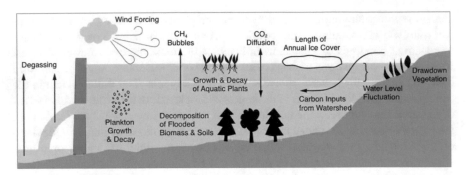

Fig. 5 The points of emission of greenhouse gases from hydro reservoirs

day, season, vegetation, and type of soil in reservoir and upstream watershed. The emissions are also affected by watershed management practices.

Dams continuously regulate water levels to produce electricity. This process also affects the amount of pollution from the reservoir that makes its way into the atmosphere. The hydrostatic pressure on submerged soils often decreases as water levels drop and allow gas bubbles to escape. Often the water column collects the methane in these bubbles, which never hits the surface. The challenge of estimating emissions also makes it hard to work out how to reduce emissions. One possible way to minimize emissions is the deliberated adjustment of water levels in the reservoir. Yet it may influence power generation, fisheries and flood control, and other operations. Protecting watersheds to avoid runoff into rivers will also minimize emissions and improve the quality of water. The experts should try to understand the different mechanisms involved in the emission of greenhouse gases from the reservoirs.

5   Run-of-River Hydropower

Hydropower plants are divided into three major groups by operation and flow type. As explained in Table 4, these groups are dammed reservoirs, run of river hydropower systems, and pumped storage hydropower systems. Run-of-river (RoR) hydroelectric systems harness energy from flowing water to produce electricity in the absence of a large dam and reservoir. These systems differ from conventional hydroelectric facilities where large dams are used to store water and run turbines. Small dams may be used for RoR hydroelectric systems to ensure that ample water goes into the penstock and some water storage for the same day use. Compared to others, the main difference in this method of electricity generation is that RoR system mainly uses the natural flow of water to produce electricity instead of the power of water dropping from a distance. In this method, the water storage or stagnation time is very small, which is a positive factor in the eutrophication process [5]. Another big difference with conventional hydropower is that RoR hydropower is used in places where water

Table 4 Classification of hydroelectric power plants

Hydropower plant type	Type of service	Impact
Dammed reservoir	Both energy and power	Modifications in flow of river and construction of reservoirs impact social setup and natural habitat
Run of river	Baseload with limited flexibility	River flow remains unchanged, limited flooding
Pumped storage	Net consumer of energy, power output only	Sometimes input is higher than output, issues related to upper reservoir
In stream	Both energy and power	Reduction of flow downstream of diversion

availability, such as in a river, is small or no. Based on their capacity, RoR systems are classified into microsystems (<100 kW), mini-systems (100 kW–1 MW), and small systems (1–50 MW).

As shown in Fig. 6, for smooth operation of RoR systems there should be a substantial flow of water either from the melting of snowpack or rainfall. Also, there should be sufficient tilt in the river to provide sufficient speed to the flowing water. The water from river is guided down to a channel. RoR systems are better implemented with relatively constant flow rates in water bodies. If RoR systems are installed in places where the flow rate for some periods of time is relatively small and then increases dramatically, a significant amount of water loss may occur during the peak flow cycle. The water loss occurs due to the fact that RoR systems are designed

Fig. 6 Design of a typical hydroelectric RoRhydroelectric system for electricity generation

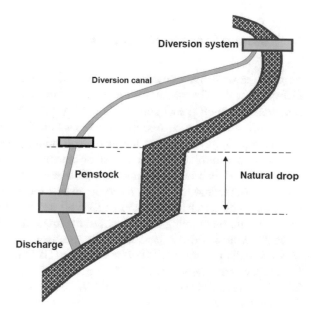

to accommodate lower flow rates and thus cannot withstand significantly higher flow rates [30].

Although RoR systems depend primarily on river flow levels to produce electricity and not on large quantities of water storage, a small dam will be enough to ensure adequate water flow through the channel. The stored water is occasionally used to make these systems more reliable by compensating the discrepancies in the water flow. The small dam is not like a ordinary hydropower reservoir, because it stores enough water for use on the same day and not for longer period [6].

There are many benefits of using RoRsystems instead of conventional dam-based hydropower systems. The construction of dams for conventional systems is time-consuming and cost-intensive. RoR systems, on the other hand, are easy to build and less expensive. These systems can be built in shorter periods of time. Many countries like Canada, which are using larger hydropower systems for electricity production, they have also developed the RoR systems in handsome number. RoR systems often prevent some of the flooding-related environmental issues, as the pondage is much smaller than the dams used by larger hydropower systems. RoR systems also produce lower amounts of greenhouse gases as compared to larger hydropower plants, gas power plants, coal power plants and oil power plants.

Despite many advantages, absence of large water storage facility means RoR plants are less reliable for production of electricity. When the water levels are reduced upstream in the winter season or due to drought, less water will pass through the channel to operate RoR system at full swing. The manipulation of river flow can adversely impact the environmental. Any diversion in the river affects the dynamics of the aquatic ecosystem, which may impact fish stocks and the overall health of the river.

6 Pumped Storage Hydropower

Pumped storage hydropower (PSH) systems are consisted of two reservoirs at different levels. The water from lower level reservoir is pumped into upper-level reservoir during off-peak hours, as shown in Fig. 7. The surplus electricity, produced during off-peak hours, is used to pump the water. The flow is subsequently reversed to produce electricity during the peak load hours. One may create two different reservoirs, one turbine, and one pump, or one may use a system running both ways. Such types of hydraulic systems are called reversible pump turbines. The idea of pumping water into the upper reservoir makes these plants net energy consumers since pumping water into the upper reservoir needs more power than that provided by PSH plant on draining water down to the lower reservoir [6].

In 2010, the worldwide installed PSH capacity was about 130 GW [31], of which 45 GW is installed in Europe, 24 GW in China, 30 GW in Japan, and 22 GW in the USA. PSH capacity increased by 24 GW during 2005–2010. The projected PSH capacity for 2050 is estimated to be about 500–600 GW. In Europe, the major contributes are Italy, Germany, France, Spain, United Kingdom, Ukraine, and Austria. At the present, a

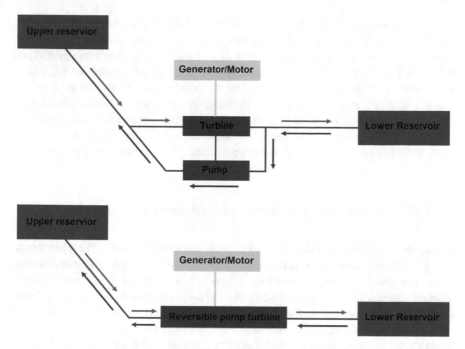

Fig. 7 An illustration of a pumped storage hydropower plant with separate pump and turbine and reversible pump-turbine

typical PSH facility in the European electricity system has a capacity of 200–300 MW with a relatively shorter storage cycle. This plant produces electricity for 4–9 h after pumping of 6–12 h. The installed capacity for electricity generation at several PSH plants is greater than that for pumping the water back into the upper reservoir. In Germany the total PSH capacity installed by 2012 was 6.8 GW for generation and 6.45 GW for pumping, respectively; and 1.9 and 1.4 GW in Switzerland. Generally, the storage capacity of the lower reservoir is kept lower than the upper reservoir, which means that the lower reservoir typically reduces the amount of energy that can be used in the storage cycles.

Europe is expected to undergo a sharp rise in installed PSH capacity in the near future. According to a market study by Ecoprog [32], in a 10-year period, about 27 GW PSH capacity would be installed in all of Europe. The major contributors will be Portugal, Spain, Germany, Switzerland, the United Kingdom, and Austria. The total capacity of under-construction projects or in licensing stage in these countries is around 18.3 GW. Considerable capacity of 3.2 GW likely to be developed in Turkey as well. New PSH facilities are being developed mostly by upgrading and extending the existing power plants with existing reservoirs. In some cases, dams are being extended to increase the storage capacity and in other cases new facilities are being built, including the building of new reservoirs.

As far as environmental impact of new PSH plants is concerned, there are two key differences in the magnitude of the environmental impact, which are creation of new reservoirs and dams and construction ofPSH plants using the existing dams and reservoirs [33]. In both cases, the construction of water tunnels, powerhouses, power grid connections, access roads have an impact on the environment. Creation of new would disturb the nature by flooding the terrestrial areas, changing the natural flow of streams, changing the land use, disrupting the river continuum, and disturbing theaquatic and terrestrial ecosystems. As discussed earlier, such impacts are similar to those that occur during the construction of dams for hydropower plants [12, 34].

7   Environment Related Issues of Hydropower Plants

Electricity use is enormously wasteful in most parts of the world. The goal should always be to increase the performance of existing energy supply and consumption before constructing new power plants. If there is a clear need for new power plants, most environmentalists support the use of wind and solar power, which is now being economically viable. Small dams, particularly in rural areas, can be a reliable and economic source of electricity. Generally, critics of big dams do not think that no new dams should be constructed. They accept that dams and other infrastructure projects can only be constructed after all relevant project information has been made public, independent experts check the arguments of project proponents about the economic, environmental, and social benefits and costs of projects, and agree that the project should be constructed with the consent of affected people. The primary cause for the worldwide opposition to major dams is the large number of people who have been evicted from their land and home to make way for reservoirs. Several million people's livelihoods still suffer as a result of the downstream impact of dams and the depletion of fisheries, polluted water, decreased quantities of water, and decreased productivity in agriculture and forests due to the loss of natural fertilizers and irrigation in seasonal floods. Dams also trigger waterborne diseases like leishmaniasis, schistosomiasis, and malaria. Environmental and climate changes influence investor choices and international financing institutions on energy projects [35]. CO_2 is the most harmful gas among greenhouse gases. Table 5 summarizes CO_2 emissions from renewable power plants in various countries, which include several tropical countries. Jungbluth et al. [36] revealed emission rates of 11 g to 13 gCO_2 per kWh from renewable power plants. To add further, Lenzen and Munksgaard [37] reported that CO_2 emission from power plants depended on load factor, lifetime, power rating, etc.

Construction of large number of dams in Brazil is deteriorating the environment in many ways [49]. Load of river sedimentation is increasing, rainfall is not following the normal cycle, flow of rivers is changing, evaporation from larger reservoirs is increasing, water supply to cities is decreasing and floods are increasing. The global impact of increasing number of dams are [10]:

Table 5 Overview of CO_2 emissions from renewable power plants in selected countries

Country	Lifetime (years)	Power rating (kW)	CO_2 per kWh (g)	Reference
Japan	20	100	123.7	[38]
USA	NA	3	NA	[39]
Denmark	20	10 × 500 8 × 500	16.5 9.7	[40]
Switzerland	20	30–800	11	[36]
Japan	NA	300	29.5	[41]
Germany	20	300/1500	250/150	[42]
Brazil	20	500	NA	[43]
Greece	20	3	104	[44]
India	20	1500	NA	[45]
Singapore	20	2.7	165	[46]
Japan	25	100	39.4	[47]
UK	20	6600	25	[48]

- Soil moisture levels will be affected by the rising temperatures due to construction of reservoirs for water storage. The air pressure and wind patterns will experience changes due to change in precipitation levels [50].
- During prolonged droughts, the reservoirs serve as a buffer in keeping with the length of the drought cycle while maintaining normal power generation capacity.
- The mean rainfall may increase with a rise in precipitation levels.
- Floodoccurence may increase due to sediment loads beyond what is expected in normal circumstances.
- Three to four times increase in average precipitation; i.e. increase in runoff.

Lehner et al. [51] studied the effect of ecological changes on Europe's hydroelectric power capacity by applying the Water-Related Assessment and Prognosis (GAP) model. The water spillover, associated atmospheric changes, and water utilization during special situations were discussed. The report revealed that the European hydroelectric power could be cut by 6% by 2070. Christensen et al. [52] assessed the impact of the atmosphere on the hydrology and water assets of the Colorado river bowl. It was revealed that precipitation is decreasing, spillover is expanding, and hydroelectric power generation is decreasing. Hydroelectric power is expected to decrease by 56% in 2039. The climate changes in different South America Megacities may raise ambient temperature from 1.7 to 6.7 °C in 2100. With a change in climate and precipitation, the temperature in central northern Brazil may also increase by 4–6 °C in 2100. In these projections, 22% reduction in rainfall in northern Brazil and 25% increase in rainfall in South America. As in any other tropical country, Brazil is highlighted as a global warming hotspot of the future and a climate-vulnerable region [53]. In a more recent report, Sorribas et al. [54] predicted the effect of climate changes on the hydrology of Amazon Basin from 2070 to 2099. As predicted, the

Table 6 CO_2 and CH_4 flow at the sediment–water interface in lakes [58, 59]

Characteristics of lakes	CO_2 emission per day (kg/km^2)	CH_4 emission per day (kg/km^2)
Eutrophic reservoirs	187.4	83.8
Oligotrophic lakes	15	3
Tropical/subtropical swamps	–	59.8
Mesotrophic lakes	114.4	33.1
Eutrophic lakes	167.2	62.4

river flow may rise by 9% in Northwest Amazon while decreasing by 15.9% in Central Amazon.

The major sources of climate change human-induced gases and trapped solar energy in the form of heat in the environment in the same way as a greenhouse [55]. CO_2 is produced under oxic and anoxic conditions in the water column and in flooded reservoir soils and sediments and is consumed by primary aquatic producers in the reservoir's euphotic zone. Nitrous oxide (N_2O) also forms along with CO_2, which causes global warming 310 times higher than CO_2. Methane (CH_4) is formed under anaerobic conditions mainly in sediments. Themethanotrophic bacteria in water and sediments, under aerobic conditions, oxidizes into CO_2. Different gases have different capacities to trap heat and impact on global warming. It is reported that global greenhouse gases have increased up to 70% since pre-industrial times between 1970 and 2004 [56]. The fossil fuel-dominated energy supply sector played central role in this huge rise in greenhouse gases. The other countable contributors are industry, population, transportation, land use and deforestation.

Both CO_2 and CH_4 gases are carbon cycle intermediates in lakes. CH_4 is produced through bacteria under anaerobic conditions while CO_2 is part of an aerobic carbon assimilation metabolism by algae and macrophytes and released as CO_2 through respiration of all species, for example, Methanobacterium and Methanococcus, especially in sediments. The methanotrophic bacteria use CO, acetic acid and H2, as end products of anaerobic fermentation to form CH_4 and are natural in aquatic systems; the production and release of methane in lakes and wetlands is a well-known cycle [57]. The strength and characteristics of the carbon cycle are determined by the lake's trophic level and nutrient-poor oligotrophic structures have low greenhouse emission levels. Table 6 shows CO_2 and CH_4 flow at the sediment–water interface in lakes. The reported data is extracted from Refs. [58] and [59].

8 Summary

Hydropower plays a significant role in meeting the global electricity demand. The developing countries are exclusively depending on hydropower, oil, and coal to deal with the growing energy crises. The emission of greenhouse gasses, disturbance in natural habitat and deforestation, however, have serious impact on the climate.

Issues of environmental degradation and climate change can have a negative impact of large-scale use of hydropower. The sustainable hydropower is called renewable energy and to manage the associated challenges, there is a dire need for proper planning and careful designing of the reservoirs and power plants. The major part of electricity worldwide is produced through coal power plants followed by gas power plants, hydropower plants, and nuclear power plants. The other sources are minor contributors to global electricity production. The electricity produced through hydropower is still small as compared to some other power sources based on solar photovoltaic panels and wind turbines. However, the major concern is the need for the vast land area to accommodate the power installations. At the same time, the construction of huge reservoirs is useful in flood regulation to avoid periodic flooding with severe impact on people in the affected areas, storage of water during periods of drought, and long-term low-cost electrical production, which is essential in many tropical and subtropical regions. Further benefits of reservoirs include recreational boating, fishing, aquaculture in net cages or tanks, water storage for intensive water-consuming industries, and irrigated agriculture production. Small dams, especially in rural areas, can be a sustainable and economic source of electricity. Generally, critics of big dams don't think that no new dams should be constructed but they argue that dams and other infrastructure can only be constructed after all relevant project information has been made public. The primary cause for the worldwide opposition to major dams is the large number of people who have been evicted from their land and home to make way for reservoirs. Several million people's livelihoods still suffer as a result of the downstream impact of dams and the depletion of fisheries, polluted water, decreased water flow, and decreased productivity in agriculture and forests due to the loss of natural fertilizers and irrigation in seasonal floods.

References

1. Bansal RC (2006) Automatic reactive-power control of isolated wind–diesel hybrid power systems. IEEE Trans Industr Electron 53(4):1116–1126
2. Bakos GC (2002) Feasibility study of a hybrid wind/hydro power-system for low-cost electricity production. Appl Energy 72(3):599–608
3. Lazarov V et al (2005) Hybrid power systems with renewable energy sources–types, structures, trends for research and development. In: Proceedings of international conference ELMA2005, Sofia, Bulgaria
4. Tørstad V, Sælen H (2018) Fairness in the climate negotiations: what explains variation in parties' expressed conceptions? Climate Policy 18(5):642–654
5. von Sperling E (2012) Hydropower in Brazil: overview of positive and negative environmental aspects. Energy Procedia 18:110–118
6. Hydropower and the environment: present context and guidelines for future action. Subtask 5 Report, Volume II: Main Report. International Energy Agency
7. da Silva RC, de Marchi Neto I, Seifert SS (2016) Electricity supply security and the future role of renewable energy sources in Brazil. Renew Sustain Energy Rev 59:328–341
8. Ramos FM et al (2009) Methane stocks in tropical hydropower reservoirs as a potential energy source. Climatic Change 93(1–2):1

9. Lumbroso D, Woolhouse G, Jones L (2015) A review of the consideration of climate change in the planning of hydropower schemes in sub-Saharan Africa. Climatic Change 133(4):621–633

10. Mukheibir P (2013) Potential consequences of projected climate change impacts on hydroelectricity generation. Climatic Change 121(1):67–78

11. Schaeffer R et al (2012) Energy sector vulnerability to climate change: a review. Energy 38(1):1–12

12. Bernai RR (2013) Renewable energy sources and climate change mitigation. Special Report of the Intergovernmental Panel on Climate Change (IPCC). JSTOR

13. Deshmukh R, Mileva A, Wu G (2018) Renewable energy alternatives to mega hydropower: a case study of Inga 3 for Southern Africa. Environ Res Lett 13(6):064020

14. Shirley R, Kammen D (2015) Energy planning and development in Malaysian Borneo: Assessing the benefits of distributed technologies versus large scale energy mega-projects. Energy Strategy Rev 8:15–29

15. Kittner N, Lill F, Kammen DM (2017) Energy storage deployment and innovation for the clean energy transition. Nature Energy 2(9):17125

16. Sokolow SH et al (2017) Water, dams, and prawns: novel ecological solutions for the control and elimination of schistosomiasis. Lancet 389:S20

17. Lele SM, Subramanian D (1988) A hydro-wood net-energy approach to hydro project design. Energy 13(4):367–381

18. Gleick PH (1992) Environmental consequences of hydroelectric development: the role of facility size and type. Energy 17(8):735–747

19. Dams WC (2000) Dams and development: a new framework for decision-making: The report of the world commission on dams. Earthscan

20. Gunkel G (2007) Contamination and eutrophication risks of a reservoir in the semi-arid zone: Reservoir Itaparica, Pernambuco, Brazil. Reservoirs and River Basins Management: Exchange of Experience from Brazil, Portugal and Germany, pp 81–95

21. Tundisi J, Matsumura-Tundisi T, Tundisi J (2008) Reservoirs and human well being: new challenges for evaluating impacts and benefits in the neotropics. Braz J Biol 68(4):1133–1135

22. McCully P (1996) Silenced rivers: the ecology and politics of large dams. Zed Books

23. Pringle CM, Freeman MC, Freeman BJ (2000) Regional effects of hydrologic alterations on riverine macrobiota in the new world: tropical-temperate comparisons: the massive scope of large dams and other hydrologic modifications in the temperate New World has resulted in distinct regional trends of biotic impoverishment. While neotropical rivers have fewer dams and limited data upon which to make regional generalizations, they are ecologically vulnerable to increasing hydropower development and biotic patterns are emerging. BioScience 50(9):807–823

24. Fearnside PM (1995) Hydroelectric dams in the Brazilian Amazon as sources of 'greenhouse' gases. Environ Conserv 22(1):7–19

25. Pringle C, Triska FJ (2000) Emergent biological patterns and surface-subsurface interactions at landscape scales

26. Goldemberg J (2000) World Energy Assessment: Energy and the challenge of sustainability. United Nations Development Programme, New York

27. Nóbrega M et al (2011) Uncertainty in climate change impacts on water resources in the Rio Grande Basin, Brazil. Hydrol Earth Syst Sci 15(2):585

28. Barros N et al (2011) Carbon emission from hydroelectric reservoirs linked to reservoir age and latitude. Nat Geosci 4(9):593–596

29. Steinhurst W, Knight P, Schultz M (2012) Hydropower greenhouse gas emissions. Conserv Law Foundation 24:6

30. Pérez-Díaz J et al (2010) A control system for low-head diversion run-of-river small hydro plants with pressure conduits considering the tailwater level variation. In: Proceedings of the international conference on renewable energy and power quality (ICREPQ 2010)

31. Deane JP, Gallachoir BP, McKeogh EJ (2010) Techno-economic review of existing and new pumped hydro energy storage plant. Renew Sustain Energy Rev 12:1293–1302

32. Ecoprog (2011) The European market for pumped-storage power plants, Ecoprog, Cologne, https://www.ecoprog.com/en/publications/energy-industry/pumpedstorage-power-plants.htm
33. Killingtveit Å. Hydropower (Hydraulic Energy)
34. Statistics, R.C., International Renewable Energy Agency (IRENA). 2016, ed.
35. Bauen A (2006) Future energy sources and systems—acting on climate change and energy security. J Power Sources 157(2):893–901
36. Jungbluth N et al (2005) Life cycle assessment for emerging technologies: case studies for photovoltaic and wind power (11 pp). Int J Life Cycle Assess 10(1):24–34
37. Lenzen M, Munksgaard J (2002) Energy and CO_2 life-cycle analyses of wind turbines—review and applications. Renew Energy 26(3):339–362
38. Uchiyama Y (1996) Life cycle analysis of photovoltaic cell and wind power plants. Assessment of greenhouse gas emissions from the full energy chain of solar and wind power and other energy sources. IAEA Working material, Vienna (Austria)
39. Haack BN (1981) Net energy analysis of small wind energy conversion systems. Appl Energy 9(3):193–200
40. Schleisner L (2000) Life cycle assessment of a wind farm and related externalities. Renew Energy 20(3):279–288
41. Hondo H (2005) Life cycle GHG emission analysis of power generation systems: Japanese case. Energy 30(11–12):2042–2056
42. Schaefer H, Hagedorn G (1992) Hidden energy and correlated environmental characteristics of PV power generation. Renew Energy 2(2):159–166
43. Lenzen M, Wachsmann U (2004) Wind turbines in Brazil and Germany: an example of geographical variability in life-cycle assessment. Appl Energy 77(2):119–130
44. Tripanagnostopoulos Y et al (2005) Energy, cost and LCA results of PV and hybrid PV/T solar systems. Prog Photovoltaics Res Appl 13(3):235–250
45. Gürzenich D et al (1999) Cumulative energy demand for selected renewable energy technologies. Int J Life Cycle Assess 4(3):143–149
46. Kannan R et al (2006) Life cycle assessment study of solar PV systems: an example of a 2.7 kWp distributed solar PV system in Singapore. Solar Energy 80(5):555–563
47. Nomura N et al (2001) Life-cycle emission of oxidic gases from power-generation systems. Appl Energy 68(2):215–227
48. Proops JL et al (1996) The lifetime pollution implications of various types of electricity generation. an input analysis. Energy Policy 24(3):229–237
49. Prado FA Jr et al (2016) How much is enough? An integrated examination of energy security, economic growth and climate change related to hydropower expansion in Brazil. Renew Sustain Energy Rev 53:1132–1136
50. da Silva Soito JL, Freitas MAV (2011) Amazon and the expansion of hydropower in Brazil: vulnerability, impacts and possibilities for adaptation to global climate change. Renew Sustain Energy Rev 15(6):3165–3177
51. Lehner B, Czisch G, Vassolo S (2005) The impact of global change on the hydropower potential of Europe: a model-based analysis. Energy Policy 33(7):839–855
52. Christensen NS et al (2004) The effects of climate change on the hydrology and water resources of the Colorado River basin. Climatic Change 62(1–3):337–363
53. Field CB (2014) Climate change 2014–Impacts, adaptation and vulnerability: regional aspects. Cambridge University Press, Cambridge
54. Sorribas MV et al (2016) Projections of climate change effects on discharge and inundation in the Amazon basin. Climatic Change 136(3–4):555–570
55. Matten SR, Frederick RJ, Reynolds AH (2012) United States Environmental Protection Agency insect resistance management programs for plant-incorporated protectants and use of simulation modeling. In: Regulation of agricultural biotechnology: the United States and Canada. Springer, Berlin, pp 175–267
56. Change OC (2007) Intergovernmental panel on climate change. World Meteorological Organization

57. Casper P et al (2000) Fluxes of methane and carbon dioxide from a small productive lake to the atmosphere. Biogeochemistry 49(1):1–19
58. Adams DD (2005) Diffuse flux of greenhouse gases—methane and carbon dioxide—at the sediment-water interface of some lakes and reservoirs of the world. In: Greenhouse gas emissions—fluxes and processes. Springer, Berlin, pp 129–153
59. Blais A-M, Lorrain S, Tremblay A (2005) Greenhouse gas fluxes (CO_2, CH_4 and N_2O) in forests and wetlands of boreal, temperate and tropical regions, in Greenhouse gas emissions—fluxes and processes. Springer, Berlin, pp 87–127

Modeling Solar Radiation in Peninsular Malaysia Using ARIMA Model

Mohd Tahir Ismail, Nur Zulaika Abu Shah, and Samsul Ariffin Abdul Karim ⓘ

Abstract The objective of this chapter is to build the ARIMA model and forecast the in-sample and out-sample daily solar radiation data in Peninsular Malaysia. Moreover, the study also investigates the stationarity, reliability, accuracy, and performance of the model. This study involves 12 states, but Perlis's data are removed because Perlis's data have the same value as Kedah. The study utilizes the Box and Jenkins methodology to develop the best model for each state. Based on the three stages of the Box and Jenkins methodology, each state can be represented by the best ARIMA model. All the ARIMA models also produced smaller error values which indicate the fitted or forecasted values follow the same trend as the actual data.

Keywords Solar radiation · ARIMA · Modeling · Forecasting

1 Introduction

The rising in global electricity consumption has led to a declining amount of fossil fuels such as coal, oil, and natural gas, which are the primary resource to generate electricity. Paul Ehrlich, in his study, once remarked that "The world will run out of oil in 2030, and other fossil fuels in 2050" Blumsack [1]. Therefore, technological advancement around the world has led to the inventions of many renewable energy resources. Solar energy is one of the renewable energy resources that is continuously

M. T. Ismail · N. Z. A. Shah
School of Mathematical Sciences, Universiti Sains Malaysia, 11800 USM, Minden, Pulau Pinang, Malaysia
e-mail: m.tahir@usm.my

N. Z. A. Shah
e-mail: zulaika12_06@yahoo.com

S. A. A. Karim (✉)
Fundamental and Applied Sciences Department and Centre for Systems Engineering (CSE), Institute of Autonomous System, Universiti Teknologi PETRONAS, 32610 Seri Iskandar, Perak Darul Ridzuan, Malaysia
e-mail: samsul_ariffin@utp.edu.my

© The Author(s), under exclusive license to Springer Nature Singapore Pte Ltd. 2021   53
S. A. Sulaiman (ed.), *Clean Energy Opportunities in Tropical Countries*,
Green Energy and Technology, https://doi.org/10.1007/978-981-15-9140-2_3

replenished and will never run out. Hence, forecasting solar radiation is essential to ensure that we have enough energy to generate electricity.

According to Abd. Aziz et al. [2], one of the countries with many energy resources, including fossil fuel and renewable energy, is Malaysia. Malaysia is a country that receives a generous amount of average daily solar radiation on average about 4500 kW/m^2. To encourage the installation of photovoltaic (PV) solar power system, the Government of Malaysia has introduced the Malaysian Building Integrated Photovoltaic Project (MBIPV) that took around five years. In order to make sure that the solar radiation is enough for 24 h demand, solar energy is stored to generate electricity at night. Thus, for the long term, a need for a forecasting process is unavoidable to maintain the demand for electricity.

Currently, the primary resource to generate electricity in Malaysia is fossil fuels. According to Rehman et al. [3], global warming is caused by fossil fuel energy usage that leads to the emission of carbon dioxide. Global warming affects the environment and disturbs the ecosystem, which consequently leads to animal extinctions. Laslett et al. [4] stated that an energy system called the renewable energy system with low greenhouse gas emission is required to prevent the calamitous climate change. As mentioned previously, Malaysia is blessed with many potential types of resources. Some of the examples of the potential resources are solar energy, wind power, biogas, biomass, and battery energy storage system. Hence, renewable solar energy should be implemented in many places in Malaysia to reduce the consumption of fossil fuels. Moreover, forecasting the amount of solar radiation is needed to maintain the future request for electricity, especially in the manufacturing companies that consume much electricity.

There are many models used to predict solar radiation. The most common model is the autoregressive—moving-average (ARMA) model since it is the simplest model of regression. If the data are not stationary, the suitable will be the autoregressive integrated moving average (ARIMA). The two researchers, Box and Jenkins in 1970, introduced a method called Box–Jenkins methodology in their book "Time Series Analysis: Forecasting and Control," and the typical model that they had discussed was ARIMA Lazim [5]. Some researches combined the other models into the ARMA process to form a new model, such as the hybrid model, to get a better forecasting result and performance with the lowest error. Nevertheless, the main objective of the present study is to forecast daily solar radiation in ten locations from ten states in Peninsular Malaysia by using the ARIMA model. The forecasting performance for in-sample and out-sample for each location is evaluated based on some error measurements.

2 Literature Review

One of the early studies using the time series model on solar radiation forecasting was by Alghoul et al. [6]. Their research took place on four stations, and they applied double differencing to remove the deterministic component. Then, they estimated the ARIMA model based on the finding of the autocorrelation function (ACF) and partial autocorrelation function (PACF) test. They found that the original data fitted to the autoregressive (AR) model. All test statistics of AR residual such as Ljung–Box and McLeod-Li showed that the value was less than chi-squared for 40 degrees of freedom for all station. Thus, the parameter AR (p) was the representation of data in this research.

The next work, performed by Perdomo et al. [7], aimed to forecast and identify the behavior of the average of daily global solar radiation in Bogota. As there were some missing values in the dataset, they used an imputation method based on a linear trend to recover the data. After that, Dickey–Fuller test showed that the series was non-stationary. Then, based on the Box and Jenkins methodology, they modeled the data using the ARIMA model. However, the results of ACF and PACF showed that the residuals of solar radiation average were out of the confidence limit. To overcome this problem, they used the Box-Cox transformation to find the model that fulfilled the stationarity conditions of the ARIMA model. As a result, the ARIMA (1, 0, 0) model fitted better with the original data and would be more useful to forecast for a long period.

Besides that, Ji and Chee [8] had discussed a novel hybrid model that combines ARMA and time delay neural network (TDNN) models to forecast the hourly solar radiation time series. In their article, both ARMA and TDNN were forecasted independently before combining them. The non-stationary trend was removed by using the ARMA model, while TDNN was conducted to forecast solar radiation. Later, a novel hybrid model was performed on the time series. ARMA showed that the best model was ARMA (1, 1) and less sensitive than TDNN. The hybrid model proved that the prediction performance was good, stable, and accurate.

However, Wu and Chan [9] proposed and compared two models to be adopted in forecasting short term solar radiation, which were ARIMA and TDNN. They claimed that the daily solar radiation of Singapore was not stable due to the weather condition. Therefore, they took the monthly average solar radiation data. By using the ARIMA model, they applied the first differencing to make the time series stationary. Then, they conducted proper forecasting by using the ARIMA model. Akaike information criterion (AIC) was then used to identify the best model. At the same time, they carried out the TDNN procedure to predict the trend of the monthly solar radiation. As a result, the forecasting performance of TDNN appeared to be better than ARIMA in every month.

On the other hand, research by Yang et al. [10] employed three different methods to forecast global horizontal irradiance (GHI). The data are collected from two weather stations in the United States of America (USA), Miami, and Orlando. Mean bias error (MBE) and root mean square error (RMSE) were employed to evaluate the approach. The result of the best approach was the third approach with the lowest RMSE of 29.73 for Miami and 32.80 for Orlando, followed by the second approach. The forecasting result in the third approach shows a small difference to the original measurements, and the accuracy improved. The first and second approaches did not produce any bell-shaped curve, therefore concluded that the third approach was the best.

Other than that, the research by Lauret et al. [11] combined two types of forecasting models, which are ARMA and neural network (NN). The combination was called Model Committee. By Bayesian model averaging, the forecasting model was evaluated based on the posterior model probabilities (PMP). This approach showed that there was an improvement in the accuracy of out-samples forecasting. The test error for Bayesian information criterion (BIC), RMSE, and MBE displayed a small value of error. Ergo, the Model Committee, showed a good performance.

A paper by Colak et al. [12] applied ARMA and ARIMA processes to ensure the stationarity of the time series data. Moreover, they also tested for the goodness-of-fit determination by log-likelihood function (LLF). The parameters would produce the observed data when the LLF value was closer to zero. The models that achieved the optimal LLF were ARMA (1, 2) and ARIMA (2, 2, 2). They employed mean absolute error (MAE) and mean absolute percentage error (MAPE) to measure the accuracy of the model. In the multi-period predictions, a one-hour (1-h) time series model was created for the solar radiation parameter. In all period tests, ARIMA (2, 2, 2) model outperformed ARMA (1, 2) model in terms of the one-period prediction of the solar radiation parameter. ARIMA showed the lowest value of MAE and MAPE hence had better performance compared to the rest.

Recently in a study conducted by Mbaye et al. [13], they predicted global solar potential in the next 24 h based on the previous data. The best model-based AIC test displayed that the ARMA (29, 0) was the best representation of data. The p-value test of white noise was 26% for a significant level of 5%. The performance of the model is verified using the RMSE, MAE, MBE, and coefficient of determination. It showed that ARMA (29, 0) was reliable. Then, Alsharif et al. [14] performed a study with the objectives of the research that were to forecast the daily and monthly solar radiation of Seoul by using the ARIMA model and built a seasonal autoregression integrated moving average (SARIMA) model. The data were collected hourly for over 37 years from Korea Meteorology Administration (KMA) and converted into daily and monthly data. The outcome for both AR and MA part showed that the ARIMA (1, 1, 2) was used to represent the daily solar radiation and seasonal ARIMA (4, 1, 1) of 12 lags were used to represent the monthly solar radiation.

Last but not least, Atique et al. [15] predicted the daily total solar energy generation, both seasonal and non-seasonal variations in Lubbock, Texas, in the USA. The differencing process is applied on a non-stationary time series to make it stationary. The data were estimated graphically by plotting ACF and PACF, and it was confirmed by using the augmented Dickey–Fuller (ADF) test. The estimation failed at the original time series (level) during the ACF plot and succeeded at first differencing after the ADF test with p-value of 0.01. The trend and seasonality were also detected during the test. Thus, a performance comparison of AIC and BIC on the seasonal and non-seasonal model was performed, and it turned out that the seasonal model performed better. Then, another performance comparison involving ARIMA models was conducted and yielded the model ARIMA (0, 1, 2) (1, 0, 1), seasonality lag of 30 with the lowest AIC. The residual analysis showed no significant correlation and followed the normal distribution. Finally, the data were forecasted by using the resulted model with a MAPE of 17.70%.

Compared to all the works of literature above, the current work differs in some ways. Firstly, the research area covers all states of Peninsular Malaysia, which constitute of twelve states. This work also involves two types of forecasting, which are in-sample forecasting and out-sample forecasting using the full step of Box–Jenkins Methodology. Based on the previous research, many researchers used MAPE in order to check the error of forecasting. According to Lazim [5], MAPE tends to produce a large forecast error when the actual data are close to zero. The data in the present work involve many values that are close to zero. For that reason, this project utilizes the Theil's inequality coefficient as one of the error evaluations of the time series. Henceforth, the accuracy and performance are checked by using the RMSE, MAE, and Theil's inequality coefficient as well as the graph comparison between the fitted value and actual value for both in-sample and out-sample data.

3 Methodology

This section discusses the data preparation and the ARIMA modeling based on Box and Jenkins methodology intensely.

3.1 Data Preparation

In this work, the daily solar radiation data are collected from Power Data Access Viewer provided by the National Aeronautics and Space Administration (NASA) Web site for almost ten years starting from January 1, 2009, until August 31, 2019 (3895 observations for one state). The solar radiation data are taken based on All-Sky Insolation Incidents on a Horizontal Surface (kWh/m^2). All-Sky Insolation Incidents on a Horizontal Surface are the average daily amount of the total solar radiation incident on a horizontal surface of the earth NASA [16]. It means that the data are not taken based on an apparent sky incident only but cover all incidents that happen on the sky that minimized the amount of solar radiation from reaching the surface of the earth, such as the presence of clouds. It is important to note that the data vary based on the condition of the sky every day.

This work also focuses on solar radiation in Peninsular Malaysia. Peninsular Malaysia has 12 states which are Perlis (Pls), Pulau Pinang (Png), Kedah (Ked), Perak (Prk), Selangor (Sel), Kuala Lumpur (Kl), Negeri Sembilan (NSe), Melaka (Mlk), Johor (Jhr), Pahang (Phg), Terengganu (Tgu), and Kelantan (Kln). Notice that this work involves only Peninsular Malaysia because of East Malaysia (Sabah and Sarawak) sunrise 30 min ahead of West Malaysia (Peninsular Malaysia) Aziz et al. [17, 18] (refer to Fig. 1). Moreover, Sabah and Sarawak received less amount of solar radiation compared with the amount of solar radiation received by Peninsular Malaysia with a 30-min gap. In order to standardize the time zone, we did not consider East Malaysia in this work. Then, we need to identify the district with the highest

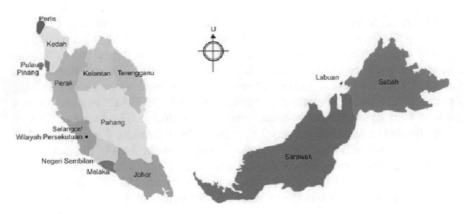

Fig. 1 States of Malaysia

solar radiation data. For this purpose, the data from all districts in each state are listed in Microsoft Excel. The district with the highest amount of solar radiation from each state is chosen based on the mean of 10-year data. Data that are provided by NASA might have a value of -999, which represents the presence of missing value. In order to fill in the missing value in the series, the data are interpolated by using linear interpolation.

However, we found that Perlis yields the same data as Kedah since they are geographically close to each other. Furthermore, the size of Perlis is outstandingly small; it is the smallest state in the country. It causes some redundancy in the collection of data, which we then ultimately decided that the data from Perlis had to be eliminated. Henceforth, by eliminating Perlis, only 11 states were observed throughout this present study (3895 observations for one state).

3.2 Basic Models of the Box–Jenkins Methodology

Box–Jenkins methodology is a three-step strategy to identify, fit, and check the model of time series data to achieve the best forecasting model. The methodology was developed by George E. P. Box and Gwilym M. Jenkins in 1976 [5]. The methodology consists of three basic models. The models are autoregressive (AR), moving average (MA), and mixed autoregressive moving average (ARMA). The choices of these three forecasting models depend on the behavior and parameter p and q estimation of the time series that we want to forecast. The parameter estimation of the time series will determine the model type, whether AR, MA, or ARMA.

According to Box et al. [19], autoregressive (AR) model is a procedure of managing the statistical time series data. It predicts the future value of time series based on past values. The current value of a series, y_t, is denoted as a function of its past value plus an error term. It means the dependent variable, y_t, has a specific lagged value used as a predictor variable. AR model can be written mathematically as:

$$y_t = c + \emptyset_1 y_{t-1} + \emptyset_2 y_{t-2} + \cdots + \emptyset_p y_{t-p} + \varepsilon_t, \tag{1}$$

where c is a constant, $\emptyset_j (j = 1, 2, \ldots, p)$ are constant AR parameters to be estimated, ε_t is an error term which assumed with mean zero and variance, $\emptyset_\varepsilon^2$, y_t is the current value of the series, and y_{t-p} is the pth order or parameter of lagged current value. There are many variations of parameters in the AR process. However, there are two most expected value of p, which are AR (1) and AR (2). AR parameter, p beyond two, is a rarely found case.

Meanwhile, in the Box–Jenkins methodology, moving average (MA) model is not just a mean of actual data, y_t. MA model in the Box–Jenkins context is defined as an average function of the error term [5]. It connects the current time series with the random error of the previous period rather than the actual time series. The formulation can be written as:

$$y_t = c + \varepsilon_t - \emptyset_1 \varepsilon_{t-1} - \emptyset_2 \varepsilon_{t-2} - \cdots - \emptyset_q \varepsilon_{t-q}. \tag{2}$$

Here, c is a constant, $\emptyset_j$, $(j = 1, 2, 3, \ldots, q)$ are the moving average parameters to be estimated, ε_{t-q} is the error assumed to be independently distributed over time, and y_t is the current value of the series. In this model, the qth order or parameter refers to the number of the lagged time period. The parameter of MA varies with different time series data used in research.

Nevertheless, the third basic of the Box–Jenkins model is the ARMA model. Lazim [5] defined that the ARMA model is a combination of AR model and MA model, which make a time series to fit better under the assumption of stationarity. Basically, the series y_t is assumed to be stationary where no differencing needed and ARMA model is written as:

$$y_t = c + \emptyset_1 y_{t-1} + \emptyset_2 y_{t-2} + \cdots + \emptyset_p y_{t-p}$$
$$- \emptyset_1 \varepsilon_{t-1} - \emptyset_2 \varepsilon_{t-2} - \cdots - \emptyset_q \varepsilon_{t-q} + \varepsilon_t, \tag{3}$$

where c is a constant, y_t is the current value, y_{t-p} is the pth order or parameter of lagged current value, $\emptyset_j (j = 1, 2, \ldots, p)$ are constant AR parameters to be estimated, $\emptyset_j$, $(j = 1, 2, 3, \ldots, q)$ are the moving average parameters to be estimated, ε_{t-q}, ε_{t-q} is the error assumed to be independently distributed over time, and ε_t is an error term. ARMA model can be expressed as ARMA (p, q) with p as AR parameter and q as MA parameter. However, there are many time series that are non-stationary where the basics model has to be reconstructed. An ARIMA model, the assumption of stationarity in the basic ARMA model, is rejected. Time series data have to be stationary before proceeding with the forecasting process. To deal with non-stationary data, Lazim [5] stated that Box and Jenkins had introduced parameter

d as the number of differencing in order to difference the data and make the model to be ARIMA (p, d, q).

ARIMA is a reconstructed model based on ARMA. The letter "I" referred to the word "integrated," which means that it merges with the ARMA model to form a new model called ARIMA. If the assumption of stationarity is rejected at the beginning of the research, the ARIMA process takes place instead of ARMA. Firstly, the data need to be differencing in order to achieve stationarity. The parameter d is introduced as the number of times the original data, and y_t have to be differentiated to obtain stationarity. Let d parameter be substituted by assuming the data is non-stationary. So, the equation of the new time series, w_t, will be formulated by introducing B as a backshift operator. A Backshift operator is an operator to produce the previous element of time series. The equation of backshift operator, B, and time series, w_t, are as follows:

$$B = y_t - y_{t-1} \tag{4}$$

$$w_t = (1 - B)^d y_t, \tag{5}$$

where w_t is a new time series after a certain number of differencing conducted, d is number of differencing, y_t is the current value, and B is the backshift operator. Next, Eqs. (4) and (5) are integrated into the ARMA formula and assume $c = 0$. ARIMA's general equation formulated as:

$$\left(1 - \emptyset_1 B + \emptyset_2 B^2 + \cdots + \emptyset_p B^p\right) w_t$$
$$= \left(1 + \emptyset_1 B - \emptyset_2 B^2 - \cdots - \emptyset_q B^q\right) \varepsilon_t.$$

Here, $\emptyset_j (j = 1, 2, \ldots, p)$ are constant AR parameters to be estimated, $0_j, (j = 1, 2, 3, \ldots, q)$ are the moving average parameters to be estimated, is backshift operator, and ε_t is an error term. The representation of the model is given by the parameter of ARIMA (p, d, q), where p is the value of the AR parameter, d is the number of differences, and q is the value of the MA parameter.

Box and Jenkins proposed a fundamental strategy or process of forecasting the time series, which consists of three main stages of modeling Lazim [5], as portrayed in Fig. 2. These are model identification, model estimation and validation, and model application. In Stage 1, model identification starts with analyzing the stationarity by using ACF and PACF for all districts. After all requirements of Stage 1 are fulfilled, the process is continued with model estimation and model validation. The Akaike information criteria (AIC) and Schwartz Bayesian information criterion (SBIC) are used to estimate possible ARMA (p, q) or ARIMA (p, d, q) model. The final stage is the model applying. Recall that the parameter d is estimated in Stage 1 for the ARIMA model. The best model is selected from the list of an estimated model in Stage 2. Stage 3 is a platform to apply the best ARMA or ARIMA model for forecasting solar radiation in 12 districts.

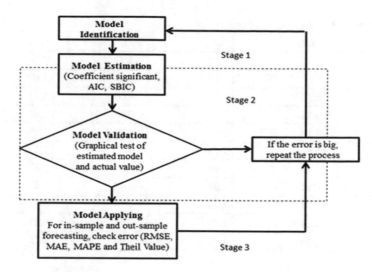

Fig. 2 Stages in ARIMA modeling

There are two types of data, which are in-sample and out-sample data. In-sample data are the sample of the current time series data used in the research, while out-sample data are the data outside the range of the current time series. Forecasting in-sample data is using the sample time series to forecast the data inside the sample. Forecasting time series outside the range of the sample is called forecasting out-sample data. Static forecasting is applied during forecasting. Static forecasting used all the value of actual data or variable as a sample to forecast time series for all districts. Thus, the forecasting result is evaluated by using root mean square error (RMSE), mean absolute error (MAE), and Thiel's inequality coefficient. A small value of RMSE, MAE, and Thiel's inequality coefficient indicates the higher accuracy of forecasting and displays a better performance in the ARMA or ARIMA model.

4 Result and Discussion

The data analyses start with the description of the data. From Table 1, it is shown that Melaka (Mlk) shows the lowest mean, while Kelantan (Kln) shows the highest mean. The lowest measure of dispersion or standard deviation (SD) of solar radiation is at Kuala Lumpur (Kl), and the highest SD is at Kelantan (Kln). It is also observed that the skewness (Skew) of the entire state shows negative values. Recall that negative skewness happens when a long tail extends to the left side of the distribution. This finding is supported by the shape of the distribution for all states shown in Fig. 3.

Kurtosis (Kurt) is the study on the peakedness of the distribution. The normal distribution kurtosis value is 3. Nevertheless, the lowest kurtosis among the district is 3.23 at Terengganu (Trg), and the highest kurtosis is 3.93 at Pahang (Phg). The

Table 1 Descriptive statistics of all states

State	Mean	SD	Skew	Kurt
Ked	4.87	1.23	−0.74	3.31
Png	5.09	1.13	−0.85	3.50
Prk	5.19	1.11	−0.86	3.47
Sel	5.11	1.19	−0.83	3.35
Kl	4.93	1.02	−0.79	3.53
NSe	5.01	1.13	−0.88	3.43
Mlk	4.72	1.10	−0.86	3.58
Jhr	4.91	1.21	−0.90	3.42
Phg	4.98	1.29	−1.16	3.93
Tgu	5.29	1.47	−0.96	3.23
Kln	5.36	1.47	−1.04	3.45

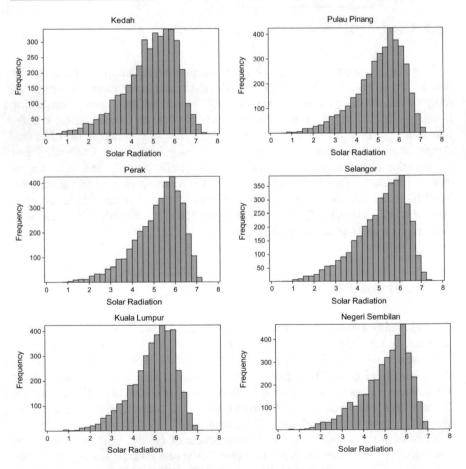

Fig. 3 Histogram shows the distribution of all states

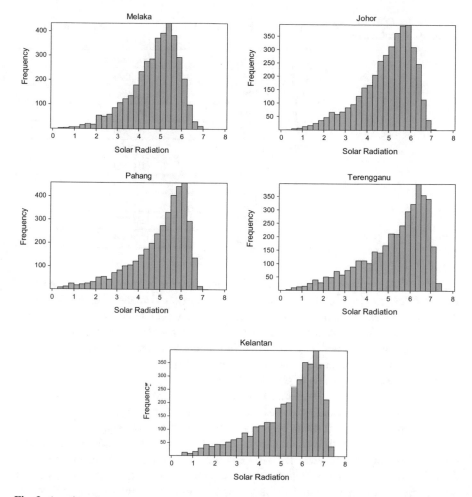

Fig. 3 (continued)

result proves that the kurtosis type is leptokurtic, which has a longer tail with kurtosis value more than 3. Again the finding is also supported by the shape of the histogram in Fig. 2.

The first step of model identification is to identify the value of d in the time series data. The value d refers to the number of differencing before the data become stationary. Noted the original time series data are the level data. If the original data are stationary at the level difference ($d = 0$), the ARMA model is being implemented. The ADF test can determine the stationarity. Table 2 shows the result of the ADF test on the level and first differencing time series data.

As displayed in Table 2, the p-value for the first difference is all significant and less than 0.05 (5% significant level). With the assumption of time series, y_t is non-stationary and having a unit root, the time series is tested on level difference.

Table 2 The ADF unit root
test results for all states

State	p-value original (level) data	p-value at first difference data
Ked	0.2132	0.0000
Png	0.2524	0.0000
Prk	0.2351	0.0000
Sel	0.2659	0.0000
Kl	0.2749	0.0000
NSe	0.3144	0.0000
Mlk	0.2645	0.0000
Jhr	0.2479	0.0000
Phg	0.1722	0.0000
Tgu	0.1602	0.0000
Kln	0.1663	0.0000

The result appears that all states are non-stationary time series since the p-value of the ADF test exceeds 0.05. Therefore, the assumption of non-stationary accepted, and a unit root exists in the original (level) time series data. However, the time series becomes stationary at the first difference ($d = 1$) with p-value of less than 0.05. Thus, the assumption is rejected, and the first differencing time series becomes stationary. Hence, the assumption of non-stationary is rejected, and the time series is stationary at first difference. The value of d is equal to one ($d = 1$).

In the second step, the baseline model used is ARIMA (1, 1, 1). In this stage, different numbers of p and q are used by increasing and decreasing the value of p and q simultaneously. The values p and q are estimated by using AIC and SBIC procedures. The first differencing time series data are used in this stage. Before choosing the lowest AIC and SBIC, the significant coefficient must be checked first for all models. If p and q values are not significant, the estimated model will be automatically rejected. Table 3 displays the best model for forecasting time series for all the states. From Table 3, the best model used for forecasting the daily solar radiation is ARIMA (1, 1, 2) used by seven states, followed by ARIMA (2, 1, 1) used by Ked and Sel while Jhr and Kln followed ARIMA (1, 1, 3).

After choosing the best model for all districts, an appropriate validation must be done in order to make sure that the best model is suitable for forecasting. The best model is validated by using the graph comparison between the actual time series data (first differencing time series) and estimated time series, based on the best ARIMA model in all states. Figure 4 presents a graph combination of actual time series data (before estimation) and estimated time series based on the best ARIMA model (after estimation) for all states. Based on Fig. 4, the blue line refers to the original time series, which is the first differencing data. The red line represents the time series after estimation of the best model.

Notice that the time series of the best ARIMA model in 11 states fit and fluctuate within the time series of actual data (first differencing time series data). Hence, it

Table 3 List of the best ARIMA model for all states

Number	State	Best model
1	Ked	ARIMA (2, 1, 1)
2	Png	ARIMA (1, 1, 2)
3	Prk	ARIMA (1, 1, 2)
4	Sel	ARIMA (2, 1, 1)
5	Kl	ARIMA (1, 1, 2)
6	NSe	ARIMA (1, 1, 2)
7	Mlk	ARIMA (1, 1, 2)
8	Jhr	ARIMA (1, 1, 3)
9	Phg	ARIMA (1, 1, 2)
10	Tgu	ARIMA (1, 1, 2)
11	Kln	ARIMA (1, 1, 3)

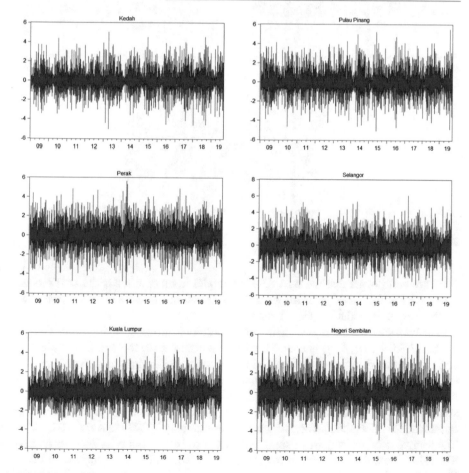

Fig. 4 Graph of actual time series (blue line) and estimated time series (red line)

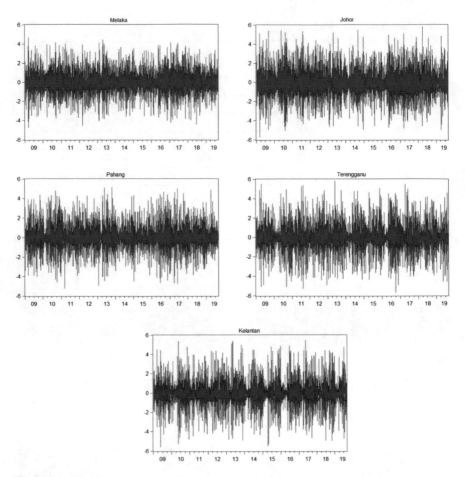

Fig. 4 (continued)

validates and proves that the best ARIMA model in all districts is the best model. The model will be used later in Stage 3 to forecast in-sample and out-sample data on the daily solar radiation for all districts.

The final stage of the Box–Jenkins methodology is model applying. In this stage, the best ARIMA (p, d, q) chosen from Stage 2 is applied to forecast in-sample data and out-sample data for all districts. The forecasting performances are evaluated by using RMSE, MAE, and Theil's inequality coefficient based on a static method of forecasting. The sample data used for forecasting are the first difference data. The smaller value of forecasting errors indicates better performance in forecasting solar radiation. The in-sample forecasting starts from January 1, 2009 to August 31, 2019. The sample used to forecast in-sample data is also from January 1, 2009 to August 31, 2019. Thus, the in-sample forecasting takes place within the sample data itself. Figure 3 obtained from Stage 2 is the result of in-sample forecasting for all districts.

Table 4 In-sample forecasting error

Number	Districts	RSME	MAE	Theil's coefficient
1	Ked	0.98	0.75	0.55
2	Png	1.00	0.77	0.51
3	Prk	1.02	0.80	0.48
4	Sel	1.11	0.88	0.48
5	Kl	0.93	0.73	0.49
6	NSe	1.05	0.82	0.48
7	Mlk	1.00	0.78	0.49
8	Jhr	1.15	0.90	0.44
9	Phg	1.12	0.85	0.53
10	Tgu	1.22	0.93	0.55
11	Kln	1.20	0.90	0.56

The error in terms of RMSE, MAE, and Theil's coefficient is calculated and listed in Table 4.

From Table 4, the values of RMSE and MAE are small with a value of less than 1.23 and 0.94, respectively. The Theil's inequality coefficient of the in-sample forecasting is small, with a value of fewer than 0.56 errors. The entire states show a small value of error, which indicates that the accuracy and performance of the best model estimated in Stage 2 is the best. The smallest RMSE is detected at Kuala Lumpur with the value of 0.93 while the highest RMSE is detected at Terengganu with the value of 1.22. On the other hand, the smallest MAE also is at Kuala Lumpur with 0.73 while the highest MAE is located at Terengganu with 0.93. The highest Theil's coefficient is located at Kelantan, with 0.56 and the lowest located at Johor with 0.44. Figure 3 displays the result of the in-sample forecasting. The red line represents the time series of forecasting of the ARIMA model, and the blue one is the actual data. The fluctuation for forecasted data fits within the actual data. The forecasted data in red line follow the same trend as the actual data. It means that the best ARIMA model chosen by estimation in all district displays an excellent performance with a small error detected.

For out-sample forecasting, the sample data from January 1, 2009 to December 31, 2018 are used for forecasting. The remaining data which are on January 1, 2019 until August 31, 2019 are reserved for forecasting. The RMSE, MAE, and Theil's inequality coefficient are listed in Table 5. From Table 5, a similar pattern of error value is detected when it is compared with in-sample forecasting in Table 4. The values of RMSE and MAE are less than 1.07 and 0.84, respectively.

The Theil's inequality coefficient obtained by out-sample forecasting is less than 0.56. All district generates a small value of error. The smallest RMSE is detected at Melaka with 0.83 while the highest RMSE is detected at Terengganu with 1.06. The smallest MAE is at Kuala Lumpur with 0.63 and Selangor with 0.83 is the highest MAE. The largest Theil's coefficient is identified at Terengganu with 0.55, and the

Table 5 Out-sample forecasting error

Number	Districts	RSME	MAE	Theil's coefficient
1	Ked	0.97	0.72	0.55
2	Png	1.00	0.74	0.50
3	Prk	0.90	0.70	0.49
4	Sel	1.03	0.83	0.48
5	Kl	0.85	0.63	0.50
6	NSe	0.85	0.64	0.49
7	Mlk	0.83	0.65	0.50
8	Jhr	0.99	0.77	0.46
9	Phg	0.83	0.65	0.53
10	Tgu	1.06	0.76	0.55
11	Kln	1.06	0.80	0.55

smallest Theil's coefficient is identified at Johor with 0.46. For out-sample time series graph, the forecasted data are compared with actual (first differencing) data for all states.

Figure 5 illustrates the performance results of the out-sample forecasting. The red line represents the fitted value of forecasting, and the blue one is the actual data. For all states, the forecasted data in the red line follow the same trend and fit within the actual data in the blue line. It is supported by the small value of RMSE, MAE, and Theil's inequality coefficient.

Hence, the values of RMSE, MAE, and Theil's inequality coefficient are small in both in-sample and out-sample forecasting. They also show excellent performance with the best model fit within actual data in Stage 2 for all states. The stationarity, reliability, accuracy, and performance, as mentioned in objectives, are achieved. For that reason, ARIMA modeling is concluded to be successful in both types of forecasting.

5 Summary

This chapter aims to investigate the performance of the ARIMA model for modeling solar radiation data in 11 states of peninsular Malaysia. Preliminary analysis indicates all states show skew- and leptokurtic-type distribution. Then, in the model identification, ADF test found that the time series appears to be non-stationary on level data. However, the first differencing is applied to build a stationary time series with value $d = 1$. In Stage 2, parameters AR (p) and MA (q) for the best model are chosen by determining the lowest AIC and SBIC. Before that, the significant coefficient is checked first from the list estimated model. The estimated model is rejected if p and q values are not significant. The best model used for forecasting solar radiation is ARIMA (1, 1, 2) by seven states, followed by ARIMA (2, 1, 1) used by Kedah and Selangor, while Johor and Kelantan followed ARIMA (1, 1, 3). In Stage 3, the best model is applied for in-samples and out-samples forecasting. The errors of forecasting are evaluated by using the RMSE, MAE, and Theil's inequality coefficient. Both in-sample and out-sample forecastings show the same pattern of error. The values of RMSE, MAE, and Theil's inequality coefficient are small. It is also supported by the comparison between the plot of the actual series and the forecast series. In conclusion, the ARIMA model is successful in modeling and forecasting solar radiation data.

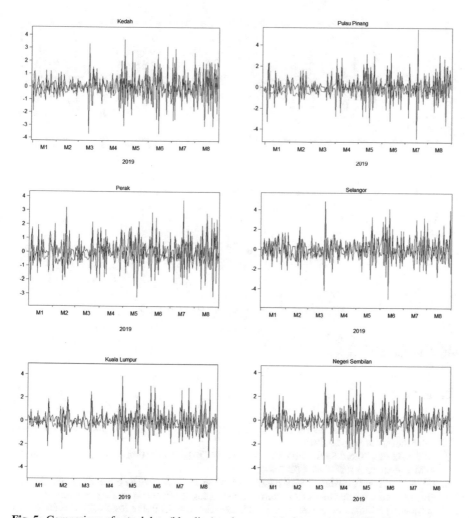

Fig. 5 Comparison of actual data (blue line) and out-sample forecasting (red line)

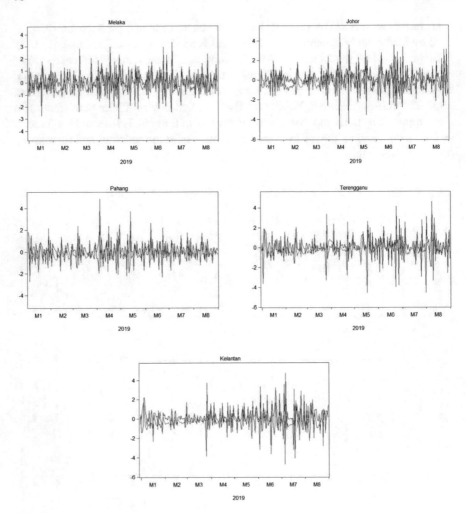

Fig. 5 (continued)

References

1. Blumsack S (2018) Are we running out of oil? https://www.e-education.psu.edu/eme801/nod e/486 Accessed Sept 29 2019
2. Abd Aziz PD, Wahid SSA, Arief YZ, Ab Aziz N (2016) Evaluation of solar energy potential in Malaysia. Trends Bioinfo 9(2):35–43
3. Rehman A, Rauf A, Ahmad M, Chandio AA, Deyuan Z (2019) The effect of carbon dioxide emission and the consumption of electrical energy, fossil fuel energy, and renewable energy, on economic performance: evidence from Pakistan. Environ Sci Pollut Res 26(21):21760–21773
4. Laslett D, Creagh C, Jennings P (2014) A method for generating synthetic hourly solar radiation data for any location in the south west of Western Australia. Renewable Energy 68:87–102
5. Lazim A (2011) Introductory business forecasting: a practical approachm 3rd edn. University Publication Centre (UPENA)

6. Alghoul MA, Sulaiman MY, Azmi BZ, Wahab MA (2006) Modeling of solar radiation in the time series domain. Int Energy J 7(4):261–266
7. Perdomo R, Banguero E, Gordillo G (2010) Statistical modeling for global solar radiation forecasting in Bogota. In: 2010 35th IEEE photovoltaic specialists conference, 2374–2379
8. Ji W, Chee KC (2011) Prediction of hourly solar radiation using a novel hybrid model of ARMA and TDNN. Sol Energy 85(5):808–817
9. Wu, J, Chan C (2012) The prediction of monthly average solar radiation with TDNN and ARIMA.2012 11th Int Conf Mach Learn Appl 2(3):469–474
10. Yang D, Jirutitijaroen P, Walsh WM (2012) Hourly solar irradiance time series forecasting using cloud cover index. Sol Energy 86(12):3531–3543
11. Lauret P, Diagne HM, David M, Rodler A, Muselli M, Voyant C (2012) A Bayesian model committee approach to forecasting global solar radiation. In: World renewable energy forum, WREF 2012, including world renewable energy congress XII and Colorado renewable energy society (CRES) annual conference, 4354–4359
12. Colak I, Yesilbudak M, Genc N, Bayindir R (2015) Multi-period prediction of solar radiation using ARMA and ARIMA models. In: 2015 IEEE 14th international conference on machine learning and applications (ICMLA), 1045–1049
13. Mbaye A, Ndiaye ML, Ndione DM, Sylla M, Aidara MC, Diaw M, Traoré V (2019) ARMA model for short-term forecasting of solar potential: application to a horizontal surface on Dakar site. OAJ Mater Device 4(1):1–8
14. Alsharif MH, Younes MK, Kim J (2019) Time series ARIMA model for prediction of daily and monthly average global solar radiation: the case study of Seoul South Korea. Symmetry 11(2):1–17
15. Atique S, Noureen S, Roy V, Subburaj V, Bayne S, MacFie J (2019) Forecasting of total daily solar energy generation using ARIMA: a case study. 2019 IEEE 9th annual computing and communication workshop and conference CCWC, 114–119
16. NASA Surface meteorology and Solar Energy—Definitions (2015) https://power.larc.nasa.gov/text/definitions.html. Accessed on 9 Sept 2019
17. Aziz H, Said S, On PWS, Salamun H, Yaakob R (2018) The change of Malaysian standard time: a motion and debate in the Malaysian Parliament. Int J Acad Res Bus Soc Sci 7(14):962–971
18. Liew-Tsonis J, Cheuk S (2012) Sustainability challenges: changing attitudes and a demand for better management of the tourism industry in Malaysia sustainable development policy and urban development tourism life science, Management and environment. IntechOpen, Chaouki Ghenai.
19. Box GEP, Jenkins GM, Reinsel GC, Ljung GM (2015) Time series analysis: forecasting and control, 5th edn. Wiley

Integration of Compound Parabolic Concentrator with Solar Power Tower Receiver

Ibrahim Alhassan Hussain, Syed Ihtsham Ul-Haq Gilani,
Hussain H. Al-Kayiem, Mohamad Zaki Bin Abdullah, and Javed Akhter

Abstract Despite the high thermal efficiency attained by evacuated tube receivers (ETR) in solar power tower (SPT) systems, optical losses due to spillage of concentrated power at the receiver are much in existent. The current work aims to minimize the concentrated solar power spillage losses at the central receiver by integrating non-imaging secondary concentrators with the evacuated tubes. A compound parabolic concentrator (CPC) with a *W*-shaped cavity at the bottom is designed based on an acceptance half-angle of 45° and coupled with an evacuated tube receiver, with 40 and 70 mm tube and glass diameters, respectively. The optical performance of the proposed scheme is analyzed using Monte-Carlo ray-tracing method. The compound parabolic concentrator is integrated with the ETR to intercept the maximum amount of solar radiation from the heliostat field and delivers it to the absorber. The distinguishing feature of this CPC is its orientation, i.e., vertically oriented, receiving solar radiation from primary concentrators (heliostats). Simulation results of the receiver with CPC have shown a recovery of 29% of the spillage losses; this improves the optical efficiency of the receiver to 75%.

Keywords Solar power tower · Spillage losses · Evacuated tube receiver · Compound parabolic concentrator · Optical performance

I. A. Hussain (✉) · S. I. Ul-Haq Gilani · H. H. Al-Kayiem · M. Z. Bin Abdullah · J. Akhter
Department of Mechanical Engineering, Universiti Teknologi Petronas, 32610 Seri, Iskandar, Perak, Malaysia
e-mail: Ibrahim_18000592@utp.edu.my

S. I. Ul-Haq Gilani
e-mail: syedihtsham@utp.edu.my

H. H. Al-Kayiem
e-mail: hussain_kayiem@utp.edu.my

M. Z. Bin Abdullah
e-mail: zaki_abdullah@utp.edu.my

J. Akhter
e-mail: javed.akhter_g03559@utp.edu.my

© The Author(s), under exclusive license to Springer Nature Singapore Pte Ltd. 2021
S. A. Sulaiman (ed.), *Clean Energy Opportunities in Tropical Countries*,
Green Energy and Technology, https://doi.org/10.1007/978-981-15-9140-2_4

1 Introduction to Solar Power Tower System

Apart from the depletion of fossil fuel reserves and the increase in fuel prices, another major concern posed by the utilization of fossil fuel is the environmental challenges such as climate change. Non-conventional energy-based systems are a better alternative to provide solutions to these challenges through the use of solar thermal energy. Solar thermal energy for concentrating solar power is a process where solar radiation is converted into heat and subsequently used for a variety of industrial applications. Among the types of solar, thermal technologies are parabolic trough collectors (PTC), linear Fresnel reflectors (LFR), parabolic dish concentrators (PDC), and solar power tower (SPT) systems. A typical solar power tower system for electricity generation is shown in Fig. 1.

Solar power tower (SPT) also known as solar central receiver (CSR) system is one of the concentrating solar power technologies for electricity generation from solar thermal energy. Solar power tower plant comprises of four main subsystems namely: central tower, energy receiver, computer-controlled mirrors called heliostats, and power conversion unit (turbine). Thermal storage is also included in some SPT plants. Heliostats intercept, reflect, and concentrate the solar radiation onto the receiver; the receiver is located on top of the central tower. The concentrated solar radiation at the receiver is then transferred to a primary working fluid (receiver coolant or heat transfer fluid) flowing within the receiver tubes. Some primary working fluids such as water double as the one employed in the power conversion system; while in some instances, the receiver coolant is a different fluid such as liquid sodium, molten salt, or synthetic oil. The heat transfer fluid (HTF) is mostly liquid in the SPT system, but gaseous working fluid such as air is also utilized recently. For liquid working fluid like water, steam is directly generated (direct steam generation) as a result of heat exchange between the receiver tubes and the water as shown in Fig. 2. For an indirect steam generation, the energy is stored in thermal storage (liquid sodium,

Fig. 1 PS10 solar tower power plant

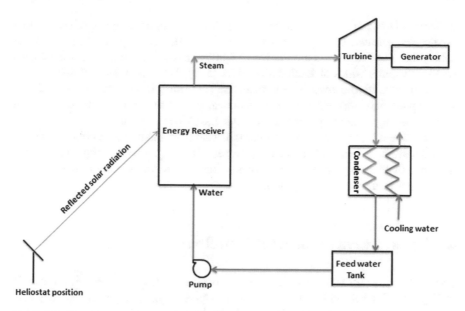

Fig. 2 Direct steam generation process in SPT system

molten salt, or synthetic oil) and is utilized later to generate steam (through heat exchange between the water and the primary working fluid). The generated steam is then passed through a steam turbine for electricity generation.

The tropical region is a very ideal location for solar thermal technology, specifically solar power tower system. The region is the closest to the equator; where solar radiation is much in abundance. Despite the abundance of solar radiation in the tropical region compared to the temperate region, most of the existing solar thermal power systems are in temperate region countries [1]. In addition to electricity generation, industries such as food processing, textiles, dairy, building, plastic, etc., can also benefit from the heat energy generated from solar power tower in the form of 'waste heat' for different industrial operations provided that they are located close to the power plant.

SPT energy receiver is the most important subsystem of the SPT plant. Highly concentrated solar energy at a high temperature is captured at the central receiver. Optical as well as thermal losses significantly affect the performance of nearly all types of receivers currently employed by the SPT system. The overall efficiency of the system depends on how well these losses are managed at the receiver. Over the years, many design modifications are proposed and investigated by many researchers [2, 3] to improve the receiver efficiency. However, most of these modifications are mainly focused on reducing thermal losses without ensuring high optical efficiency [4].

One of the optical losses that significantly affect the optical performance of SPT receiver especially ET based SPT receiver is the penetration of concentrated solar radiation through the glass tubes but miss the absorber as a result of the diameter difference between the glass cover and the absorber. Another source of optical loss is the concentrated solar power that escapes through the gaps between the glass tubes; these optical losses are known as spillage losses. Spillage losses exist even in SPT systems with conventional receiver [5–7], but it is mostly of solar power delivered outside the reach of the receiver. Therefore, in an attempt to minimize these losses and improve the receiver optical performance, a proposed design of a non-imaging secondary concentrator integrated with evacuated tube receiver to intercept the maximum amount of solar radiation from the heliostat field and deliver it to the evacuated tube absorber is presented. The proposed technique is modeled and simulated using ray-tracing software TracePro to predict some performance parameters of a solar power system.

2   Types of Conventional SPT Receivers

External and cavity receivers are the two most common types of receivers used in SPT plants. External and cavity receivers are a tubular type of receiver, and they both utilize water as heat transfer fluid [8]. The basic distinction between these types of SPT receivers is the location of the heat-absorbing tubes. The heat-absorbing surface of the external receiver consists of parallel tubes located around the outer surface of a cylindrical or rectangular receiver as can be seen in Fig. 3a. Likewise, in the cavity receiver, the heat-absorbing tubes are arranged parallel to one another and are enclosed within the inner surface of the receiver as shown in Fig. 3b. These receivers are subject to highly concentrated solar energy and are designed to absorb and transport most of the absorbed energy to a primary working fluid circulating within the tubes as mentioned earlier. Another type of receiver utilized in the SPT system which is not commonly used is the volumetric receiver; this receiver employs air as a working fluid.

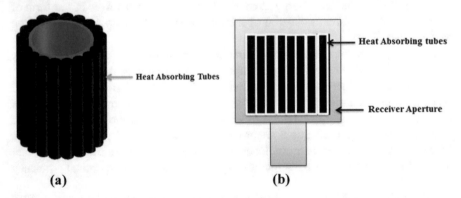

(a)   **(b)**

Fig. 3 SPT receivers **(a)** external and **(b)** cavity

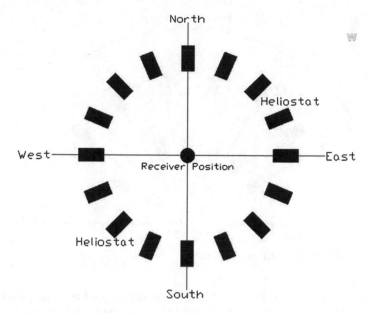

Fig. 4 Heliostats arrangement around external receiver in SPT system

3 Challenges and Limitations of Conventional Central Tower Receivers

In the external receiver design, the receiver is positioned in the center of the circular heliostats field as shown in Fig. 4. This implies that the receiver can accept reflected solar radiation from all the heliostats around the circular field. Therefore, the acceptance angle for the SPT plant with an external receiver is 360°. The disadvantage of this receiver is its large size, and the receiver tubes are unprotected from the surrounding atmosphere; which makes it prone to high thermal losses due to conduction and convection, respectively.

On the other hand, thermal losses are controlled in cavity receiver design by enclosing the heat-absorbing tubes within the inner surface of the receiver with a small cavity as the receiver aperture. It is smaller in size compared to the external receiver, but the disadvantage of this type of receiver is that it is limited to acceptance angles of 60°–120°. The heliostat field in this type of receiver is restricted to only the receiver aperture as shown in Fig. 5, unless if multiple receiver aperture is used.

4 Evacuated Tube-Based SPT (ET-SPT)

With the challenges and limitations of the two most widely used tubular receivers used in SPT plants, an evacuated tube receiver (ETR) was recently proposed for

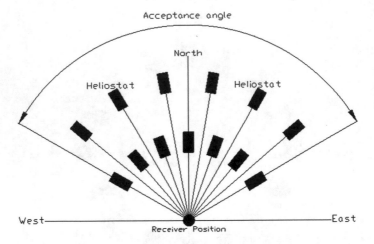

Fig. 5 Heliostat field around a cavity receiver

SPT systems [2]. The choice of ETR is based on its good thermal performance concerning convective heat losses. A unit of an evacuated tube receiver comprises an outer glass tube, a vacuum, and an inner tube with selective coating. Light rays pass through the outer transparent glass tube and get absorbed by the inner tube. The vacuum envelope between the inner (metal) and outer (glass) tube helps in retaining the collected energy inside the tube, thus, convection losses are highly suppress. ET-SPT system was designed and numerically and experimentally investigated by a few workers [2, 3, 9, 10].

Employing ERT in the SPT system proves to be thermally efficient. Receiver thermal efficiency of up to 70% is reported by Ali et al. [2]. But the challenge encountered with such a receiver is optical losses due to spillage of concentrated solar radiation at the receiver.

Concentrated solar radiation spillage losses in the ET-SPT receiver are categorized into two; within the receiver tubes and around the receiver. These two types of spillage losses make the total losses due to spillage at the receiver aperture. Spillage around the receiver is the solar radiation outside the reach of the receiver; this occurs when the rays from the heliostat are not reflected on the receiver aperture. It is either by the sides, on top, or below the receiver aperture. This type of spillage can be reduced by optimizing the size and geometry of the receiver. While the spillage within the receiver tubes is the concentrated solar radiations that escape through the gaps between tubes (Fig. 6) or penetrate the glass cover but miss the absorber (Fig. 7); this also occurs as a result of the diameter difference between the glass cover and absorbing tube. Using a secondary reflector such as a compound parabolic concentrator (CPC) could achieve a significant decrease in spillage losses within the ET-SPT receiver.

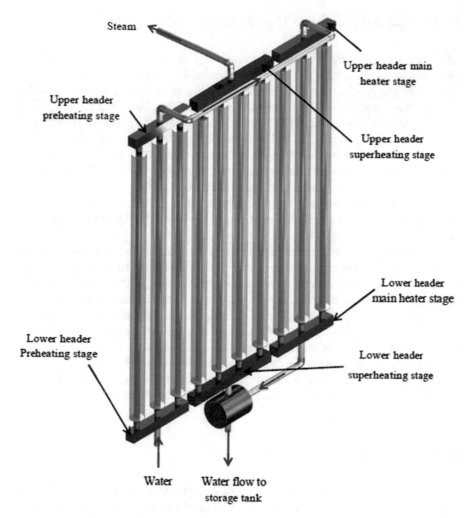

Fig. 6 Evacuated tube-based SPT receiver

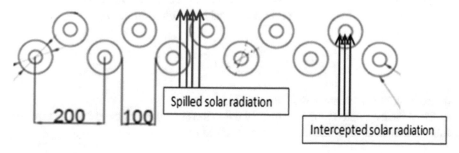

Fig. 7 Cross section of tubes zigzag arrangement in ET-SPT receiver showing spilled and intercepted radiation

5 Compound Parabolic Concentrator (CPC)

CPCs are non-imaging concentrators made by combining two parabolas in a symmetric or asymmetric way [11]. CPC can accept incoming radiation over a relatively wide range of angles and has the capability of reflecting it to the absorber placed at the exit aperture all the incident radiation within that wide range [12]. With a proper design of the reflector, CPC can reflect almost all the irradiation entering its aperture onto the receiver [13]. All these are achieved with the help of acceptance angle half ($\pm\theta_a$) phenomena. The angle through which source of the light move and yet converge at the absorber is what is termed as acceptance half-angle. For the maximum collection of reflected rays from the surface of the reflector, the absorber of the CPC has to be carefully selected. CPC is usually fitted with flat absorbers at the exit of the reflector or tubular absorbers at the combined focus of the parabola [14].

Theoretically, the CPC reflector has to touch the absorber, but in most cases, gaps exist between the reflector and the absorber to prevent heat losses by conduction. Although, the gap is provided to accommodate a cover (in case of vacuum tube absorber), differences in dimensions, tolerances for tube eccentricity, and assembly, at the same time, it is a source of solar radiation losses. Minimizing solar radiation gap losses in CPC can be achieved with a proper design of the bottom (region below the absorber tube). To eliminate these gap losses, the bottom of the reflector which is not part of the parabola can be modified and replaced with the W-shape cavity instead of involute [15]. This is done to maintain the distance between the reflector and the region below the absorber within the range of half of the radius of the absorber.

6 CPC Geometrical Model

The two most important things in CPC geometrical design are the acceptance half-angle and the CPC aperture. In this study, the acceptance half-angle was determined by the farthest heliostats to the east and west of the tower; which in turn determines the concentration ratio of the CPC. The reflector length is the same as that of the absorber, and therefore, the maximum optical concentration ratio (C_r) is given by:

$$C_r = \frac{1}{\sin \theta_a} \tag{1}$$

where θ_a is the acceptance half-angle [15].

Since the evacuated tube is to be utilized as the absorber, the height of the reflector and the radius of the glass cover were obtained by:

$$h = \frac{p}{2}\left[\frac{1}{\sin \theta_a \tan \theta_a} + \frac{1}{2} + \frac{1}{\pi \sin \theta_a} \right] \tag{2}$$

Table 1 CPC design
parameters

Parameter	Value
Acceptance angle (θ_a)	45°
Absorber diameter	40 mm
Glass cover diameter	70 mm
Absorber optical length	2.315 m

$$R = \sqrt{2}r \tag{3}$$

where 'R' is the radius of the glass cover, 'r' the radius of the absorber, 'p' the perimeter of the absorber, and 'h' is the full height of the CPC [15, 16]. All the parameters utilized in generating the CPC used in this study are given in Table 1.

The upper portion of the reflector which is the parabola is generated using the relation given by:

$$x = r \sin \vartheta - \rho(\vartheta) \cos \vartheta$$
$$y = -r \cos \vartheta - \rho(\vartheta) \sin \vartheta \tag{4}$$

where '$\rho(\vartheta)$' is the distance between two points 'jk' as shown in Fig. 5, one point on the curve of the reflector and a corresponding point on the tangent of the absorber [17, 18]. String method is used to determine the distance '$\rho(\vartheta)$' for both the parabola and the region below the absorber using the relation given by McIntire [15].

$$\rho(\vartheta) = r \frac{\left[\vartheta + \theta_a + \frac{\pi}{2} - 2\alpha - \cos(\vartheta - \theta_a)\right]}{[1 + \sin(\vartheta - \theta_a)]}$$
$$\text{for } \theta_a + \frac{\pi}{2} \leq \vartheta \leq \frac{3\pi}{2} + \theta_a \tag{5}$$

$$\rho(\vartheta) = r(\vartheta - \alpha), \quad \text{for } 2\sigma \leq \vartheta \leq \theta_a + \frac{\pi}{2} \tag{6}$$

The region below the absorber is designed by shifting the origin of the involute as in the case of ideal non-imaging reflector through angle 'α' as shown in Fig. 8 to points A and C; this implies that the top outer part of the 'W' is the starting point of the reflector curve.

For the 'W' shape cavity at the bottom of the reflector, its external section is tangent to the glass cover through the points $(-r \tan(\sigma), -r)$ and $(r \tan (\sigma), -r)$, where $\sigma = \cos^{-1}(r/(R + m))$ and $\alpha = (\tan \sigma - 2\sigma)$, while the internal section is the imaginary line from the upper tangent extended through the center [15].

Equation (3) holds for optimum reflector design of a W-shaped cavity for an absorber with radius 'r' and a gap between the absorber and the reflector 'R' without considering the cosine-weighted optical efficiency. Therefore, raising the absorber to

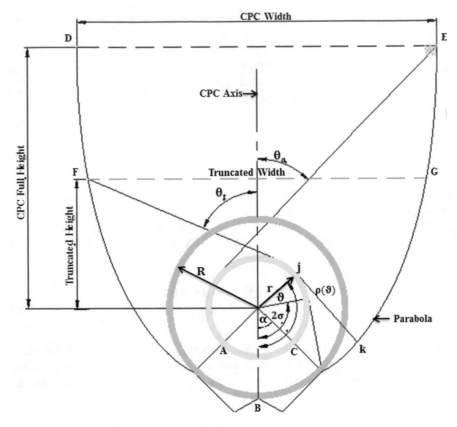

Fig. 8 Design parameters for the proposed CPC

about 5 mm from its theoretical position could improve the cosine-weighted optical efficiency. A gap of 2 mm is also provided between the reflector and the glass cover to prevent the cover from touching the reflector. These two additional distances are designated as the '*m*.' The CPC used in this research is shown in Fig. 9.

CPC as a second stage reflector is mostly used in conjunction with different types of absorbers with the sole aim of increasing the concentration ratio such as in PTC systems and Fresnel mirror field [19–21], but not for the reduction in spillage losses in an evacuated tube-based solar power system.

Using a tubular absorber in solar energy collection without a concentrator is actually de-concentrating the collected energy by a factor of $\frac{1}{\pi}$, because the heat-absorbing surface (aperture) of a tubular absorber is the width $(2r)$ of the absorber; which is the surface exposed to solar radiation. But when an ideal concentrator is used, the entire perimeter $(2\pi r)$ of the absorber is considered.

Effect of truncation on the concentrator: The height of the reflector consider-ably determines the manufacturing cost of the concentrator due to the cost of the materials. Though, the upper portion of the reflector is almost perpendicular to the

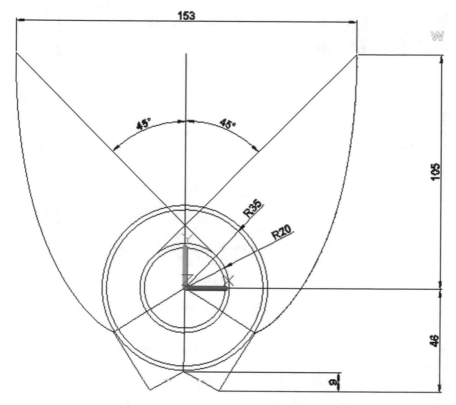

Fig. 9 Proposed CPC with *W*-shape cavity at the bottom paired with an evacuated tube receiver

concentrator aperture; so it has less influence in concentrating the incident solar radiation. Therefore, the reflector can be truncated without making much impact on the performance of the concentrator. Truncation increases the optical efficiency of CPC by increasing its acceptance half-angle for direct radiation and decreasing the number of reflections [22]. According to Ustaoglu et al. [23], optical efficiency is predicted to increase by over 20% with truncation of about 50%, and the decrease in thermal efficiency is about 1%. Truncation of the concentrator brings about savings in material cost, an increase in acceptance angle, and field view of the absorber [24], but with a slight decrease in concentration ratio. A slow decrease in the CPC width is observed at truncations below 50%, but a steep decrease is noted at the truncation of about 50% and above as shown in Fig. 10. Note that the width is proportional to the concentration ratio.

Therefore, the CPC is truncated at 50% of its full height measured from the center of the absorber; this truncation increases the CPC acceptance half-angle for direct radiation to 66° as shown in Fig. 11, and with only 6% decrease in the concentration ratio. Besides, it brings about 52% savings in the cost of the reflector material.

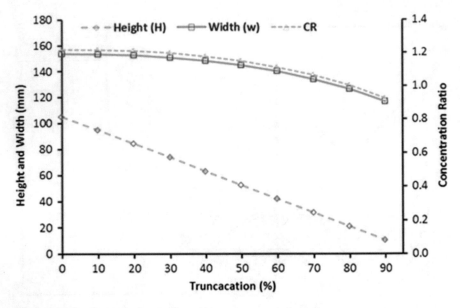

Fig. 10 Effect of truncation on height, width, and concentration ratio

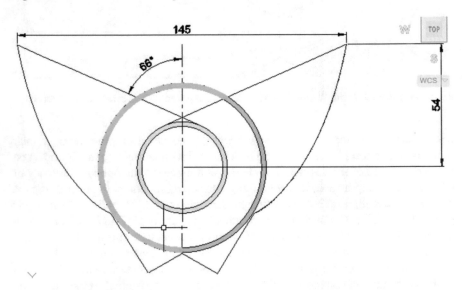

Fig. 11 Truncated CPC with W-shaped cavity

7 Integrated CPC ET-SPT Receiver

The new receiver consists of ETR and a W-shape cavity compound parabolic concentrator with an evacuated tubular absorber is mounted on top of the tower as shown in Fig. 12.

In the said system, there exist two sets of solar reflectors. The first set consists of several heliostats; which intercept, reflect, and concentrate the direct solar radiation onto the receiver aperture. The other set of concentrators are the CPCs; which intercept the reflected solar radiation from the heliostat field and reflect it onto the absorber enclosed within the glass cover. Figure 13 shows the ray-tracing simulation of one of the heliostats from the field.

Therefore, for the first concentrator, the concentrated solar power at the receiver aperture from a single heliostat in the field is given by:

$$Q_h = I_n A_h \eta_f \tag{7}$$

where Q_h is the power from single heliostat, A_h is the mirror aperture area, I_n the normal irradiance, and η_h the optical efficiency of the heliostat [25]. Also, the power

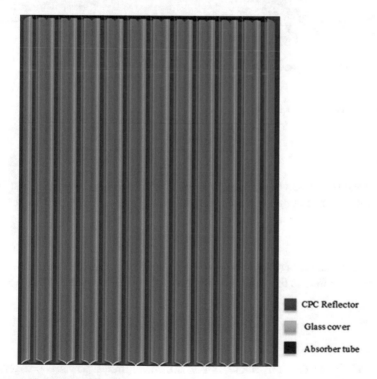

CPC Reflector

Glass cover

Absorber tube

Fig. 12 Receiver integrated with CPC

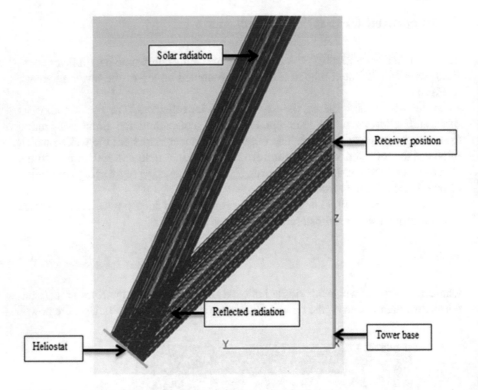

Fig. 13 Ray-tracing diagram of one of the heliostats in the SPT system

from t_1 to t_2 is obtained by:

$$Q_h = \int_{t_1}^{t_2} I_n A_h \eta_f \tag{8}$$

The optical efficiency of a heliostat is a product of multiple efficiencies that include cosine, atmospheric attenuation, shading, blocking, and reflectivity, given by:

$$\eta_f = \eta_{cos}\eta_{att}\eta_{sb}\eta_{ref}\eta_{sp} \tag{9}$$

If the heliostat field contains 'n' number of heliostats, then the total concentrated power from the field is the summation of power from individual heliostat given by:

$$Q_{Tf} = \sum_1^n \int_{t_1}^{t_2} I_n A_h \eta_f \tag{10}$$

It was observed that only about 46% of the concentrated power is present on the receiver's outer surface; this implies that 54% is wasted [3]. Out of the 54% total spillage, concentrated solar radiation spillage within the receiver tubes accounts for about 40%. It comprises of the spillage through the gaps between the glass tubes and penetration within the tube but missed the absorber, while the remaining 14% is of spillage losses outside the reach of the receiver.

Integrating a CPC in the receiver design eliminates the gap between the parallel tubes; which a source of concentrated solar radiation losses due to spillage. Also, any radiation within the acceptance half-angle of the CPC will be directed onto the absorber; preventing the passage of solar radiation from the front of the glass tube through its back; thus eliminating the second source of concentrated solar radiation spillage.

The second set of reflectors is designed to make sure that the reflected radiation from the field reaches the outer surface of the absorber. According to Akhter et al. [24], the total power at the outer surface of the observer is given by:

$$Q_{as} = I_{nT} A_c \eta_c \tag{11}$$

where Q_{as} is the power at the outer surface of the observer, A_c is the CPC entry aperture area, I_{nT} the summation of the direct normal irradiance from the field, and η_c the CPC optical efficiency. The CPC optical efficiency is obtained by:

$$\eta_c = \gamma^r \tau_g \alpha_a \tag{12}$$

where γ is the CPC reflectivity, r the average number of reflection, τ_g glass tube transmittance, and α_a absorptivity.

8 Ray-Tracing Simulations

The 3D model of the CPC was drawn using AutoCAD and then imported into TracePro. TracePro is a Monte-Carlo ray-tracing software for optical and illumination analysis. The surface and material properties of different components that made up the CPC were defined in the TracePro. A solar emulator is used to typically imitate the Sun, where the parallel Sun and 10,000 numbers of rays were assumed for the simulation. Other Sun's properties were defined together with coordinates of the location and that of the system. The coordinates of the system which comprises of heliostat base, tower base, heliostat, and tower were retrieved from the 3D model. After the system was set up, the ray-tracing was carried out for a typical day in December. The flow diagram for the simulation is shown in Fig. 14.

The heliostat was positioned due north of the tower. To keep the reflected rays on the target (receiver), relationships for azimuth elevation Sun-tracking method are used to determine the azimuth and elevation angles of the heliostat. Parameters considered are the horizontal distance from the tower, the height of the tower, latitude of the location, hour angle, Sun's azimuth, and elevation angles.

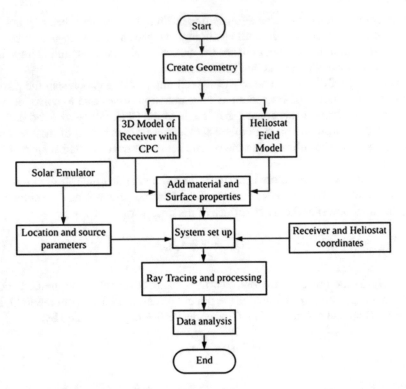

Fig. 14 Flow diagram for the simulation process

9 Results and Discussions

Ray-tracing simulation of the newly designed receiver was carried out using TracePro to determine its optical performance. The heliostat orientation is directly facing the receiver when the Sun's azimuth angle is around 180°. The heliostat intercepts and reflects the solar radiation to the target (receiver) which is 9.5 m away from the heliostat and at 9.25 m above the ground.

9.1 Radiant Flux on the Receiver with and Without CPC

The simulation was carried out for 30 min with the heliostat facing angle of 180° throughout the simulation time. Figure 12 shows intercepted radiant flux (power) on target (receiver) with CPC and without CPC. It is observed that ETR with CPC as a secondary concentrator captures over two times more solar radiation than the receiver without CPC. At the beginning of the hour, the reflected rays from the mirror were off-target because the Sun's azimuth angle is around 175°, but as the azimuth angle approaches 180°, flux is recorded on the absorber. About 30 min later, the reflected ray was off-target again because the solar azimuth angle is approaching 200° (Fig. 15).

Figure 16 shows the maximum solar power recorded at the outer surface of the absorber during the same time but on a different day using the receiver optical model. Just like in the simulation, it is assumed that there is no solar radiation recorded on the receiver aperture in the first and the last three minutes.

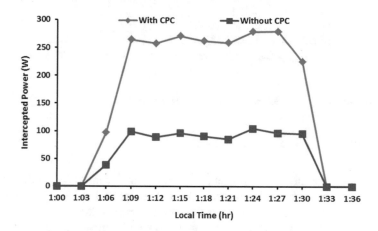

Fig. 15 Simulated intercepted power at the absorber on December 31st

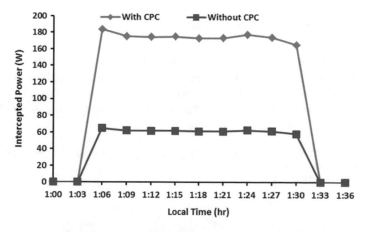

Fig. 16 Intercepted power at the absorber using the optical model on February 27th

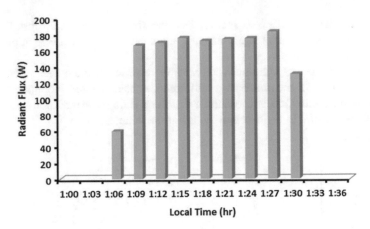

Fig. 17 Recaptured spillage losses using CPC on December 31st

9.2 Recuperative Spillage Losses

Spillage of concentrated solar power recaptured with the new receiver is shown in Fig. 17. It can be deduced from the figure that recaptured spillage losses and the amount of intercepted solar power at the receiver have a direct relationship. The more the incident solar radiation, the more the power intercepted at the absorber.

9.3 Optical Energy Distribution

The difference in the optical performance of the current (receiver without CPC) and new receiver (receiver with CPC) designs is shown in Fig. 18. Only about 46% of the total concentrated solar power at the receiver aperture is absorbed in the current receiver design, while about 75% of the total concentrated power at the receiver

Fig. 18 Optical energy distribution of ET based SPT receiver integrated with CPC

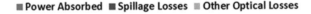

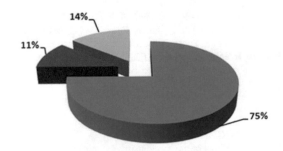

aperture is absorbed in the new receiver design. The increase in the amount of solar power absorbed is due to the presence of the secondary concentrator, which recaptures and redirects the solar power back on to the absorber.

10 Summary

This chapter presents a novel approach to minimize optical losses due to spillage of concentrated solar power at the SPT receiver and improve the optical performance of the system. The design and integration of a second stage non-imaging concentrator is presented. The optical simulation of an ET-SPT receiver integrated with a compound parabolic concentrator as secondary reflectors is carried out using the Monte-Carlo ray-tracing method to evaluate its optical performance. The simulation is carried out for 30 min between 1 and 1:30 pm on December 31st. The newly designed receiver comprising of evacuated tubes and CPC secondary reflectors has shown a better optical performance than the receiver without CPC. The concentrated solar radiation that escapes between the receiver tubes and solar radiation that pass through the glass tube but miss the absorber in an evacuated tube-based central receiver is reflected onto the absorber and captured. A significant reduction in spillage losses is achieved from 40 to 11%, and the optical efficiency is enhanced by 30%.

Acknowledgements The authors acknowledge Universiti Teknologi PETRONAS (UTP), Malaysia for providing financial support under research grant YUTP FRGS (cost center: 015LC0-209) to conduct this research.

References

1. Ogunmodimu O, Okoroigwe EC (2019) Solar thermal electricity in Nigeria: prospects and challenges. Energy Policy 128:440–448
2. Ali BH, Gilani S, Al-Kayiem HH (2016) Mathematical modeling of a developed central receiver based on evacuated solar tubes. In: MATEC web of conferences, p 02005
3. Khlief AK, Gilani SI, Al-Kayiem HH, Mohammad ST (2018) Design a new receiver for the central tower of solar energy. In: MATEC web of conferences, p 02009
4. Khullar V, Mahendra P, Mittal M (2018) Applicability of heat mirrors in reducing thermal losses in concentrating solar collectors. J Thermal Sci Eng Appl 10
5. García L, Burisch M, Sanchez M (2015) Spillage estimation in a heliostats field for solar field optimization. Energy Procedia 69:1269–1276
6. Das A, Inigo P, McGrane J, Terdalkar R, Clark M (2017) Design concepts for spillage recovery in a molten salt central receiver. J Renew Sustain Energy 9:023701
7. Ho CK, McPheeters CO, Sharps PR (2018) Hybrid CSP/PV receivers: converting optical spillage to electricity. In AIP conference proceedings, p 170006
8. Behar O, Khellaf A, Mohammedi K (2013) A review of studies on central receiver solar thermal power plants. Renew Sustain Energy Rev 23:12–39
9. Ali BH, Gilani S, Al-Kayiem HH (2014) Investigation of evacuated tube collector performance at high temperature mode using TRNSYS simulation model, Appl Mech Mater 155–160

10. Khlief AK, Gilani SIUH, Al-Kayiem HH, Mohammad ST (2017) Design and modelling of novel evacuated receiver tube of concentration solar power system by using heliostat stadium arrangement. In: MATEC web of conferences, p 02002
11. Tripanagnostopoulos Y (2014) New designs of building integrated solar energy systems. Energy Procedia 57:2186–2194
12. Kalogirou SA (2004) Solar thermal collectors and applications. Prog Energy Combust Sci 30:231–295
13. Osório T, Horta P, Marchã J, Collares-Pereira M (2019) One-Sun CPC-type solar collectors with evacuated tubular receivers. Renewable Energy 134:247–257
14. Waghmare SA, Gulhane NP (2016) Design and ray tracing of a compound parabolic collector with tubular receiver. Sol Energy 137:165–172
15. McIntire W (1980) New reflector design which avoids losses through gaps between tubular absorbers and reflectors. Sol Energy 25:215–220
16. Rabl A, Goodman N, Winston R (1979) Practical design considerations for CPC solar collectors. Sol Energy 22:373–381
17. Akhter J, Gilani S, Ali M, Gilani S (2018) Design and Optical Modeling of a Low-Profile Stationary Concentrating Solar Collector for Medium Temperature Heat Supply. J Telecommun Electron Comput Eng (JTEC) 10:65–71
18. Kim YS, Balkoski K, Jiang L, Winston R (2013) Efficient stationary solar thermal collector systems operating at a medium-temperature range. Appl Energy 111:1071–1079
19. Rabl A (1976) Comparison of solar concentrators. Sol Energy 18:93–111
20. Omer SA, Infield DG (2000) Design and thermal analysis of a two stage solar concentrator for combined heat and thermoelectric power generation. Energy Convers Manage 41:737–756
21. Collares-Pereira M, Gordon J, Rabl A, Winston R (1991) High concentration two-stage optics for parabolic trough solar collectors with tubular absorber and large rim angle. Sol Energy 47:457–466
22. Carvalho M, Collares-Pereira M, Gordon J, Rabl A (1985) Truncation of CPC solar collectors and its effect on energy collection. Sol Energy 35:393–399
23. Ustaoglu A, Okajima J, Zhang X-R, Maruyama S, (2018) Truncation effects in an evacuated compound parabolic and involute concentrator with experimental and analytical investigations. Appl Thermal Eng 138:433–445
24. Akhter J, Gilani SI, Al-Kayiem HH, Ali M (2019) Optical performance analysis of single flow through and concentric tube receiver coupled with a modified CPC collector under different configurations. Energies 12:4147
25. Stoddard MC, Faas S, Chiang C, Dirks J (1987) Solergy; a computer code for calculating the annual energy from central receiver power plants. Sandia National Labs., Livermore, CA (USA).

Nanofluids Application in Solar Thermal Collectors

Haris Naseer, Syed Ihtsham ul-Haq Gilani, Hussain H. Al-Kayiem, and Nadeem Ahmad

Abstract Depletion of conventional resources of energy and continuous elevation in average temperature of earth result in the usage of alternative energy resources, and solar energy is the biggest source in one of them. Different solar thermal collectors have been used to store heat energy from the sun and can be used in various domestic and commercial applications. Solar collector efficiency depends on the heat carrying capacity of the working fluid. Thermal effectiveness of the fluid is the combined effect of thermophysical properties of the fluid, i.e., thermal conductivity, viscosity (pressure loss and pumping power), specific heat capacity (capability of storage), and density. Conventional fluids, i.e., water, ethylene glycol (EG) and heat transfer oils, have been used in solar collectors. However, they have low thermal conductivity and high viscosities. Suspension of small concentration of nanoparticles of 1–100 nm in size in the conventional fluids has proven to display better thermal and optical properties than base fluids. Hence, increase in collector's efficiency has been reported by many researchers. This chapter aims to provide information on the calculation of concentration of nanoparticles and base fluids, preparation of nanofluids, measurements of thermal and rheological properties, and the application in different solar collectors. The major part of the chapter is allocated to the effects of nanoparticles to the base fluid's properties. Prediction models of different properties of nanofluids and correlation between the properties are also discussed in this chapter.

H. Naseer (✉) · S. I.-H. Gilani · H. H. Al-Kayiem
Department of Mechanical Engineering, Universiti Teknologi Petronas, 32610 Seri, Iskandar, Perak, Malaysia
e-mail: haris_18003305@utp.edu.my

S. I.-H. Gilani
e-mail: syedihtsham@utp.edu.my

H. H. Al-Kayiem
e-mail: hussain_kayiem@utp.edu.my

N. Ahmad
Department of Mechanical Engineering, Colonel by Hall 161, Louis-Pasture, University of Ottawa, ON K1N 6N5, Canada
e-mail: nahma081@uottawa.ca

Keywords Energy system · Solar collector · Renewable energy · Nanofluids

1 Introduction

The rise in world's population, increased industrial activities and global warming, and the depletion of conventional sources of energy like fossil fuels have motivated researchers to find alternative sources of energy that can overcome the future energy's demands [1]. There are plenty of energy resources that have been used to power up commercial and domestic sectors. These resources are shown in Fig. 1. The use of non-renewable resources for power generation, i.e., fossil fuels, natural gas, and nuclear fuels, causes the emission of CO_x, NO_x, and radioactive rays, which are responsible for global warming over a century [2]. Carbon dioxide emission in recent decades had sharply increased worldwide, from 11185.9 Mt in 1965 to 33444 Mt in 2017, and had a growth rate of 1.1% per annum from 2008 to 2018 [3]. These emissions of harmful gases have increased the average temperature of the world beyond its safe limit and have revealed the importance of developing renewable energy resources. Solar and wind energy are available worldwide, and they are clean resources of energy that can be used in various applications [2]. The total annual growth rates of solar and wind energy are 12.6% and 24.3%, respectively, in 2019 by

Fig. 1 Types of energy resources

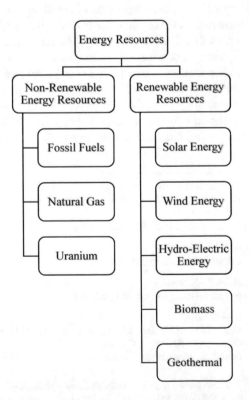

generation, which is the highest among other renewable energy resources. The Earth receives an average of 1000 W/m² of solar direct and diffused radiation that can be used in wide range of applications such as heating the fluids, electricity generation, and desalination. In electricity generation, photovoltaic panels have been used to directly convert solar energy into electricity. According to hi-Ren scenario in IEA Technology Roadmap 2014, the share of solar energy in global electricity in 2050 will be increased by 16% [4]. Thermal energy of sun can be utilized to preheat the working fluid of the power cycle, which results in reduction in the greenhouse gas emission and efficiency improvement.

1.1 Solar Thermal Collectors

Solar collector is a device that collects or concentrates the solar direct or diffused radiation onto an absorbing material. A solar thermal system consists of solar collectors, thermal storage tanks, series of pipes, valves, and sensors, i.e., temperature, pressure and fluid flow, level gauges, and pumps. The primary purpose of the thermal collectors is to preheat the working fluid.

The use of solar energy for heating of water has been traced for use in public baths in 200 BCE in Greece. This technique of solar water heater (SWH) was re-invented by Kemp in 1891 who patented his first commercial SWH but was lacking in temperature control. Bailey improved the performance of SWH by introducing a new design to control the temperature [5].

There are two methods of heating the water, i.e., direct and indirect heating. In direct heating, water is directly heated by solar thermal energy and supplied for numerous applications. In indirect heating, thermal storage tank is used which is installed with the system that stores the heat energy of the working fluid and can be used in night or cloudy days. The solar collectors have many applications [6] in washing clothes, cleaning purposes, heat treatment processes, preheating of boiler feed water, central heating system, etc.

Various types of solar collectors are used for direct heating or indirect heating. Direct absorber solar collectors, compound parabolic collectors, evacuated tubes, and flat plate solar collectors are examples of solar collectors. The classification of solar thermal collectors is shown in Fig. 2.

Concentrating collectors are generally characterized as medium concentration as they reflect direct and diffused radiations to focal point or focal line, e.g., parabolic or compound parabolic concentrator. They can be designed either to a fixed device or a rotating reflector to capture the maximum solar radiation and to work effectively. These collectors can achieve a higher temperature range of 150–500 °C as compared to the non-concentrating solar collectors, which can only achieve a temperature range of 100–200 °C [6]. The schematic of a solar thermal collector is shown in Fig. 3.

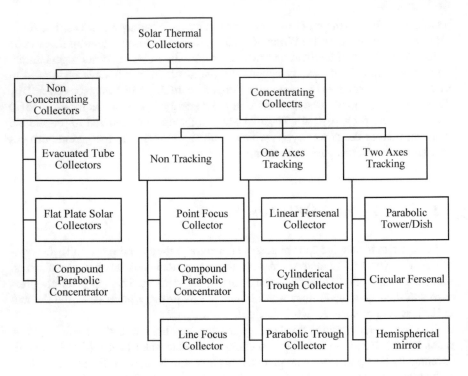

Fig. 2 Types of solar collectors

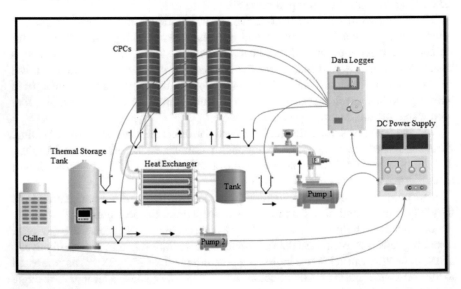

Fig. 3 Schematic of a solar collector

1.2 Thermal Performance Modeling of Solar Collectors

The energetic efficiency of a solar collector is calculated based on the useful heat absorbed by the nanofluids [7], which is given by:

$$\eta_{\text{eng}} = \frac{Q_u}{A_c G_t} \tag{1}$$

where Q_u is the useful heat gain. Here, A_c and G_t are the collector's area and total solar radiation in the collectors, respectively. The denominator, $A_c G_t$, is the available solar energy. The available solar energy is determined by:

$$Q_u = \dot{m} C_p (T_{\text{out}} - T_{\text{in}}) \tag{2}$$

$$Q_u = A_c F_R [I_0(\tau\alpha) - U_L(T_{in} - T_a)] \tag{3}$$

$$Q_u = A_c \left[S - U_L \left(T_{\text{pm}} - T_a \right) \right] \tag{4}$$

where I_o and are total incident solar radiation, and product of transmissivity of the glass cover and absorptivity of the nanofluids, respectively. The effective radiation on the collectors and its optical efficiency is represented by S. The heat removal factor, F_r, is determined by:

$$F_R = \frac{Q_u}{Q_{u(T-T_i)}} \tag{5}$$

For most of the solar collector application, $F_{r,}$ and U_L are constants. U_i is the overall heat loss from the collector, which determines how successful the collector is in preventing heat loss from the top, back (U_b), and edge (U_e) insulation and can be expressed by:

$$U_L = U_i + U_b + U_e \tag{6}$$

1.3 Heat Transfer Criteria of Solar Collector

The collector's heat transfer coefficient is given by:

$$h_f = \frac{Nu.k_{nf}}{D_i} \tag{7}$$

where Nusselt number, Nu, depends on Reynolds number, Re, for determination of the flow regime and the Prandtl number, Pr. Once the regime of fluid flow is determined, the estimation of Nusselt number can be established. The Reynolds and Prandtl numbers are given by:

$$\text{Re}_{nf} = \frac{\rho_{nf} \cdot V \cdot D_i}{\mu_{nf}} = \frac{4 m_{\text{riser}}}{\pi \cdot D_i \cdot \mu_{nf}} \tag{8}$$

$$Pr_{nf} = \frac{\mu_{nf} \cdot C_{p,nf}}{k_{nf}} \tag{9}$$

It can be observed from the equations above that the thermal properties of nanofluids play an important role by changing the thermophysical properties of the fluid flow and influencing the values of dimensionless heat transfer coefficients.

1.4 Nanoparticles and Properties

The particles of 1–100 nm in size, which were introduced by Choi in 1995 [8], have better thermos-physical properties and therefore can be mixed with base fluids to increase the thermal behaviors. There are a wide range of metallic nanoparticles (Zn, Cu, Al, Ag, Au, etc.) and metallic oxides (TiO_2, CuO, Al_2O_3, SiO_2, MgO, etc.) that can be used for this purpose. The selection of nanoparticles is done on the basis of their applications, i.e., temperature ranges, density of base fluids, economical and environmental view of points.

Thermal properties of nanoparticles depend on the shape, size, pH, viscosity of fluids, specific heat, and volume fraction. Properties of different nanoparticles are shown in Table 1. These nanoparticles are hydrophilic (water loving) or hydrophobic (water resistant) in nature. But their nature can be changed by functionalizing with different medias (attach hydroxyl or carboxylic group on the particles) or by using

Table 1 Properties of different nanoparticles

Nanoparticles	Density (kg/m^3)	Thermal conductivity (W/m K)	Specific heat capacity (kJ/kg.K)
CuO [9]	6000	33	0.551
TiO_2 [9]	4230	8.4	0.692
Al_2O_3 [10]	3890	36	0.880
SiO_2 [9]	3970	3.2	0.765
Cu [11]	8933	400	0.385
MgO	3580	30	0.880
MWCNT [12]	2100	1500	0.710
AlN [13]	4175	8.4	0.692

the surfactants, e.g., sodium dodecyl sulfate, sodium dodecyl benzene sulfate, cetyl trimethyl ammonium bromide, etc.). Surfactants are long organic molecules that possess both hydrophilic and lipophilic groups and used to cancel the repulsion or attraction forces. It is the most economical method in large-scale production.

1.5 Base Fluids and Properties

Water is used as a base fluid for most of the applications as it is easily available and very low in cost. It has good thermophysical properties and can be prepared as nanofluids easily. In addition, water has no hazards on environment and health like other chemicals. Most of the nanoparticles are hydrophilic in nature and are easily soluble in water. The main drawback of the water is that it has low boiling point and may not be feasible for use in medium- or high-temperature applications such as in compound parabolic concentrator. Thermal oils and glycols have promising heat transfer properties, high-temperature boiling points, low-pressure applications and can be widely used in applications such as aerospace, solar collectors, power plants, and automobiles. Apart from heat transfer properties, thermal oils are used as coolants and lubricants in bearings, moving parts of engines, and also for cleaning.

Since the introduction of nanofluids, many researches have been conducted to increase the thermal effectiveness behaviors of the base fluids (water, oils, ethylene glycol (EG), etc.). Further to that, many scientists have been attempting to increase the heat transfer properties of base fluids by using mono or hybrid nanoparticles. In recent years, researchers are combining two or more nanoparticles of different properties to make hybrid nanofluids, which have enhanced intermediate properties and stability (Table 2).

Table 2 Properties of different base fluids

Fluid name	Density (kg/m^3)	Thermal conductivity (W/m K)	Specific heat capacity (kJ/kg K)	Viscosity (Pa s)	Boiling point ($^\circ$C)
Water [11]	997	0.613	4.179	0.00089	100
Therminol-55 [14]	865	0.127	1.945	0.02000	390
Ethylene glycol [13]	1115	0.249	2.664	0.00300	198
Therminol VP-1 [15]	1056	0.135	1.575	0.00328	257

2 Methodology

2.1 Selection of Nanoparticles and Base Fluids

The nanoparticles are selected according to their applications. For solar collectors, the nanofluids are selected on the basis of three factors:

a. Thermal conductivity
b. Density
c. Specific heat capacity
d. Viscosity

2.2 Nanofluids Preparation

Nanofluids can be prepared by two methods: one-step method and two-step method. In the one-step method, the nanoparticles are simultaneous made and dispersed in base fluids. The stability of fluid is increased because the processes of drying, storage, transportation, and dispersion of nanoparticles are avoided. The main drawback of using this method is that the chemical reactions cannot be controlled, that is, either it is completed or not, in terms of impurities in the final form, and the concentration of the nanoparticles in the base fluids (Table 3).

In the two-step method, the nanoparticles are prepared separately by using chemical or physical reactions. Then, the particles disperse in the base fluids by using magnetic force (stirring), ultra-sonication, high shear mixing, and homogenizing. Different researchers used different techniques to prepare nanofluids, as shown in Table 4.

Table 3 Preparation methods used by different researchers

References	Nano particles	Base fluid	Preparation method
Aberoumand et al. [16]	Ag	Oil	Single step
Botha et al. [17]	Silver–silica	Transformer oil	Single step
Aberoumand and Amin [18]	Cu	Engine oil	Single step
Farbod et al. [19]	CuO	Engine oil	Two step
Karami et al. [20]	CNTs	Water	Two step
Alim et al. [9]	Al_2O_3, SiO_2, CuO, TiO_2	Water	Two step
Moghadam et al. [21]	CuO	Water	Two step
Gupta et al. [22]	Alumina	Distilled water	Two step
Asadi [23]	MWCNT–ZnO	Engine oil	Two step
Baojie et al. [24]	$SiC–TiO_2$	Diathermic oil	Two step
Esfe and Abbasian [25]	$MWCNT–SiO_2$	Engine oil	Two step
Gulzar et al. [26]	$Al_2O_3–TiO_2$	Therminol oil	Two step

Table 4 Thermal conductivity prediction models

Model	Assumptions	Equation
Maxwell [29]	Spherical-shaped, micro- or millimeter-sized particles	$\dfrac{k_{eff}}{k_f} = \dfrac{k_p + 2k_f + 2\phi_p(k_p - k_f)}{k_p + 2k_f - \phi_p(k_p - k_f)}$
Hamilton and Crosser [29]	N = empirical shape factor: For spherical shape, $n = 3$ For cylindrical shape, $n = 6$	$\dfrac{k_{eff}}{k_f} = \dfrac{k_p + (n-1)k_f - (n-1)\phi_p(k_f - k_p)}{k_p + (n-1)k_f + \phi_p(k_f - k_p)}$
Extendable form of Maxwell equation [31]	For hybrid nanofluids	$\dfrac{k_{hnf}}{k_{bf}} = \dfrac{\left(\frac{(\phi_{p1}k_{p1} + \phi_{p2}k_{p2})}{\phi_{tot}} + 2k_{bf} + 2(\phi_{p1}k_{p1} + \phi_{p2}k_{p2}) - 2\phi_{tot}k_{bf}\right)}{\left(\frac{(\phi_{p1}k_{p1} + \phi_{p2}k_{p2})}{\phi_{tot}} + 2k_{bf} - (\phi_{p1}k_{p1} + \phi_{p2}k_{p2}) + 2\phi_{tot}k_{bf}\right)}$

2.3 Nanofluids Concentration

There are a few bases for representing the concentration of nanofluids: volume fraction and weight fraction. Volume fraction is defined as the ratio of the volume of the nanoparticle to the volume of all the nanoparticles and base fluids. The volume fraction and weight fraction are given as:

$$\varphi_{wt.} = \left[\frac{m_{NP.1} + m_{NP.2}}{m_{NP.1} + m_{NP.2} + m_{BF.1} + m_{BF.2}} \right] \times 100 \qquad (10)$$

$$\varphi_{v} = \left[\frac{\left(\frac{m}{\rho}\right)_{NP.1} + \left(\frac{m}{\rho}\right)_{NP.2}}{\left(\frac{m}{\rho}\right)_{NP.1} + \left(\frac{m}{\rho}\right)_{NP.2} + \left(\frac{m}{\rho}\right)_{BF.1} + \left(\frac{m}{\rho}\right)_{BF.2}} \right] \times 100 \qquad (11)$$

where m and ρ are the mass and true density of the nanoparticles and base fluids, respectively. Subscripts NP and BF represent nanoparticle and base fluids, respectively.

3 Results and Discussion

3.1 Nanofluids Thermophysical Properties

Prior to employing nanofluids in different applications, it is necessary to examine the thermophysical properties to estimate the best use. The experimentation methodology is shown in Fig. 4. However, the presently available theoretical prediction methods

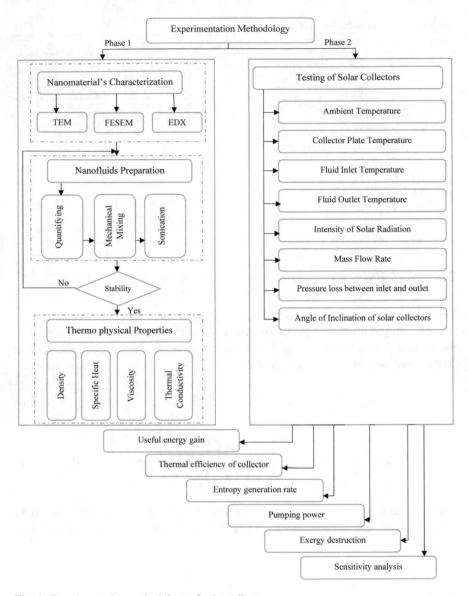

Fig. 4 Experimentation methodology of solar collectors

have failed to give appreciable results and could not predict the properties in high concentration [27].

It has been proven that adding nanoparticles of different thermal properties compared to conventional fluids (water, oils, ethylene glycol (EG), etc.) affects the thermophysical properties of fluids. Generally, the thermal conductivity of the nanofluids is greatly enhanced, while the specific heat is decreased comparable to that of the base fluid. The literature on the dynamic viscosity shows that adding more nanoparticles to the base fluids increased the dynamic viscosity, but it also shows the decreasing trend with increase in temperature. Nanofluids with high dynamic viscosity affect the pumping power and cause pressure loss, thus decreasing efficiency of the solar collectors [28].

3.2 Thermal Conductivity

Thermal conductivity of nanofluids depends on two mechanisms. The first one is the Brownian motion, which is due to motion of solvent molecules. The effective thermal conductivity enhances by the convection caused by the Brownian movement of nanoparticles. The second one is the interfacial nanolayer structure which is formed near by the surface of a solid particle. The effective thermal conductivity is the sum of the static thermal conductivity and Brownian motion thermal conductivity [29]. Thermal properties analyzer KD2 Pro KS-1 was used to measure the thermal conductivity of pure oil and nanofluids. The device works on the principle of transient hot wire method with a maximum deviation of 5.0% and accuracy of ±0.01 W/m °C [30]. There are different models developed by the researchers, which are used to predict the thermal conductivity at different volumes or weight concentrations. These models are shown in Table 5. Although these models do not predict the exact thermal conductivity, they do not consider the important factors such as shape and size of nanoparticles, pH of the nanofluids, agglomeration, and temperature effect on the Brownian motion.

Karami et al. [20] investigated the CNTs behavior in water on seven different volume fractions and with and without surfactants. He concluded to achieve 32% thermal conductivity by adding 150 ppm nanoparticles and also not to use surfactants in CNTs. Asadi et al. [23] calculated the thermal conductivity of MWCNTs-ZnO (15–85%) in engine oil (10W40) at different temperatures, and the maximum enhancement in thermal conductivity was 40% at 1.0% weight concentration. Detailed elaborations of thermal conductivity enhancement of hybrid nanofluids are shown in Table 6.

3.3 Viscosity

The viscosity of nanofluids is very important factor because it affects the pumping power, pressure drop, overall efficiency of solar collectors, and the stability. The viscosity depends on the volume fraction and particle diameter. Viscosity has inverse relationship with the temperature and is directly proportional to the concentration of nanoparticles. As nanofluids' temperature increases, its viscosity decreases due to particle Brownian motion. Hence, the nanoparticles suspended in the fluids start settling down. It has been observed by many researchers that by suspending the

Table 5 Outcomes of thermal conductivity of nanofluids

References	Nanoparticle	Base fluid	Volume fraction (%)	Outcomes
Asadi [23]	MWCNT-ZnO (15–85%)	10W40 (Engine Oil)	0.125, 0.25, 0.5, 0.75, 1	Thermal conductivity increased up to 40% at 1% concentration. Also, small concentration has no effect on pumping power
Vafaei et al. [32]	MWCNTs-MgO	EG	0.05, 0.1, 0.15, 0.2, 0.4, 0.6	New correlation has been proposed by using ANN method with maximum deviation of 0.8%
Esfe et al. [33]	SWCNT-MgO (60–40%)	EG	0.015, 0.025, 0.05, 0.1, 0.2, 0.35, 0.45, and 0.55	Increasing concentration and temperature increased the TCR up to 22.2% at 1.94% VF
Esfe [34]	MWCNT-SiO$_2$ (15:85)	EG	0.05–1.95%	Thermal conductivity increased up to 22.2% for 1.94%
Asadi et al. [10]	MWCNT Al$_2$O$_3$ (15–85)	Thermal Oil	0.125–1.5%	Dynamic viscosity increased with increasing the solid concentration. Maximum enhancement in thermal conductivity is 45%
Abed Moradi et al. [35]	MWCNT-TiO$_2$	EG-Water	0.0625, 0.125, 0.25, 0.5, 0.75, 1.0	34.31% increase in thermal conductivity at 1.0%

(continued)

Table 5 (continued)

References	Nanoparticle	Base fluid	Volume fraction (%)	Outcomes
Balaga et al. [36]	fMWCNT-Fe$_2$O$_3$	DI Water	0.01, 0.02, 0.03 wt %	13.53% increase in thermal conductivity at 0.02%

Table 6 Effective viscosity prediction model

References	Model
Einstein Model [29]	$\mu_{nf} = (1 + 2.5\varphi)\mu_{bf}$
Batchelor [37]	$\mu_{nf} = \left(1 + 2.5(\varphi) + 6.2\left(\varphi^2\right)\right)\mu_{bf}$
Wang et al. [30]	$\mu_{nf} = \left(1 + 7.3(\varphi) + 123\left(\varphi^2\right)\right)\mu_{bf}$

nanoparticles in the base fluids, affects the nature of the fluid's behavior; for example, it changes the Newtonian behavior to non-newtonian behavior of the fluid [29]. Many viscosity prediction models have been proposed by researchers, as listed in Table 6. Thermo-physical properties of nanofluids at different temperature and concentration are shown in Table 7.

3.4 Application in Solar Thermal Collectors

The assessment of a solar collector can be performed by the collector surface area, measured solar irradiance, mass flow rate, heat capacity and temperature of heat transfer fluids at collector inlet and outlet. Alim et al. [9] analyzed the entropy generation and pressure drop by using Al$_2$O$_3$, SiO$_2$, CuO, and TiO$_2$ in water with different concentration and volume flow ranges for flat plate solar collectors. They found that CuO had reduction in the entropy generation by 4.34% and increased the heat transfer coefficient to 22.15% as compared to water.

The stability of nanofluids is very important factor when used in solar collectors as sedimentation can cause the blocking of channels, pipes, pumps, and this will increase the maintenance cost. The issue of sedimentation of nanofluids in solar flat panel collectors was investigated by Colanelo et al. [38]. They used Brinkman's model to calculate the nanofluid's viscosity by modifying the cross section of header pipe to keep a constant velocity profile. They prepared nanofluids of spherical 45 nm Al$_2$O$_3$ and spherical 30 nm Fe$_2$O$_3$ with varying concentration and found 6.7% enhancement in thermal conductivity and 25% improvement in convective heat transfer coefficient. The efficiency of solar collectors directly depends on the heat transfer properties of the fluid. The efficiency of flat plate solar collector was calculated by using CuO with water in 0.4% volume fraction and mass flow rate of 1 to 3 kg/min by Moghadam et al. [21]. The collector efficiency increased to 21.8% with 1 kg/min of mass flow

Table 7 Thermophysical properties of nanofluids

References	Nanoparticle	Base fluid	Volume fraction	Surfactant	Sonication time	Stability	Outcomes
Alarifi et al. [39]	MWCNT-TiO$_2$ (80–20%)	5W50 (engine oil)	0.25, 0.5, 0.75, 1, 1.5, 2	No	1 h	14 Days	Dynamic viscosity increases with increasing concentration. The proposed co-relation has been developed
Asadi [23]	MWCNT-ZnO (15–85%)	10W40 (engine oil)	0.125, 0.25, 0.5, 0.75, 1	No	1 h	10 Days	Thermal conductivity increased up to 40% at 1% concentration. Also, small concentration has no effect on pumping power
Moldoveanu et al. [40]	Al$_2$O$_3$–SiO$_2$	Water	0.50 vol.% Al$_2$O$_3$ + 0.50 vol.% SiO$_2$, 0.50 vol.% Al$_2$O$_3$ + 1.00 vol.% SiO2, 0.50 vol.% Al$_2$O$_3$ + 1.50 vol.% SiO$_2$, 0.50 vol.% Al$_2$O$_3$ + 2.50 vol.% SiO$_2$	No	1 h	10 Days	Hybrid nanofluids have more thermal conductivity than mono nanofluid. Silica's concentration is proportional to the thermal conductivity till certain value

(continued)

Table 7 (continued)

References	Nanoparticle	Base fluid	Volume fraction	Surfactant	Sonication time	Stability	Outcomes
Esfe [33]	SWCNT-MgO (60–40%)	EG	0.015, 0.025, 0.05, 0.1, 0.2, 0.35, 0.45, and 0.55	No	5.5 h		Increasing concentration and temperature increased the TCR up to 22.2% at 1.94% VF
Esfe [34]	MWCNT-SiO$_2$ (15:85)	EG	0.05–1.95%	No	7 h		Thermal conductivity increased up to 22.2% for 1.94%
Madhesh et al. [41]	Cu–TiO$_2$	Water	0.1–2.0%	No	2 h		Nusset number increased up to 49% at 1.0% volume fraction. But marginal decrement was observed after 1.0% VF
Baojie et al. [24]	SiC–TiO$_2$	Diathermic oil	0.1–1.0%	Oleic Acid	2 h	10 Days	Thermal conductivity is linear to the volume fraction

(continued)

Table 7 (continued)

References	Nanoparticle	Base fluid	Volume fraction	Surfactant	Sonication time	Stability	Outcomes
Asadi et al. [10]	MWCNT-Al$_2$O$_3$ (15–85)	Thermal oil	0.125–1.5%	No	1 h	7 days	Dynamic viscosity increased with increasing the solid concentration. Maximum enhancement in thermal conductivity is 45%
Hussein [13]	Alumina nitride	EG	1–4%	No	2.5 h	2 months	Heat transfer enhancement factor and friction factor is dependent on the volume fraction and temperature
Esfe et al. [42]	MWCNT-CuO (1:9)	10W40 (engine oil)	0–10%	No	3 h		Variation in rheological behavior is more dependent on temperature than concentration
Mwesigye et al. [43]	SWCNTs	Therminol VP-1	0.25–2.5%	No			Thermal efficiency depends on thermal conductivity as well as specific heat
Esfe and Abbasian [25]	MWCNT-SiO$_2$ (40–60)	5W50 (engine oil)	0–1%	No			Artificial neural network and mathematical modeling has done to predict the results

(continued)

Table 7 (continued)

References	Nanoparticle	Base fluid	Volume fraction	Surfactant	Sonication time	Stability	Outcomes
Gulzar et al. [26]	Al_2O_3–TiO_2(60–40)	Therminol-55	0.05–0.5%	No	2 h	7 days	Temperature decreased the viscosity of the fluid up to 15% at 145 °C

rate. Gupta [22] used alumina in 0.005% volume fraction in distilled water with 1.5, 2, 2.5 LPM and observed enhancement in collector efficiency to 8.1% for 1.5 LPM and 4.2% for 2 LPM. Direct absorption solar collector was studied experimentally and numerically by Otanicar et al. [44]. Efficiency enhancement up to 5% was observed by using CNTs, graphite and silver. The comparison between experimental and numerical data revealed that 3%, 5%, and 6% efficiencies could be achieved by using graphite's nanoparticles, by using silver nanoparticles, and when the particle size was halved, respectively. An almost 24% enhancement at 0.05 wt. % of water-based Cu nanofluids in flat plate collector was found by Tajik et al. [45]. They used 0.01 wt. % and 0.1 wt. % and analyzed the efficiency variation.

Nanofluids' viscosity was evaluated by using Einstein model of viscosity, and Sint et al. [46] studied theoretically the effect of tilt angle on performance of flat plate solar collector using water-based CuO nanofluids. The viscosity of nanofluids increased with increase in volume concentration and particle size. The overall heat loss coefficient decreased with increasing volume concentration of nanoparticles of up to 2%. At above 2% concentration, an increase in heat loss coefficient was observed. The effect of nanoparticle size was negligible on the efficiency of collector.

Mogadham et al. [21] observed the effect of mass flow rate on the performance of flat plate collectors. They applied 40 nm CuO dispersed in water with a volume fraction of 0.4%. They concluded that collector's efficiency was the function of mass flow rate. The better performance was observed at low mass flow rate. Hybrid mixtures of water and propylene glycol at five concentrations of 0%, 25%, 50%, 75%, and 100% were used as working fluid in flat plate collector by Shojaeizadeh et al. [2] They found that the maximum efficiency was achieved when the ratio was from 25 to 75% was; at beyond 75%, the efficiency decreased by 8.3%. In a recent research, Sundar et al. [2] analyzed the effects of alumina nanofluids in a flat plate collector under turbulent flow. Thermal conductivity and viscosity were measured by using Maxell and Brinkman's models. An enhancement in the collector's heat transfer of up to 49.75% at Re of 13,000 and 0.3% volume was observed. Meanwhile, a thermal effectiveness enhancement by 76% and 58% was observed by using nanofluids. The brief application of nanofluids in solar collectors is shown in Table 7 (Table 8).

4 Summary

The recent advances in the field of applied nanotechnology, the more precise calculation and analyzing technique, and advanced simulation tools have enabled researchers to find out the real-time applications of nanofluids in solar collectors. In this chapter, the detailed preparation, characterization, and experimentation methods are discussed along with models developed by different researchers. Due to modern era of green, environment friendly and sustainable energy, solar collectors have gained attention from industrialists and researcher to improve further and make them more feasible and practical for use in industries and in remote areas.

Table 8 Application of nanofluids in solar collectors

References	Solar collectors	Nanoparticles/base fluid	Outcomes
Karami [20]	Direct sunlight absorber	f-CNTs/water	Thermal conductivity improvement up to 32% by adding only 150 ppm functionalized CNT. Surfactants are not suitable for CNT in DASC
Michael [47]	Flat plate solar water heater	Copper oxide/water	Thermosiphon has higher efficiency than forced. Improved the thermal performance up to 6.3%. Efficiency increases in forced circulation of 0.1 kg/s, but it can be improved by adding more power
Alim [9]	Flat plate solar collector	Alumina, copper oxide, titanium oxide, silicon oxide/water	By increasing the nanoparticles' volume percentage, the heat transfer features improves. CuO nanofluid increases the heat transfer coefficient by 22.15% and reduces the entropy generation by 4.34%. It also has small effect on pumping power by 1.58%
Moghadam [21]	Flat plate solar water heater	Copper oxide/water	Collector efficiency increases by 21.8% with 1 kg/min of mass flow rate
Kasaeian [48]	Parabolic trough collector	MWCNTs/oil	The copper absorber tube, coated with black chrome has the highest absorptivity (0.98) and thermal conductivity. The vacuumed absorber tube has averagely 11% higher efficiency than bare absorber tube, due to convection losses
Gupta [22]	Direct absorption solar Collector	Alumina/Distilled Water	Collector efficiency enhances to 8.1% for 1.5 LPM and 4.2% for 2 LPM

References

1. Hassan A et al (2020) Thermal management and uniform temperature regulation of photovoltaic modules using hybrid phase change materials-nanofluids system. Renew. Energy 145:282–293
2. Wahab A, Hassan A, Qasim MA, Ali HM, Babar H, Sajid MU (2019) Solar energy systems—potential of nanofluids. J Mol Liq 289
3. BP (2020) Statistical review of world energy globally consistent data on world energy markets, and authoritative publications in the field of energy. The statistical review world of world energy and data on world energy markets from is the review has been providing, p 66
4. IE Agency (2011) Technology roadmap. Springer Reference
5. Bellos E, Said Z, Tzivanidis C (2018) The use of nanofluids in solar concentrating technologies: a comprehensive review. J Clean Prod 196:84–99
6. Jesko Z (2008) Classification of solar collectors. Eng Rural Dev 22–27
7. Sahin AZ, Uddin MA, Yilbas BS, Al-Sharafi A (2020) Performance enhancement of solar energy systems using nanofluids: an updated review. Renew Energy 145:1126–1148
8. Choi SUS (1995) Enhancing thermal conductivity of fluids with nanoparticles. Am Soc Mech Eng Fluids Eng Div FED 231:99–105
9. Alim MA, Abdin Z, Saidur R, Hepbasli A, Khairul MA, Rahim NA (2013) Analyses of entropy generation and pressure drop for a conventional flat plate solar collector using different types of metal oxide nanofluids. Energy Build. 66:289–296
10. Asadi A, Asadi M, Rezaniakolaei A, Aistrup L, Afrand M (2018) International Journal of Heat and Mass Transfer Heat transfer efficiency of Al_2O_3-MWCNT/thermal oil hybrid nanofluid as a cooling fluid in thermal and energy management applications: an experimental and theoretical investigation. Int J Heat Mass Transf 117:474–486
11. Usman M, Hamid M, Zubair T, Haq RU, Wang W (2018) International Journal of Heat and Mass Transfer Cu-Al_2O_3/Water hybrid nanofluid through a permeable surface in the presence of nonlinear radiation and variable thermal conductivity via LSM. Int J Heat Mass Transf 126:1347–1356
12. Moghaddam MA, Motahari K (2017) Experimental investigation, sensitivity analysis and modeling of rheological behavior of MWCNT-CuO (30–70)/SAE40 hybrid nano-lubricant. Appl Therm Eng 123:1419–1433
13. Hussein AM (2017) Thermal performance and thermal properties of hybrid nanofluid laminar flow in a double pipe heat exchanger. Exp Therm Fluid Sci 88:37–45
14. Akhter J, Gilani SI, Al-kayiem HH, Ali M, Masood F (2019) Characterization and stability analysis of oil-based copper oxide nanofluids for medium temperature solar collectors. Materwiss Werksttech 50(3):311–319
15. Therminol VP-1 datasheet, downloaded from https://www.therminol.com/pages/products/vp-1.asp, pp 1–5
16. Aberoumand S, Jafarimoghaddam A (2017) Experimental study on synthesis, stability, thermal conductivity and viscosity of Cu–engine oil nanofluid. J Taiwan Inst Chem Eng 71:315–322
17. Botha SS, Ndungu P, Bladergroen BJ (2011) Physicochemical properties of oil-based nanofluids containing hybrid structures of silver nanoparticles supported on silica. Ind Eng Chem Res 50(6):3071–3077
18. Aberoumand S, Jafarimoghaddam A, Moravej M, Aberoumand H, Javaherdeh K (2016) Experimental study on the rheological behavior of silver-heat transfer oil nanofluid and suggesting two empirical based correlations for thermal conductivity and viscosity of oil based nanofluids. Appl Therm Eng 101:362–372
19. Farbod M, Kouhpeymaniasl R, Noghreh Abadi AR, Morphology dependence of thermal and rheological properties of oil-based nanofluids of CuO nanostructures,. Colloids Surf A Physicochem Eng Asp 474:71–75
20. Karami M, Akhavan Bahabadi MA, Delfani S, Ghozatloo A (2014) A new application of carbon nanotubes nanofluid as working fluid of low-temperature direct absorption solar collector Sol Energy Mater Sol Cells 121:114–118

21. Moghadam AJ, Farzane-Gord M, Sajadi M, Hoseyn-Zadeh M (2014) Effects of CuO/water nanofluid on the efficiency of a flat-plate solar collector. Exp Therm Fluid Sci 58:9–14
22. Gupta HK, Das Agrawal G, Mathur J (2015) Investigations for effect of alumina nanofluid flow rate on the efficiency of direct absorption solar collector. Case Stud Therm Eng 5:70–78
23. Asadi A (2018) A guideline towards easing the decision-making process in selecting an effective nano fluid as a heat transfer fluid. 175:1–10
24. Wei B, Zou C, Yuan X, Li X (2017) International journal of heat and mass transfer thermophysical property evaluation of diathermic oil based hybrid nanofluids for heat transfer applications. Int J Heat Mass Transf 107:281–287
25. Hemmat Esfe M, Abbasian Arani AA (2018) An experimental determination and accurate prediction of dynamic viscosity of MWCNT(%40)-SiO$_2$(%60)/5W50 nano-lubricant. J Mol Liq 259:227–237
26. Gulzar O, Qayoum A, Gupta R (2019) Experimental study on stability and rheological behaviour of hybrid Al$_2$O$_3$-$_{TiO2}$ Therminol-55 nanofluids for concentrating solar collectors. Powder Technol 352:436–444
27. Asadi A et al (2019) Recent advances in preparation methods and thermophysical properties of oil-based nanofluids: a state-of-the-art review. Powder Technol 352:209–226
28. Asadi A et al (2018) Heat transfer efficiency of Al$_2$O$_3$-MWCNT/thermal oil hybrid nanofluid as a cooling fluid in thermal and energy management applications: an experimental and theoretical investigation. Int J Heat Mass Transf 117:474–486
29. Ilyas SU, Pendyala R, Shuib AS, Marneni N (2014) A review on the viscous and thermal transport properties of nanofluids. Adv Mater Res 917:18–27
30. Ilyas SU, Pendyala R, Narahari M (2017) Rheological behavior of mechanically stabilized and surfactant-free MWCNT-thermal oil-based nanofluids. Int Commun Heat Mass Transf 87(August):250–255
31. Karabay H (2019) International journal of heat and mass transfer comparison of a theoretical and experimental thermal conductivity model on the heat transfer performance of Al$_2$O$_3$-SiO$_2$/water. 140:598–605
32. Vafaei M, Afrand M, Sina N, Kalbasi R, Sourani F, Teimouri H (2017) Evaluation of thermal conductivity of MgO-MWCNTs/EG hybrid nanofluids based on experimental data by selecting optimal artificial neural networks. Phys E Low-Dimens.l Syst Nanostruct 85:90–96
33. Hemmat M, Esfandeh S, Kiannejad M, Afrand M (2019) A novel applicable experimental study on the thermal behavior of SWCNTs (60 %)-MgO (40 %)/ EG hybrid nano fluid by focusing on the thermal conductivity. Powder Technol 342:998–1007
34. Esfe MH, Behbahani PM, Akbar A, Arani A, Sarlak MR (2017) Thermal conductivity enhancement of SiO$_2$-MWCNT (85: 15 %)– EG hybrid nanofluids. J Therm Anal Calorim 128(1):249–258
35. Moradi A, Zareh M, Afrand M, Khayat M (2020) Effects of temperature and volume concentration on thermal conductivity of TiO$_2$-MWCNTs (70–30)/EG-water hybrid nano-fluid. Powder Technol. 362:578–585
36. Balaga R, Ramji K, Subrahmanyam T, Babu KR (2019) Effect of temperature, total weight concentration and ratio of Fe$_2$O$_3$ and f-MWCNTs on thermal conductivity of water based hybrid nanofluids. Mater Today Proc 18:4992–4999
37. Batchelor BGK (1977) The effect of Brownian motion on the bulk stress in a suspension of spherical particles. 83
38. Colangelo G, Favale E, De Risi A, Laforgia D (2013) A new solution for reduced sedimentation flat panel solar thermal collector using nanofluids. Appl Energy 111:80–93
39. Alari IM, Bader A, Ali V, Minh H Asadi A (2019) On the rheological properties of MWCNT-TiO$_2$/oil hybrid nano fluid : an experimental investigation on the effects of shear rate, temperature, and solid concentration of nanoparticles. 355:157–162
40. Madalina G, Huminic G, Adriana A, Huminic A (2018) International journal of heat and mass transfer experimental study on thermal conductivity of stabilized Al$_2$O$_3$ and SiO$_2$ nanofluids and their hybrid. Int J Heat Mass Transf 127:450–457

41. Madhesh D, Parameshwaran R, Kalaiselvam S (2014) Experimental investigation on convective heat transfer and rheological characteristics of Cu-TiO$_2$ hybrid nanofluids. Exp Therm Fluid Sci 52:104–115
42. Hemmat Esfe M, Zabihi F, Rostamian H, Esfandeh S (2018) Experimental investigation and model development of the non-Newtonian behavior of CuO-MWCNT-10w40 hybrid nano-lubricant for lubrication purposes. J Mol Liq 249:677–687
43. Mwesigye A, Yılmaz İH, Meyer JP (2018) Numerical analysis of the thermal and thermo-dynamic performance of a parabolic trough solar collector using SWCNTs-Therminol®VP-1 nanofluid. Renew Energy 119:844–862
44. Otanicar TP, Phelan PE, Prasher RS, Rosengarten G, Taylor RA (2010) Nanofluid-based direct absorption solar collector. J Renew Sustain Energy 2(3)
45. Jamal-Abad MT, Zamzamian A, Imani E, Mansouri M (2013) Experimental study of the performance of a flat-plate collector using cu-water nanofluid. J Thermophys Heat Transf 27(4):756–760
46. Khin N, Sint C, Choudhury IA, Masjuki HH, Aoyama H (2017) Theoretical analysis to determine the efficiency of a CuO-water nanofluid based-flat plate solar collector for domestic solar water heating system in Myanmar. Sol Energy 155:608–619
47. Michael JJ, Iniyan S (2015) Performance of copper oxide/water nanofluid in a flat plate solar water heater under natural and forced circulations. Energy Convers Manag 95:160–169
48. Kasaeian A, Daviran S, Azarian RD, Rashidi A (2015) Performance evaluation and nanofluid using capability study of a solar parabolic trough collector. Energy Convers Manag 89:368–375

Plate Fin Heat Sink for the Application in Solar Thermal Power

Adeel Tariq, Khurram Altaf, Ghulam Hussain, and Muhammad Reezwan

Abstract Solar power is a renewable energy source and has the potential to be use as a source of electricity around the globe. Various types of solar power plants include parabolic troughs, parabolic dishes, and solar towers. Solar power is generally used in solar collector, solar water heater, solar chimney, and photovoltaic cell. In most of the applications of solar power, heat exchangers are used and for effective exchange of the heat, fins are used. In this chapter, analytical investigation is made for three-dimensional fluid flow and convective heat transfer for designed fin. The aim is to achieve the highest rate of convective heat transfer by changing the relative attack angle of fin. For analytical investigation, simulation software is used to save time and cost of the project by running multiple simulations on different attack angle of the fin. The results are compared with other simulation running with different velocity magnitude. The designed fin with perfect relative attack angle can be used in future industry that needs heat transfer application as it is more reliable and cost-effective.

A. Tariq (✉)
Universiti Teknologi Petronas, House 9, Bandar U24 Seri Iskandar, Perak, Malaysia
e-mail: adeeltariq.238@gmail.com

K. Altaf · M. Reezwan
Department of Mechanical Engineering, Universiti Teknologi Petronas, Block 17, 32610 Seri Iskandar, Perak, Malaysia
e-mail: khurram.altaf@utp.edu.my

M. Reezwan
e-mail: muhammad_22705@utp.edu.my

G. Hussain
Mechanical Engineering Department, GIK Institute of Engineering Sciences and Technology, Topi 23460, Pakistan
e-mail: gh_ghumman@hotmail.com

© The Author(s), under exclusive license to Springer Nature Singapore Pte Ltd. 2021
S. A. Sulaiman (ed.), *Clean Energy Opportunities in Tropical Countries*,
Green Energy and Technology, https://doi.org/10.1007/978-981-15-9140-2_6

1 Solar Power

Solar power is generated from the Sun through solar radiation. Although solar power represents a small percentage of global electricity production, vast amounts of solar energy are being used by many countries. China and the USA lead the world in overall solar electricity, while Germany is a nation with a substantial share of solar power [1]. A range of approaches are used to gain the capacity to harness sunshine to produce power with that energy.

Steam has been used for decades in engineering research. One of the most common instruments used to transform steam into mechanical operation is probably the steam locomotive engine. Growing modern steam engine conducts a comparable cycle at higher energy transmission efficiency. Most steam turbines are used due to their good efficiency to convert steam energy into kinetic rotational energy. Such rotational energy can also be used to drive an electric motor or any other device involving mechanical energy. Historically, the steam required for such processes has been derived from the burning of fossil fuels such as coal or natural gas, though it has used solar thermal energy experimentally for over a century. The steam produced by alternative methods (such as solar radiation) is identical to the steam created by burning coal to heat water, and the processes of converting solar energy into mechanical and electrical energy are basically like those used in combustion systems. Focusing solar thermal systems are best designed for generating high temperatures under higher loads, satisfying the demands of large-scale turbines, which also require a significant amount of high-quality steam.

The solar energy collected and converted to heat by the collector device is transported by the thermal fluid to the storage, and then to a boiler where steam is generated. Additionally, steam is supplied in the heat engine to a generator, where it is converted into mechanical energy, while some energy is discharged. If the electric power is necessary as output, a generator is supplied with the mechanical energy, where it is transformed to electricity. Due to non-100 percent performance, we should assume certain losses at each conversion stage. One of the problems here is that solar collector output decreases with that operating temperature, while heat engine efficiency rises at a higher temperature. Therefore, to choose the working conditions of the system, optimization is important. Usually, the temperatures given by the flat-plate collectors are too small for heat engines to be efficient; thus, the concentration of collectors (e.g., parabolic systems) or evacuated tubular collectors is preferred options.

2 Types of Solar Power Plants

Although there are several different types of solar thermal power plants, they are the same as they are used to absorb sunlight, only at one stage, mirrors concentrate photons. At this stage, the solar power is absorbed and converted into thermal energy, which creates steam to drive a turbine, which produces energy.

2.1 Parabolic Troughs

These troughs, also recognized as line focus collectors, contain a huge parabolic-shaped reflector that focuses on a pipe running down the trough in the incident sunlight. Sometimes the collectors use a single-axis solar tracking system to trace the Sun across the sky as it passes from east to west to ensure that the mirrors also have the maximum exposure to solar power. The receiver pipe in the middle will exceed temperatures above 400 °C, as Sun is focused at 30–100 times its normal intensity [2].

Such troughs are arranged in rows into a solar field. The temperature of the heat transfer fluid increases, as it flows through the tubing into the parabolic trough. The fluid then flows in a central location to heat exchangers, where the heat is transferred to water, creating superheated steam under high pressure. This steam then moves a turbine to power a generator to generate electricity. After the exchange of heat, the fluid cools down and spreads out into the solar field [2].

2.2 Parabolic Dishes

These are large parabolic dishes which track the Sun using motors. This means they can generate the highest possible amount of incoming solar radiation which they then reflect on its focal point. Such dishes can absorb sunlight even better than parabolic troughs, and the fluid that passes through them can radiate heat above 750 °C [2].

Throughout such devices, a stirling engine adds heat to mechanical energy by compressing working fluid when cold and allowing the hot fluid to expand in a piston or pass through a turbine. A generator then converts the mechanical energy into electricity [2].

2.3 Solar Towers

Solar power towers are big buildings which serve as main solar energy recipients. They all concentrate sunlight on a spot in the center of a vast array of mirrors in the network. This many flat mirrors are called heliostats which face the sky. On the pole is mounted a heat exchanger, where heat exchange fluid is heated. Perhaps 1500 times as strong as incident sunlight [2], the heat focused on this level. The hot fluid is then used to produce steam and create electricity to run a turbine and generator.

One downside of these towers is that in order to be productive, they must be very large.

2.4 Benefits and Drawbacks of Solar Power

The ability of these instruments to produce steam at extremely high temperatures causes an effective conversion of heat energy to electricity. These plants also get around the issue of not being able to efficiently conserve energy by being able to store heat instead. Storing heat is more efficient and more cost-effective than storing electricity.

Such plants will also produce dispatchable baseload electricity, which is important because it means that these plants have a constant amount of energy that can be turned on or off at will to satisfy the demands of society [3]. In addition, solar thermal power plants are a type of technology for producing electricity that is cleaner than fossil-fuel-generated power production. These are also two of the cleanest solutions for producing electricity. Nonetheless, these plants do have similar environmental implications as a full life-cycle analysis would show the sources of carbon dioxide used in the creation of these plants. However, emissions are already considerably lower than those linked with fossil-fuel industries.

Some of the drawbacks include the huge area of land needed to operate successfully at these plants. The water requirement of such plants may also be a problem, because enough steam production requires significant amounts of water [4].

The negative impact of these plants on birds is one more potential risk of using large focused mirrors. Birds which travel in the way of Sun's intense rays can be incinerated. Some records of bird deaths at power plants like these amounts to approximately one bird per two minutes [5].

3 Applications of Solar Power Plants

3.1 Solar Collector

A solar collector is a device that collects and/or concentrates solar energy from the Sun. These systems are used mainly for active solar heating and provide for personal water heating [6]. Usually, these collectors are installed on the roof and must be very robust because they are subjected to extreme weather conditions [6].

The use of these solar collectors offers an alternative to traditional home water heating using a water heater, which potentially decreases the energy cost over time. Many such collectors may be installed in an array as well as in suburban settings and used in solar thermal power plants for producing electricity.

There are various kinds of solar collectors, but all rely on the same general concept. A certain material is typically used to capture and channel energy from the Sun to heat up water. The most of these structures use a black film that protects pipes from which water flows. The black material reflects the solar light very well and heats the water it covers. It is a very rudimentary style, but it can be very complex for collectors. Absorber plates can be used where a high temperature rise is not needed but typically devices that use reflective materials to absorb sunlight result in a higher temperature rise.

3.2 Solar Water Heating

Solar water heating (SWH) is the process by which sunlight is converted into energy that can then be used to heat household water. Such hot water is used for laundry rooms, radiant floor heating, or running pools [7]. Solar water heating systems use a wide variety of techniques, and they can be used nearly everywhere in the world. In each of the systems-storage tanks and solar collectors, there are two main components. There are also two major types of solar water heating systems: active and passive, pump-free systems with rotating pumps [8].

Solar water heating systems normally consist of two major parts: storage tanks and solar collectors. The collectors help to absorb heat from the Sun and preserve it. Once the Sun heats the collected water, the heat is transferred on to a liquid known as the heat transfer fluid. When this fluid becomes hot, the collector heat Is removed, and the liquid water is released for use or storage. When this fluid is not hot enough, heat exchangers are used to transfer heat from the heat exchange fluid to a home's water supply [9]. Inactive heating devices, pumps are also used to control the temperature at which the water enters and how rapidly it flows.

3.3 Solar Chimney

A solar chimney is a type of passive solar heating and cooling device which can be used to change the temperature of a building as well as to provide ventilation. Solar chimneys, like a Trombe wall or solar panel, are a means of achieving the buildings' energy-efficient architecture. Solar chimneys are fundamentally hollow containers which connect the interior of the building with the exterior of the building [10].

The system used to heat a room using a solar chimney is relatively easy. The air layer inside the chimney is heating up when the solar light hits the chimney wall. Once the top external vents of the chimney are closed, the hot air is pushed down into the living room. This makes for a form of heating of convective air. For reheating of the air, it is pulled back into the solar chimney as the air in the room cools [11].

While solar chimneys are used for heat, they work in a similar fashion to Trombe walls. Cooling a space with a solar chimney is somewhat distinct from cooling down with a Trombe wall. A roof overhand cannot be installed alongside a solar chimney, although there are two extra vents available. The first vent was talked about, the one at the chimney peak [12]. The second is at the opposite end of the building, providing the space to allow the passage of air between the structure and the outside. Once solar radiation hits the chimney wall, the air layer within the chimney heats up again. The vent at the chimney's top is left open so it does not get trapped in this hot dust. This warm air is pulled up and out of the chimney, drawing new air from the outside and creating a sort of "wind" that brings cool, fresh air to the building [11].

3.4 Photovoltaic Cell

A photovoltaic (PV) cell is a power conversion device which converts solar energy into usable electricity by means of a process called the photovoltaic effect. Many different types of PV cells use semiconductors to interact with incident photons from Sun and generate an electrical current. While silicon (Si) is the most popular commercial building material for solar cells, others include gallium arsenide (GaAs), cadmium telluride (CdTe), and copper indium gallium selenide (CIGS). Solar cells can be formed either from brittle crystalline structures (Si, GaAs) or as compact thin-film cells (Si, CdTe, CIGS). Crystalline solar cells can be further classified into two groups, called monocrystalline and polycrystalline. As the names suggest, monocrystalline PV cells are made of a fixed or single crystal lattice while polycrystalline cells create various or differing crystal structures. Additionally, the number of solar cell layers or "p-n junctions" can be classified. Many consumer PV cells are single junction only, but even multi-junction PV cells that have improved efficiency at a low cost have been developed.

4   Overview of Heat Transfer

An analytical investigation is performed for three-dimensional fluid flow and convective heat transfer for modified plate fins. This chapter aims to achieve a higher rate of convective heat transfer by changing the relative attack angle between the heat sink and fluid flow. For the analytical investigation, simulation software is used to save time and cost of the project by running multiple simulations on different attack angle of the fin. The simulations are performed at different airflow velocities. The modified plate fins with perfect relative attack angle can be used in solar thermal power applications for augmentation of heat transfer while staying cost-effective.

Heat transfer can be described as the flow of heat due to temperature difference. It is a process of heat flow from high temperature toward the low temperature. In terms of thermodynamics, heat transfer is the movement of heat along the system's boundary due to the temperature difference between the surroundings and the system. The heat will continue to flow across the boundary until reaching the same temperature, which is called as thermal equilibrium.

5 Modes of Heat Transfer

There are three modes of heat transfer; conduction, convection, and radiation. Conduction occurs when two solids, having different temperatures, contact with each other. The heat is transferred when adjacent atoms vibrate against one another. The heat transfers from the high temperature region to lower temperature region. The heat will continue to flow until both solids achieve thermal equilibrium, at a point which both solids have the same temperature. Heat can be transferred in three modes which are conduction, convection, and radiation. Conduction can be represented by:

$$Q = \frac{kA(T_2 - T_1)}{d} \tag{1}$$

where Q is amount of heat transferred, k is thermal conductivity of the material, A is the heat transfer area, $T_2 - T_1$ is the temperature gradient (the difference between two temperatures), and d is material thickness.

Convection is a transfer of heat by the mass movement of a fluid when the heated fluid is caused to move toward the opposite of the heat's source, carrying energy with it. Heat convection is often the primary mode of energy transfer in liquids and gases. Convection occurs because the particles of hot air expand and become less dense, which will rise and replaced by the cold air. It created circulation patterns in a system, causing convection currents to transport energy within the air. The convection heat transfer is presented by:

$$Q = hA(T_2 - T_1) \tag{2}$$

where Q is the amount of heat transferred, h is convective heat transfer coefficient, A is the heat transfer area, and $T_2 - T_1$ is the temperature gradient (the difference between two temperatures).

The radiation heat transfer is expressed as:

$$Q = \sigma A(T_1^4 - T_2^4) \tag{3}$$

where Q is the amount of heat transferred, σ is Stefan–Boltzmann constant (5.6703×10^{-8}), A is the heat transfer area, T_1 is hot body absolute temperature, and T_2 is the cold surroundings absolute temperate.

In this chapter, the important equation to be considered is the convective heat transfer equation. In convective heat transfer, a major role in the overall energy transfer process is the bulk fluid motion of the fluid. There are two types of convective heat transfer, which are forced convection and free convection. In forced convection, the fluid is forced to flow over a surface by an external force, while for free convection the fluid motion is caused by buoyancy effect.

6 Heat Sinks

The rate of convection heat transfer can be increased by increasing the temperature difference between the surface and fluid or increasing the exposure or contact area between the surface and fluid. One of the ways to increase the surface area is to use heat sinks (fins). The geometry of fins can be modified to achieve a higher rate of heat transfer as compared to the plain fins. Also, adjusting the attack angle of the fin can increase the rate of convection heat transfer.

Fin efficiency can be interpreted as the ratio of actual heat transfer through the base of fin divided by highest possible heat transfer rate through the base of the fin, which can be obtained if the entire fin is at base temperature. It is very useful in designing a system or in the estimation of the system's performance if we know the fin efficiency. The fin efficiency can be derived from uniform heat transfer coefficient, constant fluid temperature, and heat conduction in one dimensional of the fin.

6.1 Geometrically Modified and Inclined Heat Sinks

According to Jasim and Soylemez [13], the introduction of perforated fin could improve thermal performance of the heat sink. Perforated fin is designed to increase the area of heat transfer so that the amount of heat transferred can be increased. Hence, it can increase the heat transfer efficiency of the fin. Slightly changed the inclination orientation of the perforated fins can increase the amount of heat transferred compared to parallel orientation. Jasim and Soylemez found that few advantages with inclined perforation can be achieved by increasing the inner convection area, increasing the external perforated area, and change the inclination angle. The result of these modifications led to a change in the effectiveness of heat transfer. Improving the thermal performance of pin fin by 65% and decreasing thermal resistance were the results of applying inclined perforation fins.

In the work of Rasel et al. [14], the forced convection applied to the rectangular body with staggered arranged circular fins affected the flow characteristics. When the Nusselt number, Nu, and heat transfer coefficient increased, the velocity increased. As the velocity increased, the heat transfer rate from fin decreased and the base increased. As the velocity increased, the heat transfer for base increased and the heat transfer for fin decreased.

According to Kadbhane et al. [15], the convective heat transfer of fin depended on various factors such as geometrical and operating parameters, number of fins, attack angle of the fins, and material of the fins. Due to the limitation to increase the size of the fins, optimal fins spacing were found as the most important factor to enhance convective heat transfer. Plus, forced convection had better enhancement in convective heat transfer compared to natural convection.

Narato et al. [16] reported that the inclined angle of pin arrays could affect the heat transfer characteristics. They made research and experiment for the case of the angle at 60°, 90°, 120°, and 135° at constant Reynolds number Re of 32,860. They used simulation software ANSYS Workbench to gain the data. Similarly, Yuan [17], Shaeri [18], Ibrahim [19], and Sonawane [20] concluded that geometric modifications and inclinations could enhance the heat transfer rate of the heat sinks.

7 Nusselt Number

The Nusselt number is the ratio of convective to conductive heat transfer at the boundary in the fluid. The higher the Nusselt number, the higher the convective heat transfer. The Nusselt number can be calculated by:

$$Nu = \frac{\text{convective heat transfer}}{\text{conductive heat transfer}} = \frac{hL}{k} \tag{4}$$

where k is thermal conductivity of the fluid, L is characteristic length, and h is convective heat transfer coefficient.

8 Numerical Analysis of Fins

8.1 CAD Modeling

The model geometry of fin design was produced with the assistance of ANSYS SpaceClaim. It is an intuitive 3-D modeling software that enables the user to create, edit, and repair geometry in the workbench. Several conditions were considered while designing the geometry, considering design for manufacturing principles. The heat sink geometry was generated by creating three slots in the plate fins resulting in the conversion of plate fins into rectangular pin fins. The dimensions of the heat sink geometry are described in Table 1. Figures 1 and 2 show the model geometry of the fin design. The material used for this fin design was aluminum.

The relative attack angles of fin used in this project were 15°, 30°, 45°, and 60° from the heat source as shown in Figs. 3 and 4. The designing of the domain was necessary because it acted like an adiabatic wall to avoid heat loss throughout the system starting from inlet to outlet.

Table 1 Heat sink
dimensions

Properties	
Length Unit	Millimeters, mm
Solid/fluid	Solid (defined by geometry)
Base	34 mm × 24 mm × 25 mm
Fin dimension	7 mm × 8 mm × 234 mm
Number of fins	12

Fig. 1 Front view of the
heat sink

Fig. 2 Top view of the heat
sink

8.2   CFD Analysis at Multiple Fin Inclinations

Computational fluid dynamics (CFD) requires discretization or meshing of a model
that continuous in time and space originally. The meshing of a model impacts the
accuracy of the result at the end of the simulation process. It also influences the

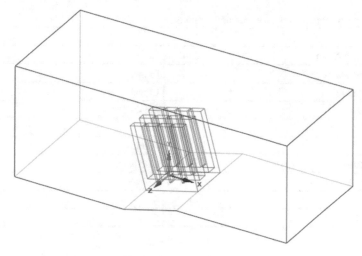

Fig. 3 Isometric view of the heat sink with domain

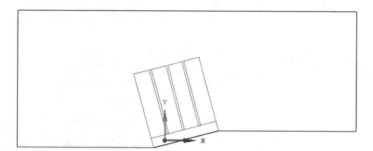

Fig. 4 Rear view of fin design with domain

convergence and speed of the solution when running the calculation. In the computer-aided engineering simulation process, any computational cells such as prisms, tetrahedrons, or hexahedral are few examples of meshing an integral part. A higher number of mesh's cells will consume a long time for the solver to compute. The properties of mesh generated in this case are shown in Table 2 Mesh properties Thus, mesh independence is an important step to reduce the time taken throughout the whole project. The heat sink and domain consist of multiple faces, as shown in Table 3.

8.2.1 Boundary Conditions

Initial and boundary conditions are needed to solve any computational fluid dynamics problems. Boundary conditions are an essential component of a mathematical model. In ANSYS fluid flow (FLUENT), it is equipped with standard boundary conditions

Table 2 Mesh properties of heat sink and air domain

Sizing	
Physics preference	CFD
Solver preference	Fluent
Element order	Linear
Element size	0.3 mm
Growth rate	1.2
Max size	10 mm
Size function	Curvature
Relevance center	Coarse
Smoothing	Medium
Transition	Slow

Table 3 List of the named sections of the heat sink and air domain

Part name	Inlet	Outlet	Heat source	Solid interface	Fluid interface
Geometry (faces)	1	1	1	52	52

Table 4 Type of boundaries of the domain

Domain	Boundaries	
Main body (solid)	Boundary: Air inlet	
	Type	Velocity-inlet
	Boundary: Air outlet	
	Type	Outlet-vent
	Boundary: Heat source	
	Type	Wall

such as inlet, outlet, heat source, wall and medium of the system, as shown in Tables 4 and 5. For the present work in this chapter, several boundary conditions were set up to calculate the results.

8.2.2 Solution Method

In this stage, the computation of the solution is accomplished in ANSYS FLUENT. The objective is to generate a technique of simulating temperature and pressure distribution in the system precisely. The accuracy and speed of the solver depend on geometry configuration and meshing quality. Then, the results can be obtained from the computation to analyze further.

For pressure–velocity coupling solution method, SIMPLE algorithm was used as it is a steady-state calculation, as shown in Table 6. This scheme is referred to

Table 5 Boundary condition for heat sink and air domain

Part: Air inlet	
Parameter	Value
Velocity specification method	Magnitude, normal to boundary
Reference frame	Absolute
Velocity magnitude (m/s)	2/4/6 (each attack angle run with different velocities)
Thermal temperature	24 °C
Part: Air outlet	
Parameter	Value
Velocity specification Method	Magnitude, normal to boundary
Reference frame	Absolute
Initial gauge pressure (Pa)	0
Thermal temperature	24 °C
Part: Heat source	
Parameter	Value
Thermal conditions	Heat flux
Heat flux (W/m^2)	8578
Heat generation rate	0
Material name	Aluminum

Table 6 Pressure–velocity coupling spatial discretization

Spatial discretization	
Gradient	Least squares cell-based
Pressure	Second order
Momentum	Second-order upwind
Turbulent kinetic energy	First-order upwind
Turbulent dissipation rate	First-order upwind

as a pressure-based segregated algorithm. The SIMPLE algorithm is the default for pressure–velocity coupling solution. It is suitable for one or more skewness correction schemes up to a value of 0.7 and for simple flow involving turbulence or laminar physical model only.

Standard initialization was used, and the computation was from the inlet. The number of iterations for calculation was 50, with reporting interval and profile update interval of 1.

9   Thermal and Flow Behavior of Inclined Fins

The results and outcomes obtained based on simulation calculation are exported in visualization and data to acquire the anticipated information. The ANSYS results were exploited in the form of contour plots and probe tool for temperature and pressure value at certain points. In Fig. 5, this is one example of the contour distribution of temperature and pressure from ANSYS. The temperature distribution is portrayed at the solid fin, while the pressure distribution is portrayed at the plane along the domain starting from inlet to outlet.

9.1   Temperature Distribution in Heat Sink

Under the applied boundary conditions, thermal behavior of the heat sink can be observed by looking at Table 7, which describes the temperatures at the inlet, outlet and heat sink base, whereas Fig. 6 represents the trend of the base temperature of the heat sink at different inclination angles and different airspeeds.

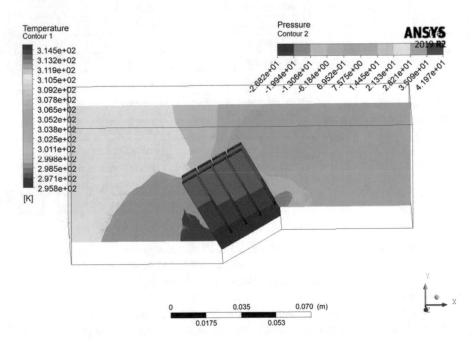

Fig. 5 Temperature distribution in heat sink and air domain

Table 7 Temperature distribution at different sections of heat sink and air domain

Angle (degree)	Velocity (m/s)	Temperature (°C)		
		Inlet	Base	Outlet
15	2	24.00	50.13	26.93
	4	24.00	40.04	25.34
	6	24.00	38.66	24.79
30	2	24.00	59.55	26.38
	4	24.00	47.51	25.33
	6	24.00	42.20	25.01
45	2	24.00	58.21	25.09
	4	24.00	45.94	25.45
	6	24.00	42.07	24.96
60	2	24.00	59.65	26.07
	4	24.00	47.50	25.40
	6	24.00	42.31	24.73

Fig. 6 Base temperatures of the heat sink at different inclinations and various airspeeds

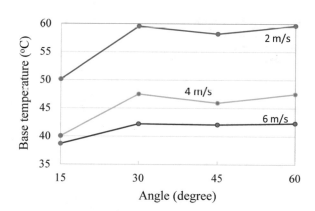

9.2 Pressure Distribution Across Heat Sink

The pressure drop across the heat sink is calculated by subtracting the outlet pressure from inlet pressure of the domain, as shown in Table 8. The pressure drops across the heat sink at different inclinations and different airspeeds are shown in Fig. 7.

10 Summary

From the results obtained, it can be concluded that changing the relative attack angle of the fin can affect the convective heat transfer. The highest convective heat transfer

Table 8 Pressure distribution around the heat sink

Angle (degree)	Velocity (m/s)	Pressure (Pascal)		
		Inlet	Outlet	ΔP
15	2	1.75	0.00	1.75
	4	7.04	−0.01	7.05
	6	15.44	−0.01	15.45
30	2	3.08	0.00	3.08
	4	11.92	−0.02	11.94
	6	26.69	−0.02	26.71
45	2	5.16	−0.01	5.17
	4	20.21	−0.03	20.24
	6	43.74	−0.06	43.80
60	2	9.32	−0.01	9.33
	4	36.83	−0.06	36.89
	6	82.56	−0.20	82.76

Fig. 7 Pressure drop of the heat sink at different inclinations and various airspeeds

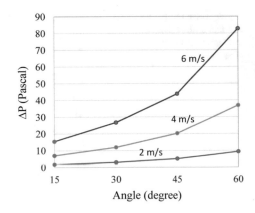

efficiency has been achieved by fin with an angle of 30° and 60°. The lowest to achieve high convective heat transfer efficiency is the fin with an angle of 15°. The velocity magnitude affects on base and outlet temperature, as the higher the velocity magnitude, the lower the final temperature. This is due to the forced convective heat transfer from inlet to outlet, passing the fin. The fin with the highest pressure drop in this project is for the fin with an angle of 60° at 6 m/s. The pressure drop is influenced greatly by the velocity magnitude. All results with velocity magnitude of 6 m/s have a high pressure drop. Changing the relative attack angle of the fin can affect the convective heat transfer. A deep study needs to be carried out to understand how suitable angle with specific design affects the convective heat transfer. This study would be helpful for other engineers to design fins for a system for world industry for the future.

References

1. Shahan Z (August 19, 2015). Top solar power countries (Online). Available: http://cleantech nica.com/2012/06/12/top-solar-power-countries-per-capita-per-gdp-per-twh-of-electricity-produced-in-total/2/
2. EIA (August 18, 2015) Solar thermal power plants (Online). Available: http://www.eia.gov/Energyexplained/?page=solar_thermal_power_plants
3. Visions of Earth (August 18, 2015) Solar thermal power (Online). Available: http://www.visionofearth.org/featured-articles/solar-thermal-power/
4. Wisions (August 11, 2015) Solar power tower (Online). Available: http://www.wisions.net/technologyradar/technology/solar-power-towers
5. Fecht S (August 11, 2015) Solar power towers are vaporizing birds (Online). Available: http://www.popsci.com/solar-power-towers-are-vaporizing-birds
6. Boyle G (2004) Renewable energy: power for a sustainable future, 2nd edn. Oxford University Press, Oxford
7. CANSIA (August 17, 2015) What is solar water heating (Online). Available: http://www.cansia.ca/solar-energy-101/what-solar-water-heating
8. Energy.gov (August 17, 2015) Solar water heaters (Online). Available: http://energy.gov/energysaver/articles/solar-water-heaters
9. Home Power (August 17, 2015) What is solar water heating? (Online). Available: http://www.homepower.com/articles/solar-water-heating/basics/what-solar-water-heating
10. Greenzly (August 7, 2015) How do solar chimneys work? (Online). Available: http://www.greenzly.org/article/how-does-solar-chimney-work/277
11. Autodesk Sustainability Workshop (August 10, 2015) Stack Ventilation and Bernoulli's Principle (Online). Available: http://sustainabilityworkshop.autodesk.com/buildings/stack-ventilation-and-bernoullis-principle
12. Bansal N, Mathur R, Bhandari M (1993) Solar chimney for enhanced stack ventilation. Build Environ 28:373–377
13. Jasim HH, Soylemez MS (2018) The effects of the perforation shapes, sizes, numbers and inclination angles on the thermal performance of a perforated pin fin. Turk J Sci Technol 13(2):1–13
14. Rasel MAJ, Islam MZ, Hasanat A (2016) Analysis of heat transfer characteristics under forced convection in a rectangular body with circular fins. Am J Eng Res (AJER) 5(10):311–316
15. Kadbhane SV, Palande DD (2016) Review of convective heat transfer from plate fins under natural and mixed convection at different inclination angle. Int Res J Eng Technol (IRJET) 3(2):467–473
16. Narato P, Wae-hayee M, Vessakosol P, Nuntadusit C (2017) Effect of inclined angle of pin arrays on flow and heat transfer characteristics in flow channel. In: Materials science and engineering. Songkhla, Thailand
17. Yuan ZX, Zhao LH, Zhang BD (2007) Fin angle effect on turbulent heat transfer in a parallel-plate channel with flow-inclining fins. Int J Numer Meth Heat Fluid Flow 17(1):5–19
18. Shaeri MR, Yaghoubi M (2009) Numerical analysis of turbulent convection heat transfer from an array of perforated fins. Int J Heat Fluid Flow 30:218–228
19. Ibrahim TK, Mohammed MN, Mohammed MK, Najafi G, Sidik NA, Basrawi F, Abdalla AN, Hoseini SS (2018) Experimental study on the effect of perforations shapes on vertical heated fins performance under forced convection heat transfer. Int J Heat Mass Trans 118:832–846
20. Sonawane R, Palande DD (2016) Heat transfer enhancement by using perforation. Int Res J Eng Technol (IRJET) 3(4):2624–2629

An Overview of Biomass Conversion Technologies in Nigeria

Hadiza A. Umar, Shaharin A. Sulaiman, Mior Azman Said, Afsin Gungor, and Rabi K. Ahmad

Abstract In Nigeria, maintaining steady power supply to the teeming population of over 200 million people has become a challenge. The country stands to be the most populous and yet having the highest economy in the African region. The present electricity generation capacity of 12,552 MW is not enough to cater for the rural and urban dwellers, and this constitutes a great challenge to the country's development. As such there is a need for sustainable and clean energy to cater for the ever increasing energy demand. In this regard, this chapter looks into the available biomass conversion techniques in the country, their technology levels and the benefits of adopting biofuel technology. It further dwells on the challenges impeding their implementation which include lack of workable policies, research funding, public awareness and loan and incentive schemes. The chapter provides recommendations to address the issues through proper policy and framework, research, training and development, public enlightenment and award of loans. The study identifies anaerobic digestion as the most promising conversion technology in the Nigerian scenario for a clean and sustainable energy although other processes like gasification and pyrolysis are also promising.

H. A. Umar (✉) · S. A. Sulaiman · M. A. Said · R. K. Ahmad
Department of Mechanical Engineering, Universiti Teknologi PETRONAS, 32610, Seri Iskandar Perak, Malaysia
e-mail: hadiza_16000717@utp.edu.my

S. A. Sulaiman
e-mail: shaharin@utp.edu.my

M. A. Said
e-mail: miorazman@utp.edu.my

R. K. Ahmad
e-mail: rabi_17000319@utp.edu.my

A. Gungor
Department of Mechanical Engineering, Faculty of Engineering, Akdeniz University, 07058 Antalya, Turkey
e-mail: afsingungor@hotmail.com

Keywords Biomass · Nigeria · Gasification · Anaerobic digestion · Transesterification · Combustion

In the modern contemporary world, there is no doubt that electricity generation has become mandatory for socio-economic development. However, in Nigeria maintaining steady power supply to the teeming population of over 200 million people has become a challenge. The country stands to be the most populous and yet having the highest economy in the African region, whose economy depends largely on fossil fuel reserves. The present electricity generation capacity of 12,552 MW is not enough to cater for the rural and urban dwellers, and this constitutes a great challenge to the country's development. As such there is a need for sustainable and clean energy which can be cheaply sourced to cater for the ever increasing energy demand. In this regard, this chapter looks into the available biomass conversion techniques in the country, their technology levels and the benefits of adopting biofuel technology. It further dwells on the challenges impeding their implementation which include lack of workable policies, research funding, public awareness and loan and incentive schemes. The chapter is concluded based on recommendations to address the issues through proper policy and framework, research, training and development, public enlightenment and award of loans. The study identifies anaerobic digestion as most promising conversion technology in the Nigerian scenario for a clean and sustainable energy due to its high potential, even though other processes like gasification and pyrolysis are also promising.

1 The Nigerian Energy Scenario

The present scenario of energy in Nigeria is a very complex issue. Although the country is endowed with abundant natural resources and fossil fuel deposits as shown in Fig. 1. The rapid energy demand increase has been due to the growth in technologies, economies and also the human population. Nigeria has the capability in the use of renewable energy sources like biomass, wind, hydro, tidal energy and solar as the major energy source, yet it depends highly on non-renewable energy. The Nigerian populace suffer the problem of power shortage. Nigeria is the tenth largest exporter of crude oil globally, with a fossil fuel reserve of about 37 million barrels, 182 trillion cubic feet natural gas and 209 million short tons of coal [1]. On the other hand, high oil prices before the COVID-19 pandemic greatly affected Nigeria's export revenue. However, the country's electricity generation solely depends on natural gas and coal, and to some extent the hydropower. Alas, the total installed electricity capacity generated in Nigeria is only 12,522 MW, which is less than the present demand of 98,000 MW throughout the country. Less than 45% of the country's population has direct access to electricity [2] and 50% of the households in the country are connected to gridline [3]. Most of the private industries, urban and rural household rely on self-generated electricity by operating diesel/petrol-fuelled generators. Such

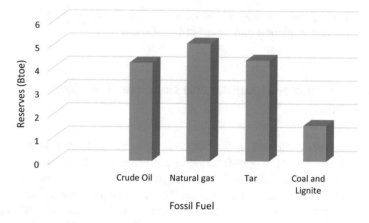

Fig. 1 Nigerian fossil fuel reserves

problem has been lingering for the past twenty years, ranging from power generation, distribution and transmission. This is due to reasons like deficiencies in electricity production, losses on electrical network infrastructure, the challenge of energy access [4], and failed transformers and transmission lines resulting from substandard products. However, the rural areas suffer transmission problems due to poor road network and terrain. Huge amount of funds is needed for installation and distribution as the villages are far away from the grid infrastructure. Other problems are identified as poor voltage regulation and low load factor [5]. There exists a huge gap between the demand and supply of electricity thereby crippling most of the country's industries which results to economic crisis [1]. In the absence of reliable performance, the Nigerian power sector was reformed in 2005 (power sector reform act). The reform agenda constituted the establishment of National Electricity Regulatory Commission (NERC) as a stand-alone agency to overlook the electric power sector. The reform further looked into reviewing tariffs and implementing subsidies to promote different energy mix including renewable energy. Despite the massive monetary expenditure to improve the power sector, misappropriation of the funds retarded the expected development [5]. Up-to-date urban areas suffer the habitual load shedding and hence are left for hours without electricity on a daily basis.

Amongst the existing renewable energy resources available in the country, hydropower is the only source that generates sufficient amount of power about 2000 MW [5] that is incorporated into the grid. This is being generated from about 200 dams across the country with a total capacity of 15 billion cm^3/year [6]. Other renewable sources that are utilized include small-scale solar systems which are utilized for street lighting, household and schools, and water pumping in rural communities. The biomass is also largely utilized for cooking and heating purposes through combustion.

For a progressive economy, there is no doubt that steady, clean and sustainable power is required which is possible through renewable energy sources. The fossil fuel reserves are decreasing with time, as such have to be replaced with alternative fuels. Biomass stands to be a good choice due to its abundance. It is characterized as sustainable, non-depleting and environmentally safe.

2 Trend of Energy Demand and the Drive for Biomass in Nigeria

The trend of energy demand in the Nigerian residential sector both in rural and urban households has been spontaneous in the last one to two decades. This has been triggered by the increasing per capita growth domestic product (GDP), rise in what population and improved lifestyle. The main energy sources on which this sector depends are majorly wood, charcoal, electricity, kerosene and LPG. The total commercial energy consumption of the sector has been changing in the past years with increment at a compound annual growth rate (CAGR) of 1.93% from 101 peta joules (PJ) in 1990 to 148 PJ in 2010 [7].

Households utilize energy to accomplish different tasks like cooking, heating, lightning and appliances operation such as refrigerators, air conditioning systems and television sets. In the rural areas, majority of households depend on fuel wood for cooking, some few use kerosene while the smallest percentage about 3.4% use modern cooking fuels. In terms of electrification of the households, only 41% have been electrified. However, in the urban areas, electricity access has improved tremendously over the years with almost every house hold connected to the grid except for uncompleted buildings. In terms of modern cooking fuels, most households utilize kerosene, only 8.5% have access to clean cooking fuels.

The energy requirements for cooking vary from charcoal, kerosene, fuel wood, LPG and electricity. Fuel wood dominates the rural areas as it is easily collected from forests close to the rural areas. In urban areas, it is made available by vendors who sell it at cheap rates compared to LPG and kerosene. It is worth noting that use of LPG and electricity for cooking is adapted by very few households in the urban areas which may be attributed to the level of income of the household. However, other minute factors like availability and cultural norms influence the choice of cooking fuels. Recently, there have been discussions on plans and efforts by the government and civil societies to provide all Nigerian households with modern cooking equipment by 2030 [8]. In terms of lighting, Nigerian households rely on electricity, dry cell battery, kerosene and candles, but at national level, electricity and kerosene serves as the major sources of lighting for the larger percentage of the households [8].

The industrial sector has also witnessed some form of transitions in terms energy supply and distribution. Initially, industries depended on the government-owned utility company for power supply. The electric grid was initially characterized by frequent temporary interruptions which later developed into consistent power shortages which led to abrupt process halting, process restarts and product wastage. These had negative impacts on the industries productivity, and hence sparked the need for self-generated power. Many industries ventured into the self-generated power scheme; unfortunately, many industrial plants that have invested in this way are currently underutilizing their power plants, with utilization factor as low as 50% in some cases [4]. Some manufacturers are addressing the utilization challenge by opting for scalable solutions for their electrical power plants. However, the need to reduce related cost of underutilized power plant capacity and the need for most manufacturers to concentrate more on their core business interest lead to industrial energy outsourcing. Outsourcing is seen as a means of curbing capital investments. This system allows an independent service provider to supply energy functions such as electricity and steam generation. However, there is a need for increased energy investment, efficiency and conservation to further reduce related costs which leads to transition to industrial energy conservation era. This era is attributed to emphasis on energy efficiency and conservation measures, which assist in terms of waste and cost reduction, hence increasing competition amongst the industries [4].

The problem of unstable power supply poses a great challenge to both residential and industrial sectors in Nigeria, as such both sectors heavily depend on petrol and diesel generators to complement the electricity from the grid. This has impacted negatively on the cost of economy in the rural areas, manufacturing and exports. The Nigerian households and small business spend more than twice on kerosene, petrol and diesel then they do on grid electricity. Studies have shown that about 15,000 MW installed capacity of generators exist in the country with the generator sets ranging from 0.5 kVA for small traders to 75 kVA/60 kW serving residential estates and industries across the country.

In the effort of the Nigerian government to secure a stable electricity supply, a program sustainable energy for all (SE4All) was launched. It features three major ambitious target by 2030;

1. Access to electricity for all

The sustainable energy for all action agenda targets 90% of both urban and rural areas to be fully electrified by 2030.

2. Renewable energy share of 30%

According to the SE4All agenda and the renewable energy and energy efficiency plan (REEEP), Nigeria intends to generate about 9000 MW of on grid power from renewable energy sources by 2030. This is intended to be 30% of the country's energy mix.

3. Climate emissions target

In compliance with the Paris Agreement on climate change, Nigeria is set to reduce its carbon emission from the power and economic sectors by 2030. The target for emissions reduction is to keep it below 300 Mt CO_2e (million tons of carbon dioxide equivalent) by 2020. Hence, in order for Nigeria to achieve these targets, diversification of power sources into renewable energy especially the biomass becomes mandatory.

3   Biomass as a Source of Energy in Nigeria

Biomass is a natural occurring matter which is composed of chemical energy that is capable of producing energy carriers in the form of solid, liquid or gaseous fuels and is sourced from plants and animals. This includes forest resources such as grass, trees, shrubs, wastes from animals, industrial, municipal and agricultural residues [9]. Located in the sub-Saharan West African region, Nigeria is characterized by thick dense forest which houses huge stem trees, shrubs and grasses in the southern part of the country. Nigeria is a country blessed with arable land of 34 million ha [10]. The country has an environment and geography that supports the growth of plants. It has different plantations which lead to abundant biomass resources [11]. In Nigeria, agricultural residues are highly important sources of biomass fuels for both the domestic and industrial sectors. The availability of primary residues for energy application is high but due to difficulty in collection and knowledge of converting it to useful biofuel, biomass is extensively used to provide heat and electricity, fertilizers and animal feeds as well as biofuel for the rural community [2].

Nevertheless, secondary residues are in plentiful quantities at the processing site. They can be used as a captive energy source for processing plants that require minimum transportation and handling cost. Biomass usage helps in waste management, increases revenue generation and improves productivity for industries. Biomass has the potential to supply more energy than solar and wind energy. Figure 2 shows the forest types found in Nigeria based on data from the literature [12].

The country has a total of 36 states, with its land area reaching to 92,376,000 ha out of which agriculture covers 71 million ha. About 90% of the harvested wood is used as fuel wood for cooking and heating by low-income households and the rural populace. Nigeria possesses sufficient biomass potential of about 144 million tons/year, from which most Nigerians utilize which accounts for 80% of the basic

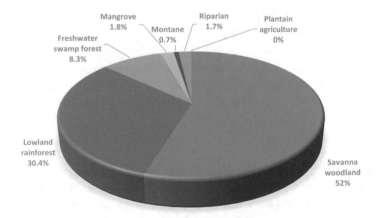

Fig. 2 Nigerian forest types

Table 1 Various agricultural produces and their biomass residue

Crop	Residue	Crop	Residue
Cocoa	Pods, husk	Cassava	Peelings
Paddy	Straw, husk, stalk	Bean	Stalk
Groundnut	Shells, stalk	Yam	Peelings
Coconut	Husk, shell	Palm fruit	Palm kernel shell
Maize	Cob, stalk	Sugarcane	Bagasse
Wheat	Straw, husk	Sorghum	Stalk
Soy bean	Stalk		

consumed energy [13]. The development of biomass energy technologies will not only significantly increase Nigeria's electricity capacity but also ease power shortages in the country [2]. This can be achieved through substantial investment, development of indigenous technologies, stakeholder cooperation and community awareness. Africa's most populous country needs more than 10 times its current electricity output to guarantee supply for its 198 million people where nearly half of whom have no access to electricity at all. Table 1 shows the different types of biomass used in Nigeria and their corresponding residues while Fig. 3 shows photographs of selected residues. The average annual crop residue and their corresponding energy potential in the six geopolitical zones of the country are displayed on Figs. 4 and 5, respectively. The geopolitical zones are North Central (NC), North East (NE), North West (NW), South East (SE), South (SS) and South West (SW).

4   Major Biomass Classification in Nigeria

(a) Energy crops: Crops that have high energy content that are capable of producing biofuel are referred to as energy crops. Such crops include sugarcane, cassava, corn and sorghum from which bioethanol is produced, and others like soy bean, jatropha, oil palm, cotton and coconut for biodiesel [14]. The country has huge potential for the growth of these types of crops due to abundant arable lands and water; however, such energy crops are used solely as food and not for energy production in order to ensure food security. The potential of utilizing such crops for biofuel and biodiesel is shown in Figs. 6 and 7, based on data from the literature [1], in order to estimate bioenergy production capacity. Images of some energy crop growing farms are shown in Fig. 8

(b) Agricultural resources: These are residues that are generated during harvesting and processing of agricultural products. At the point of harvesting, residues obtained include leaves, stalks and straws, while at the processing site cobs of corn/millet, cocoa pods and peels of cassava/yam are generated. All these residues have potential of generating bioenergy since they have remarkable heating values, as shown in Table 2, based on data from the literature [13].

Sugarcane bagasse   Groundnut shells

Corn cobs   Coconut shells

Rice husk   Cassava peels

Fig. 3 Some Nigerian agricultural biomass residue

(c) Forest residues: Forest residues are the unused biomass part of forest resources, usually obtained from wood processing during carpentry. This includes saw dust, veneer rejects, wood scrapings, edgings, slabs and trimmings which can be used to generate heat, electricity, solid and liquid fuel [14].

(d) Wastes generated by human activities: The Nigerian environment is extremely polluted by the numerous wastes generated daily classified as municipal solid waste (MSW), food waste, animal waste and industrial waste. These wastes have the potential to be utilized in biofuel generation especially biogas through the process of anaerobic digestion [13, 15]. Municipal solid waste generates as a result of human activities in commercial, residential and industrial sectors. These wastes are majorly biodegradable due to few industrial activities in the country and so making it suitable for biogas generation. On a daily basis, about 74,428 ton of MSW is generated which is capable of producing 2.01 million m^3 of biogas [12]. However, despite this advantage MSW is usually burnt or used for landfills. Food wastes are generated by hotels, restaurants and food

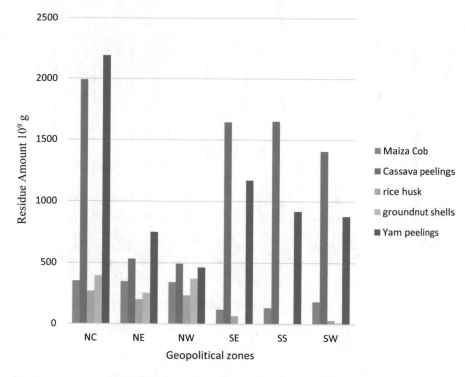

Fig. 4 Average annual residue of major crops in the different geopolitical zones

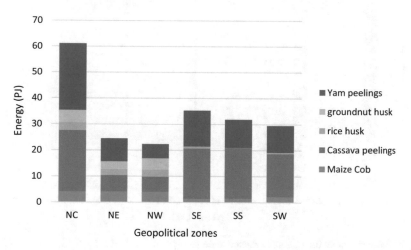

Fig. 5 Energy potential of major crop residues for the different geopolitical zones in Nigeria

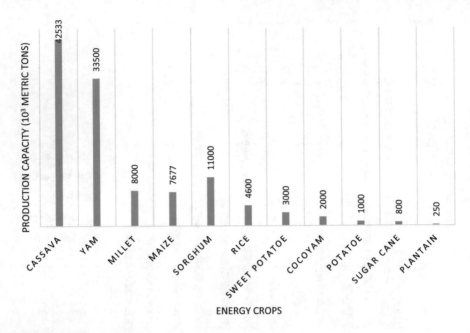

Fig. 6 Biofuel potential of some energy crops

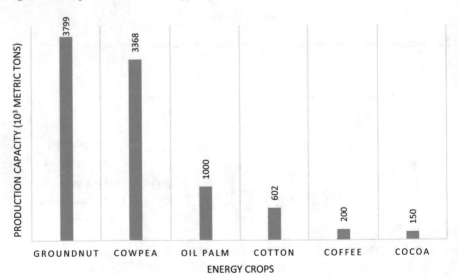

Fig. 7 Biodiesel potential of some selected energy crops

Fig. 8 Energy crops plantation in Nigeria

Table 2 Some major crop residues and their energy contents

Crop	Residue type	Total residue (tons)	LHV (MJ/kg)	Energy potential (PJ)
Rice	Straw	7.86	16.02	125.92
	Husk	1.19	19.33	23.00
Cowpea	shell	9.77	19.44	95.06
Sugar cane	Bagasse	0.14	18.1	1.99
	Leaves	0.14	15.81	2.21
Cassava	Stalks	85.07	17.5	297.68
	Peelings	127.6	10.61	812.3
Maize	Cob	15.35	19.66	211.35
	Husk	2.10	16.28	34.19
	Stalks	1.54	15.56	14.35
Groundnut	Shells	1.81	15.66	28.35
	Straw	8.74	17.58	76.83
Millet	Straw	9.05	12.38	89.63
Sorghum	Straw	8.93	12.38	88.39

processing industries which may contain fruits vegetables, dairy products and food remnants. The content of the food waste may determine the type of biofuel to be generated example, oily food wastes may be used for biodiesel production, starch and sugary wastes may be used for biogas. Animal wastes are produced from livestock which are reared in the farm lands. In the northern part of Nigeria, cattle, sheep and goat rearing is very predominant, that of pig dominates the southern part. The rearing of poultry is also a common feature of most households both in rural and urban areas. These wastes generated by the animals are excellent source of biogas. On a daily basis, Nigeria generates 227,500 tons of fresh animal waste capable of producing 6.8 million m^3 of biogas [13]. The conversion methods in Nigeria are majorly thermochemical (gasification and combustion) and biochemical (fermentation and anaerobic digestion).

5   Application of Biomass Conversion Technologies in Nigeria

Conversion of biomass involves its transformation into different fuel types, and this depends on several factors such as the method applied, the type of feedstock, moisture content, end use and the economy of the process [13]. In Nigeria, few conversion technologies are applied and are yet to reach full commercialization scale.

(a)   Gasification

Gasification is a thermochemical conversion process that takes place within a gasifier which involves the application of high temperature 700–1000 °C to a feedstock. The end product is syngas, a mixture of CH_4, CO, CO_2 and H_2 [16]. In Nigeria, gasification plants are not very common due to financial and technical constraints. In the south-eastern part of the country, Ebonyi state, a rice husk gasification plant with 25 MW electric capacity was put in place through the joint effort of the state government and United Nations agency. The plant provides electricity to street lights, primary healthcare centre, information and communication centre, and an existing palm kernel processing plant. Another gasification plant has also been established in the same state with the aim to supply electricity to the small scale rice millers around the area, Ebonyi agro-rice mill, local traders, Oferekpe mega water scheme [5].

(b)   Anaerobic digestion

Anaerobic digestion is a biochemical process of producing energy from high moisture biomass which may be agricultural or other biodegradable residue. The high moisture biomass of about 90% is acted upon by microorganisms in an oxygen-deficient condition. This results in CO_2, CH_4 rich gas (biogas) and traces of other gases like H_2S. The digestate obtained after the process is utilized as fertilizer [13]. In Nigeria, biogas research was initiated since 1982, and since then ongoing research has been carried out in the sector at pilot scale. Currently, numerous biogas plants with capacity 10–20 m^3 exist which include [17]:

1. Zaria Prison, Kaduna State
2. Ojokoro, Lagos State
3. Mayflower School Ikene, Ogun State.
4. Usman Danfodio University Sokoto, Sokoto State.

Other Nigerian universities that have also embarked on the biogas project through the construction of biogas digesters include Obafemi Awolowo University, University of Illorin, Ladoke Akintola University of Technology, Ogbomoso, Ambrose Ali University, Nnamdi Azikwe University, Awka, Ibrahim Badamasi Babangida University, Lapai. Non-governmental organizations like the United Nations Development Program (UNDP) also introduced the low technology biogas system which was operated in some selected northern states that include Kano, Yobe and Jigawa States. Since then, Kwachiri community of Kano state depended on the biogas plant which was operated on cow dung for their cooking needs. The UNDP has also set up similar projects in market abattoirs of some selected northern states [1].

Another potential biomass for biogas generation is the municipal solid waste. It has a high potential of biogas generation due to its moisture content, and it is also readily available about 25 million tonne is generated annually in Nigeria. According to Odetoye [18], the recent trend in biogas research is on yield increment and quality enrichment to improve the purity level of the gas.

(c) Combustion

Combustion is also a thermochemical conversion whereby feedstock is burned in the presence of excess air to result in high-temperature heat. Combustion produces energy in the form of heat and steam, which is used to generate electricity. Direct combustion is the most common biomass conversion procedure in Nigeria, as most household's combust agricultural residues for the purpose of cooking and heating. The native cooking procedure which employs the use of fire woods has low efficiency (5–10%) and also poses threats of health hazards. In order to minimize the adverse effects, the Energy Commission of Nigeria (ECN) developed customized cooking stoves with the following features [1]:

- Enclosed hearth (box-shaped), in which wood combustion takes place within, thereby preventing heat lost due to convection and radiation.
- Multi-pot design, which is able to accommodate more than a pot at a time, hence utilizing recovered heat that may otherwise have been lost.
- Chimney for fresh air inflow and passing out of smoke to make the kitchen more convenient.
- Briquetting

This is a physical conversion method where by an extrusion machine transforms sawdust and husk of rice and millet to briquettes. These briquettes are later used as fuel for power generation. There are quite a number of briquetting machines that have been developed by the Nigerian tertiary and research institutes. In Sokoto Energy Research Centre, Usman Dan Fodio University Sokoto, a briquetting machine was developed which is capable of producing 13 kg/h of briquettes. Other machines exist in Obafemi Awolowo University, Ile-Ife, University of Ibadan and Abubakar Tafawa Balewa University (ATBU), Bauchi Nigeria. The briquetting machine in ATBU was developed by the university's centre for Industrial Studies (CIS) in collaboration with Raw Material Research and Development Council (RMDC) Abuja. The machine is capable of producing 40 kg/h of sawdust/agricultural waste briquette. It is worth noting that small-scale companies that produce and market these briquettes do exist in Kaduna and Ogun states [1].

(e) Transesterification

Transesterification is a process in which biodiesel is produced through the reaction of glycerides and alcohol specifically methanol or ethanol in the presence of a catalyst, to produce fatty acids, alkyl esters and alcohol. Odetoye [18] reported several researches carried out in Nigeria, on the production of biodiesel using different feed stocks such as palm kernel oil, mango seed oil, coconut oil, tobacco seed oil, jatropha oil, pinari oil, sheabutter oil and milk bush oil. Alamu et al. [19] reported the production of palm kernel oil biodiesel using sodium hydroxide as catalyst. The fuel exhibited properties that were in agreement with the American Society for Testing and Materials (ASTM) standards and hence could be used as fuel in biodiesel engines. In another research by Umaru et al. [20], mango seed oil was used for biodiesel production which yielded over 80% biodiesel at optimal temperature of 60 °C. The biodiesel was in good agreement with the threshold standard values given by ASTM for biodiesel and fossil diesel.

6   Benefits and Prospects of Biofuel Industrialization in Nigeria

The industrialization of biofuel in Nigeria can lead to numerous benefits as described in Fig. 9, which can be broadly classified into three;

(a) Environmental

The utilization of biomass resources and production of biofuel offsets the adverse effects of the released greenhouse gases, which causes pollution of both land and aquatic systems. The implementation of biomass technology helps in controlling waste utilization which aids maintaining a clean healthy environment and also promotes erosion control thereby protecting the soil. When efficient biofuel technologies are kept in place, deforestation becomes minimum, thereby conserving the forest and wild life simultaneously. Other major problems like global warming and acid rain will also be controlled.

(b) Socio-economic

Biofuels processing could have impact on the livelihood of rural dwellers. It will provide new means of employment to both the youth and peasant farmers, thereby minimizing rural to urban migration. Also, the rural settlements will have access to modern amenities such as good roads, health facilities, education and training centred, and skill acquisition centres. Also, agricultural development, technological

Environmental
- Land Management
- Water requirements
- Fertilizer use
- Soil protection
- Reduced greenhouse gas emission
- Reduction of deforestation
- Biodiversity concerns

Socio-economic
- Improved living standard
- Gender aspects
- Food security
- Access to water and sanitation
- Job/ income generation
- Affordable financing
- Technology availability
- Economic feasibility

Clean energy
- Sustainable
- source of heat, electricity and biofeul
- Better air quality
- Fuels are less toxic and biodegredable
- Enhancement of energy reliability and security

Fig. 9 Benefits of biofuel technology adoption in Nigeria

advancements and the general well-being of both rural and urban sectors will be enhanced. There would be inflow of exchange currency due to biomass feedstock export.

(c) Clean energy

Affordable reliable, sustainable and clean energy will be made available. Biomass is capable of producing solid, liquid and gaseous fuels through different conversion processes like pyrolysis, gasification, transesterification, anaerobic digestion and fermentation. There is assurance of better air quality and presence of domestic energy source that is capable of running continuously.

7   Challenges Facing the Technologies and the Way Forward

The production of biofuel in Nigeria is still at infancy [13] stage as commercialization of the processes still has a long way to go. In order to progress in the quest for industrialization and commercialization of biofuel, there is need to develop the research aspect. There is an urgent need of reviving research institutes as they are the primary means of nurturing local technologies, and hence developing the immediate community. Existing biofuel researches are not well coordinated and require adequate funding coupled with centres of excellence for conducting the research [18]. There are other hindrances associated with the implementation of biomass technologies in Nigeria, such problems and their potential solutions were enumerated by Babalola et al. [1] as illustrated in Fig. 11. There is presently no comprehensive and well-articulated renewable energy policy regarding development of biomass technologies. There is lack of awareness as to the benefits of these renewable energy technologies amongst majority of Nigerians. This may be due to the literacy level of most people living in the rural areas, and these technologies are seen to be used with discomfort or sacrifice. Another factor is lack of proper funding to support these technologies. Loans should be made available to farmers to embark on biofuel crops cultivation. Funding of researches and training has to be sustained as well. Improvement in the logistics facility, as most of the biomass resources are unevenly distributed within

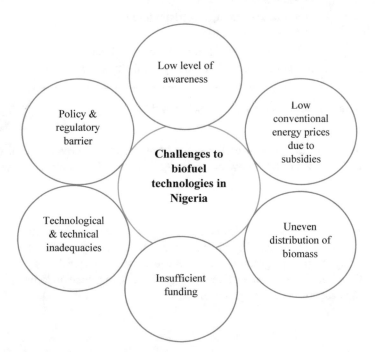

Fig. 11 Challenges facing biofuel technology in Nigeria

the country, hence developing conversion centres close to the biomass depots will reduce transportation problems.

In order to overcome such problems, the government should put in place workable policies and regulatory framework which will ensure stakeholder participation and application of the acquired technology practices. The government should be committed to funding research development, training and technological support for biofuel sector development. The masses should be encouraged to embark into small-scale agricultural practices that would generate biomass for biofuel conversion. Awareness campaign should be embarked upon through mass media on the importance of biofuel utilization. Loans and incentives should be given to farmers to encourage cultivation of energy crops like jatropha.

8 Summary

The renewable energy technologies available in Nigeria include hydropower, solar energy, biomass, wind, wave and tidal. Biomass energy is tapped through sources that include energy crops, forest residues and human wastes (MSW). The implementation of the available biofuel technologies that is gasification, anaerobic digestion, combustion and briquetting is majorly at the pilot level, with some few reaching up to commercial scale.

Nigeria has abundant biomass resources as a result of its diverse vegetation zones. The country can diversify its fuel supplies thereby reducing dependence on it fossil fuel reserves. Most of the renewable energy technologies require long-term financial preparations for the technology to be properly put in place and to penetrate to both the urban and rural populace. So, a simpler and quicker approach for the Nigerian government to embark on renewable energy in the current scenario will be to embark on biogas production through anaerobic digestion. This can be a very promising renewable energy source in Nigeria at the moment, due to the raising tons of municipal waste that is accrued on daily basis resulting from the teaming population. It will serve as a means of environmental sanitation and as well as a means of clean and sustainable energy.

This study has contributed to the body of literature in biomass conversion technology research in Nigeria. It has provided an insight of the energy scenario in Nigeria, the available conversion techniques that are implemented and the challenges impeding their implementation. The study has offered useful insights for the government to play a major role in ensuring the implementation of these technologies commercially, since abundant resources exist.

References

1. Babalola P, Oyedepo S (2019) Bioenergy technology development in Nigeria -pathway to sustainable energy development. Int J Environ Sustain Dev 18:175–205
2. Posibi T (2020) Biomass energy in Nigeria: an overview. Bio Energy Consult. Available: https://www.bioenergyconsult.com/biomass-energy-in-nigeria/
3. Mohammed YS, Mokhtar AS, Bashir N, Saidur R (2013) An overview of agricultural biomass for decentralized rural energy in Ghana. Renew Sustain Energy Rev 20:15–25
4. Edomah N (2019) Governing sustainable industrial energy use: energy transitions in Nigeria's manufacturing sector. J Clean Prod 210:620–629
5. Mohammed YS, Mustafa MW, Bashir N, Ibrahem IS (2017) Existing and recommended renewable and sustainable energy development in Nigeria based on autonomous energy and microgrid technologies. Renew Sustain Energy Rev 75:820–838
6. Ogbonnaya C, Abeykoon C, Damo UM, Turan A (2019) The current and emerging renewable energy technologies for power generation in Nigeria: a review. Therm Sci Eng Prog 13:100390
7. Energypedia.info. Nigeria Energy Situation (Online). Available: https://energypedia.info/wiki/Nigeria_Energy_Situation#cite_note-Energy_Commission_of_Nigeria_.28ECN.29.2C_.E2.80.9CDraft_National_Energy_Master_Plan_.5B2014.5D.E2.80.9D-19
8. Dioha MO, Kumar A (2020) Exploring sustainable energy transitions in sub-Saharan Africa residential sector: the case of Nigeria. Renew Sustain Energy Rev 117:109510
9. Demirbas MF, Balat M, Balat H (2009) Potential contribution of biomass to the sustainable energy development. Energy Conv Manage 50(7):1746–1760
10. . Nigeria Agricultural Land (2020) Trading economics. Available: https://tradingeconomics.com/nigeria/arable-land-hectares-wb-data.html
11. Owoeye F (2017) Biomass: a preferred alternative source of energy generation in Nigeria. Nairametrics. Available: https://nairametrics.com/2017/11/12/biomass-a-preferred-alternative-source-of-energy-generation-in-nigeria/
12. Giwa A, Alabi A, Yusuf A, Olukan T (2017) A comprehensive review on biomass and solar energy for sustainable energy generation in Nigeria. Renew Sustain Energy Rev 69:620–641
13. Ben-Iwo J, Manovic V, Longhurst P (2016) Biomass resources and biofuels potential for the production of transportation fuels in Nigeria. Renew Sustain Energy Rev 63:172–192
14. Agbro EB, Ogie NA (2012) A comprehensive review of biomass resources and biofuel production potential in Nigeria. Res J Eng Appl Sci 1(3):149–155
15. Akorede M, Ibrahim O, Amuda S, Otuoze A, Olufeagba B (2017) Current status and outlook of renewable energy development in Nigeria. Niger J Technol 36(1):196–212
16. Basu P (2010) Biomass gasification and pyrolysis: practical design and theory. Academic press
17. Akinbomi J, Brandberg T, Sanni SA, Taherzadeh MJ (2014) Development and dissemination strategies for accelerating biogas production in Nigeria. BioResources 9(3):5707–5737
18. Odetoye T, Ajala E, Ogunniyi D (2019) A review of biofuels research in Nigeria. Arid Zone J Eng Technol Environ 15(1):153–162
19. Alamu O, Akintola T, Enweremadu C, Adeleke A (2008) Characterization of palm-kernel oil biodiesel produced through NaOH-catalysed transesterification process. Sci Res Essays 3(7):308–311
20. Umaru M, Mohammed IA, Sadiq M, Aliyu A, Suleiman B, Segun T (2014) Production and characterization of biodiesel from Nigerian mango seed oil. In: Proceedings of the world congress on engineering 1:5
21. Simonyan K, Fasina O (2013) Biomass resources and bioenergy potentials in Nigeria. Afr J Agric Res 8(40):4975–4989

The Potential of Coconut Shells Through Pyrolysis Technology in Nigeria

Rabi K. Ahmad, Shaharin A. Sulaiman, Sharul Sham Dol, and Hadiza A. Umar

Abstract The outer hard shell that encloses the coconut fruit is known as the coconut shell. It is available in plentiful quantities and utilizes as an energy source throughout the tropical countries worldwide. The current world production of coconut fruits and the availability of its biomass wastes have the potential to generate power for low emissions in different applications. Coconut shells are among the untapped energy source from agricultural residues. Rural small-scale farmers in Nigeria are not familiar with the technological process of utilizing the coconut shells for sustainable biofuel production. Usually, they use the shells for open burning charcoal production and as organic fertilizer. Future sustainable and eco-friendly thermochemical technology like pyrolysis produces the entire three biofuels products (solid, liquid, and gas) based on the biomass type and its availability. The products are in the form of charcoal/biochar, bio-oil, and biogas fuels used as a substitute for fossil fuels. This chapter highlights the potentials of utilizing the coconut shells by the thermochemical conversion method, specifically the pyrolysis method. The major contribution it will generate to the economic and industrial development of the country. Industries have the potentials in contributing to this sector when biomass wastes are utilized as raw materials in industrial applications. The use of biofuel obtained from biomass

R. K. Ahmad (✉)
Department of Agricultural and Environmental Engineering, Bayero University Kano, 3011 Kano, Nigeria
e-mail: rkahmad.age@buk.edu.ng

S. A. Sulaiman
Department of Mechanical Engineering, Universiti Teknologi Petronas, 32610 Seri Iskandar, Perak, Malaysia
e-mail: shaharin@utp.edu.my

S. S. Dol
Department of Mechanical Engineering, Abu Dhabi University, 59911 Abu Dhabi, United Arab Emirates
e-mail: sharulshambin.dol@adu.ac.ae

H. A. Umar
Department of Mechanical Engineering, Bayero University Kano, 3011 Kano, Nigeria
e-mail: ummihadiza@gmail.com

© The Author(s), under exclusive license to Springer Nature Singapore Pte Ltd. 2021
S. A. Sulaiman (ed.), *Clean Energy Opportunities in Tropical Countries*,
Green Energy and Technology, https://doi.org/10.1007/978-981-15-9140-2_8

thermochemical conversion has more advantages over the use of raw biomass, in terms of clean energy and environmental sustainability concerns.

1 Overview

The chapter describes the basic principles of the thermochemical conversion processes of coconut shell biomass to useful fuel products. Nigeria is among the tropical region that is blessed with coconut trees. Coconut production and its biomass residues play a major role in contributing to the country's economy. Biomass is a renewable resource obtained from agricultural residues, industrial wastes, animal material, etc., for power generation. The major biomass conversion technologies are chemical, biochemical, thermochemical, and electrochemical processes. The thermochemical technology concepts will be discussed in this chapter. It involves the conversion of biomass residues to useful fuel in the form of solid, liquid, and gases based on the input process variables, and they include torrefaction, pyrolysis, gasification, combustion, and liquefaction. The chapter starts with the general need for alternative energy, coconut production, and its availability in Nigeria, biomass as renewable resources, and potentials of the coconut shells biomass. The basic principles of thermochemical conversion routes, the steps involved, and the opportunities are discussed toward the end of this chapter.

2 The Need for Renewable Energy

The search for renewable and carbon-neutral energy sources is the motivation of many countries [1]. The world's energy depends substantially on non-renewable fuels like coal, petroleum, and natural gas. The demand for energy is on the increase with the emerging new technologies, increase in oil prices, and population size [2]. This has driven the search and use of biomass fuels in addition to the fossils fuels back in action by the industries toward increasing the energy demand and reduction in emissions by fossil fuels [3, 4].

Biomass is the largest renewable source of energy-containing carbon that is naturally available and can partly substitute fossil fuels [5]. It refers to any plant-derived organic matter which composes of lignocellulosic materials and industrial residues available on a renewable basis [6, 7]. The usage of biomass for energy relative to the use of fossil fuels has global significant environmental benefits. This depends greatly on the biomass conversion efficiency to energy. The technology should be able to reduce the net carbon dioxide emissions that contribute to the greenhouse gas effects [6]. Biomass technologies are now available in abundant, and increasingly cheap, which could enable deep cuts in carbon without likely to endanger economic growth.

The use of alternative fuels from renewable energy more especially the use of biomass has several great potentials for both the developed and developing countries. It reduces the energy demand and reliance on conventional fossil fuels [3, 4]. Biomass as a source of carbon also reduces the greenhouse gases emissions and enhances economic and sustainable development [8]. The development of an efficient conversion technology is the key factor to be considered when using biomass as a source of energy to compete with fossil fuels based systems [9]. The energy from biomass is more economical when compared to other energy forms. Biomass production is equivalent to one-eighth of the total annual world consumption of all other forms of energy sources. The recent technologies, technical development, and process conditions can offer a suitable eco-friendly process that will increase the use of biomass as an energy resource [10]. Considering the reliance on fossil fuels, the search for another energy raw material is very important for the future energy demand in Nigeria. The developing and developed world is quitting gradually from the use of fossil fuels for energy to sustainable energy resources (biomass); Nigeria should also take part in the race. Only 19% of the total electricity generated in Nigeria is from renewable energy, where the biomass contribution is extremely small. The total potential of these renewables (almost 68,000 MW) is more than the current power output [11].

3 Biomass as a New Technology

Bioenergy is the energy obtained from biomass that is available as a renewable source. It is termed as an alternative feedstock for sustainable future energy. Biomass residues are the wastes that remain or generated when the useful raw products are extracted for further processing [8]. The need for biomass wastes recycling and energy-saving technologies is increasing because of economic and environmental conditions. Thus, it leads to the development of diverse technologies to solve the issue. This technology involves the conversion of wastes into numerous forms of useful fuel, which can be used to supply power [12]. National energy policy and strategy focus more on modern technologies with advanced pollution-free, renewable sources, and its conservation, and this brings the importance of using biomass fuel. Biomass produces significantly less amount of carbon dioxide, nitrogen oxides, and sulfur dioxide when compared to fossil fuels [5, 6]. Biomass usage does not produce heavy metals, as such biomass energy does not contribute anything harmful to the environment, it lies very consistently with environmental protection policies [5]. Biomass energy conversion has several sustainable technologies and process options that are available for its processing. The transesterification technique is divided into thermochemical conversion methods and biochemical conversion methods. The methods are among the major pollution-free biomass conversion techniques to useful energy. The thermochemical method involves thermal decomposition of the biomass whereas the biochemical method involves the use of microorganisms or enzymes [12]. In recent years, one of the effective approaches used as environmental-friendly exploitation and conversion of biomass waste into useful biofuels is the thermochemical method. It is further subdivided into pyrolysis, gasification, liquefaction, and combustion.

Biomass is generally considered as an energy source completely CO_2-neutral. The underlying assumption is that the CO_2 released in the atmosphere is matched by the amount used in its production. This can be possible if biomass energy is consumed sustainably. This may not be the case in many developing countries. The reduction of CO_2 emissions applies to electricity production with biomass as the source of energy [5]. Therefore, utilization of biomass energy would be practical because it does not contribute to global warming unlike the non-renewable energy resource [13]. Ethanol and methanol are some of the examples of liquid biomass fuels that emit less CO, hydrocarbons, and carcinogenic compounds than the petroleum fuels. The use of biomass fuels in a well-managed and sustainable way will not contribute to carbon dioxide levels that cause global warming, unlike the fossil fuel combustion [5].

3.1 Biomass Usage in Developing World

The extent of biomass usage in other countries is quite diverse, with most developing countries having biomass as the paramount energy source for households use, and large group activities. Biomass is used for thermal, electricity, and mechanical power generation. It possesses certain advantages as energy resources because it is relatively inexpensive, and it provides an effective source of low sulfur fuel and reduces environmental hazards if processed properly [14]. Enhancement of several societal needs would also be addressed by the use of biomass because it is the bountiful renewable carbon-neutral source obtained from plants and animals for the production of bioenergy and biomaterials [7, 15, 16].

The developing countries uses biomass as their primary source of energy [5]. In the near future, biomass will be the cost-effective, sustainable, and promising method for electricity production and energy supply to users. Its usage can help countries in meeting the target of GHGs reduction. Biomass development projects are aimed at identifying the possibility of utilizing biomass as an energy resource that can substitute a portion of conventional fuel use [5]. In view of the plentiful within easy reach supply of agricultural residues worldwide, they are the most potential biomass, but yet they are not fully utilized in harnessing energy [13]. The use of biomass for energy applications in developing countries is increasing day by day because most of their economies depend predominantly on forestry and agriculture; this can probably tackle the issue of biomass waste disposal if it serves as an energy source [8].

4 Coconut Production and Coconut Shells Availability

A coconut tree is an important smallholder palm of the tropics [17]. The coconut tree (Cocos nucifera) is the only species of the genus Cocos, and it is a member of the palm family. Coconuts are known for their flexibility ranging from food supplements (a regular part of the diets in the tropics and subtropics regions) to cosmetics. The palm of coconut trees produce coconuts up to 13 times a year, and although it takes a year for the coconuts to mature, a fully blossomed tree can produce between 60 and 180 coconuts in a single harvest [18]. The coconut palm consists of the trunk, frond, leaves, and the coconut fruit. The total coconut fruit comprises the husk, shell, meat, and water with a percentage of 50%, 15%, 25%, and 10%, respectively [19] as shown in Fig. 1 [20].

Coconut is among the world's major oil crops; therefore, it is produced in large quantities [21]. They are mostly cultivated by the coastal and nearby regions throughout the whole world [19]. The major coconut areas lie between 20°N and 20°S on both sides of the equator. Though it is found beyond this region as far as 28°N and 27°S, cultivation in these extreme regions has not been successful, and the palm does not fruit in cooler climate [22]. Coconut is termed as the eco-friendly small-holder palm of the tropical environment produced in 92 countries worldwide.

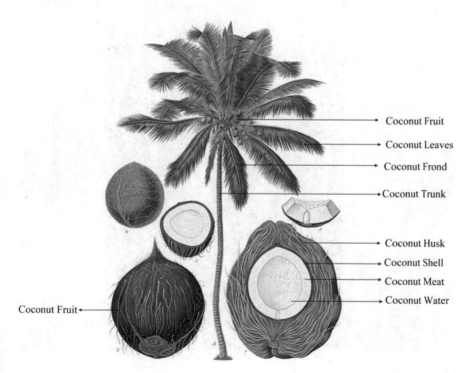

Fig. 1 A coconut palm tree and the cross-sectional view of the coconut fruit illustrating the layers

It has an annual production of 64.3 billion nuts which covers 12.28 million hectares in 90 countries globally. Asia and the Pacific's regions dominate the production of coconut. 75% of the world coconut production is from three countries (Indonesia, Philippines, and India) with Indonesia being the largest producer [17, 21]. The leading producer among the South Pacific countries is Papua New Guinea, and in Africa, Tanzania is the largest producer while Brazil accounts for nearly 43% of the coconut area in Latin America [22]. Global coconut smallholders earn USD$ 7.73 billion per annum from coconut production [17].

Coconut plantations has a sustainable, and renewable nature as shown in Fig. 2 [23] is related to energy crop plantations and also serve as a major source of a wide variety of products.

Apart from the traditional products of coconut in the form of coconut oil and meal, it has a unique advantage of producing a substantial variety of products from foods to eco-friendly non-foods. The biomass wastes from the perennial energy harvest in the coconut plantation include the husks, shells, trunks, fronds, and leaves [23, 24] as illustrated in Fig. 3.

The total coconut fruit comprises the husk, shell, meat, and water with a percentage of 50%, 15%, 25%, and 10%, respectively [19]. The shell encloses the Copra also

Fig. 2 Coconut plantation

Husks   Shells   Trunks

Fronds   Leaves

Fig. 3 Coconut plantation residues

known as the coconut meat. Coconuts are distinct from other fruits for their applications. It serves as a drink from its potable coconut water, milk, and when mature, oil is extracted from the coconut meat as shown in Fig. 4 [23]. The hard shells are used for coconut shells charcoal [24]. Coconut fruit is an excellent source of minerals that are used in the production of many products [25]. Due to the fruit economic importance and potentials, the development of coconut processing industries rapidly emerged [21]. Annually, a high amount of coconut fruits are processed for different purposes [26], such as foods, health and beauty, beverages, energy and cooking oil, fresh coconut milk, coconut water, and powdered milk [27, 28]. The coconut shells are obtained in large quantity as a by-product from these industries.

These valuable green energy sources obtained from the plantation and industries must be utilized, because the energy harvest is regular throughout the year and the supply of the energy is uniform. The energy-rich material is a low cost yet has the potential for high-value green products [17] such as coconut shell charcoal,

Coconut water   Coconut milk   Coconut oil

Fig. 4 Coconut water, milk, and oil extracted from coconut meat

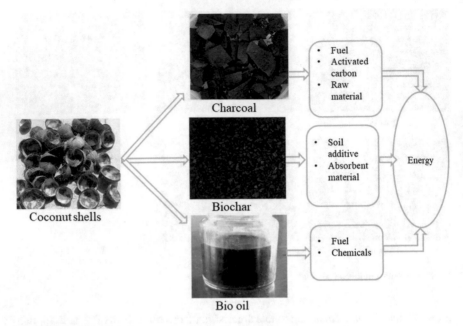

Fig. 5 Coconut waste to energy

biochar, and bio-oil shown in Fig. 5 [8]. In some coconut growing countries, the coconut is the main or only origin of earnings of foreign exchange. Research must be directed toward harnessing the potential of this energy crop, because the coconut shell, husk, and leaves have exciting prospects as energy sources and align with the global sustainability and climatic change aspirations and projects.

Coconut fruit is a non-indigenous crop in Nigeria. The country is known for its arable rain forest fertile for any type of crop. It is among the tropical countries with a large production of coconut fruits with the northern part of the country dominating the production. Nigeria is the 18th coconut fruit producer in the world ranking with 265,000 metric tons of coconut production. Hence, the nation can participate positively in the coconut production market. The prime coconut producing states includes: Kano, Jigawa, Bauchi, Kaduna, Niger, Katsina, Kebbi, Zamfara, Sokoto, Adamawa, Yobe, Borno, Taraba, Plateau, Nasarawa, Lagos, and Ogun states [19]. Nigeria's capital can be maximized based on the availability of renewable resources and reduce reliance on its notable crude oil. The vast amount of biomass is being generated from the country's agricultural activities and agro-industries where coconut shells belong.

With the high production of coconut fruit and its benefit, there is a high coconut demand which makes its shells available and discarded in plentiful quantities everywhere as a biomass waste in tropical countries worldwide. The coconut shells contain 34% cellulose 36% lignin, 29% pentosane, and 1% ash [19]. The coconut shells can be obtained from coconut processing industries and the local coconut hawkers in the country. In many countries with Nigeria inclusive, coconut shells are subjected to open burning which contributes remarkably to GHGs emissions [21, 29, 30]. Instead of open burning, the biomass can be processed through thermochemical conversion methods. Over 90% of coconuts fruits are grown by small income farmers in Nigeria, who do not have access to electricity. The coconuts are processed locally in small processing plants, and therefore, the energy of the unutilized biomass can be used through thermochemical conversion methods to generate power to the processing plant and the nearby community [22].

5 Thermochemical Conversion of Biomass

The thermochemical conversion processes involve the use of heat to convert biomass into useful energy in the form of solid, liquid, and gases. The efficient pathways for biomass conversion to bioenergy are subdivided into pyrolysis, torrefaction, gasification, combustion, and liquefaction [6, 31]. The processes listed above can use any biomass type for power generation based on biomass availability. The reactors involved the factors that are responsible for the complete process, and the final by-products yield at each process for these pathways are highlighted with further elaboration with more focus on the pyrolysis method.

Torrefaction involves subjecting the biomass to thermal heating at a temperature range of 200–300 °C in an air-free environment. The process is carried out to remove some part of the volatile matters and to retain the maximum solid yield of the biomass [31, 32]. The reactions involve during biomass torrefaction are the devolatilization process, depolymerization process, and carbonization process of hemicellulose, lignin, and cellulose. The factors influencing the torrefaction process are the residence time, temperature, heating rate, feedstock moisture content and particle size, gas flow rate, and the ambient pressure [31]. The types of the reactors used for biomass torrefaction include torbed, rotary drum, belt dryer, moving type compact bed, screw conveyor, microwave, and multiple hearth furnace fluidized bed reactors [33]. Solid mass reduction, rise in energy content, and chemical compositions are some of the physicochemical changes that the torrefied biomass materials undergo during the torrefaction process. At the end of torrefaction, the end product is a brown/black uniform material that has 70% mass solid product, and 30% is the lost mass with regards to condensable and combustible gases. The process improves product properties such as moisture content, carbon content, hydrogen content, energy content, and particle shape and size. The carbon and energy value of the torrefied biomass increases by 15–25 wt% than the raw biomass, and the moisture content decreases to less than 3 wt% [34]. The energy content of oil palm frond and *leucaena leucocephala* increases from 17.67 MJ/kg and 17.93 MJ/kg to 25.16 MJ/kg and 24.92 MJ/kg when torrefied at 300 °C, respectively [34]. Torrefaction product produces a high-grade smoke-free fuel which can be used in residential and commercial combustion, as well as in the gasification process [31].

Pyrolysis involves the thermal conversion of biomass into solid, liquid, and gases at a specific process condition in a controlled or oxygen-free environment depending on the final product [6]. The three types of pyrolysis method include slow/carbonization mainly for solid charcoal and biochar, fast that produces bio-oil, and lastly flash pyrolysis that involves biogas production. The methods are carried out using different types of reactors. This method will be described further in later sections.

Gasification is another conversion method that produces biogas as the major fuel. It involves the partial oxidation of biomass carried out in the presence of a medium and temperature to produce vapors, CO_2, CO, CH_4, hydrogen, chemicals, and other gaseous hydrocarbons [31]. The process medium for reaction can be gas or supercritical water that include air, oxygen, subcritical steam, or there mixture [6]. The series of reactions to understand the process flow of biomass conversion and production of syngas is through dehydration, pyrolysis, cracking, reduction, and combustion processes [31]. These reactions are carried out in a gasifier. The major gasifiers include downdraft and updraft, moving bed, fixed bed, fluidized bed, and entrained gasifiers. A survey on the gasifier's availability and usage revealed that 75% are downdraft, 20% are fluidized beds, 2.5% are updraft, and 2.5% for various other designs. The resulting biogas, a mixture of other gases, is utilized as an energy source for space and household heating and electricity [5]. CO and H_2 from biomass gasification are used as transportation fuels, such as gasoline, synthetic chemicals, and methane for power generation [6].

From time immemorial, the forest woods are used directly for space heating and cooking. Combustion is the oldest technology for the utilization of biomass. It is an exothermic reaction that involves the direct high-temperature conversion of biomass to produce CO_2 and steam (H_2O) in excess air [5, 6]. Conventional combustion technologies are available for the combustion of biomass that includes fluidized bed combustors, bubbling, and circulating fluidized bed combustors [35].

Lastly, liquefaction is the production of liquid biofuels from biomass. It involves the decomposition of large feedstock molecules to liquids with smaller molecules in a liquid medium at high temperatures and pressures. It is similar to the pyrolysis process in the production of liquid. Pre-drying of biomass is not needed here because biomass with a high amount of moisture is preferable in this process [36]. Liquefaction of solid biomass feedstock can be carried out through pyrolysis, gasification, and hydrothermal process. Catalysts are used in liquefaction of biomass to biofuels [6].

6 Pyrolysis

There are several barriers to be overcome for technological advancement and commercialization to progress. One of such is the conversion of high-quality fuel from biomass through a thermal conversion like pyrolysis for energy utilization [8]. Pyrolysis is one of the relatively simple, inexpensive, and robust most promising thermochemical technologies that convert biomass into bioenergy in the form of useful products (gas, liquid, and solid). The biomass is processed at a temperature range of

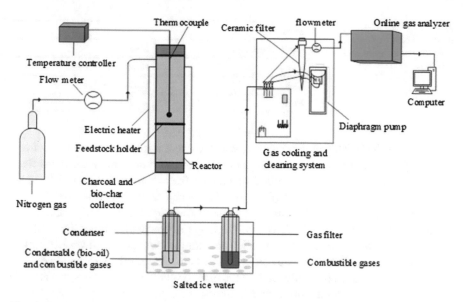

Fig. 6 Pyrolysis process for biomass conversion using downdraft pyrolysis reactor

200–650 °C in an inert atmosphere [7, 37] as shown in Fig. 6. The yield and the proportions of the final products (gas, char, and liquid) are subject to different factors like biomass properties, operational process conditions (temperature, pressure, particle size, residence time, and heating rate) [6, 8]. Pyrolysis methods are classified into three based on the two operational factor temperature and vapor residence time as slow, fast, and flash pyrolysis [2].

Biomass pyrolysis is the fundamental chemical reaction for both the combustion and gasification processes which occurs naturally in the first two seconds. The products of biomass pyrolysis include biochar, bio-oil, and gases (CH_4, CO, CO_2, and H_2) [6]. Depending on the thermal environment and the final temperature, pyrolysis will yield mainly charcoal or biochar at low temperatures, less than 450 °C, when the heating rate is quite slow, and bio-oil at an intermediate temperature and under relatively high heating rates. At high temperatures greater than 800 °C, with rapid heating rates, the biofuel product mainly gases [2].

6.1 Type of Pyrolysis Process

The pyrolysis process is categorized based on the operating conditions such as temperature, residence time, and desired final product into a slow, fast, and flash process in an inert atmosphere. Based on the medium, pyrolysis is further divided into hydrous-pyrolysis and hydro-pyrolysis using water and hydrogen environment, respectively [10]. Table 1 shows the type of pyrolysis methods.

Table 1 Types of pyrolysis method based on the operating conditions and medium

Pyrolysis type	
Operating conditions	Medium
Slow	Hydrous-pyrolysis
Fast	Hydro-pyrolysis
Flash	–

a. Slow pyrolysis: This process is known as the major traditional method practiced for decades in the production of charcoal (carbonization method). It involves heating the biomass in a limited or oxygen-free environment to obtain solid carbon as the main product [31]. At a low temperature less than 400 °C and long residence time, the biomass is heated slowly to obtain mainly solid as the by-product in the absence of oxygen. The long residence time allows the condensable vapor to be converted into solids and combustible gases. In ancient times, the biomass is allowed to stay in the process for several days to maximize the formation of solids [6]. Solid char is the main product of the process and is characterized by low temperature and long residence time [38], which can yield a good quality charcoal [10]. The overall process results in 35% char, 30% bio-oil, and 35% gases [6].

b. Fast pyrolysis: This process is used for the production of bio-oil. Fast pyrolysis aims to maximize liquid or bio-oil products. During the process, the biomass residues are heated so rapidly in the absence of oxygen to reach a temperature of almost 650 °C to decompose at a high heating rate (1000–10,000 °C/s) in some seconds. The complete process yields about 60% bio-oil, 20% biochar, and 20% syngas [6, 38]. Fast pyrolysis unlike the slow pyrolysis uses higher heating rates for liquid products [31]. The important factors that increase the bio-oil yield include: high temperature in the range of 425–600 °C or greater, short residence time less than 3 s, and a very high heating rate with rapid quenching of the product gas [6]. The maximum liquid is obtained around 500°C [39]. Fast pyrolysis also converts biomass to phenol oil [3]. Some conventional fast pyrolysis process converts the biomass to yield the entire pyrolysis products (solid char, liquid, and gas) at a moderate residence time, heating rate, and temperature of approximately 600 °C [6].

c. Flash pyrolysis: This is known as the up-gradation of a fast pyrolysis process that is characterized by the reactive flash volatilization method. The process involves thermal decomposition of the non-volatile solids and liquids at higher processing conditions of heating rate, and temperature and lower residence time to volatile compounds [31]. Usually, in flash pyrolysis, liquids and gaseous are the end product. The biomass is heated at a modest temperature from 450 °C up to 600 °C within a short residence time yielding a product containing condensable and combustible gases, which quits the pyrolysis reactor system as bio-oil and biogas [39]. However, if the production of gas is the primary aim of interest, the peak temperature will be raised to 1000 °C. From the total pyrolysis biofuels from flash pyrolysis, the yield of bio-oil ranges from 70 to 75% [6].

6.2 Pyrolysis Reactor Systems

Many pyrolysis systems are designed for biomass conversion into useful products. The yields of these products depend on the reactor configuration and operational conditions. The ablative, fixed bed, bubbling and circulating fluidized bed, vacuum, rotating disk, rotating cone, plasma, microwave, and auger reactors are the types of the rectors used for pyrolysis process [6, 29].

(a) Ablative: This type of reactor uses high pressure. It involves creating high pressure in between the biomass particle and the reactor wall. The pressure is created by a centrifugal force or a mechanical means [6], where the biomass is pressed and melts to produce a bio-oil yield of 70% [31]. This employs short residence time and high heat transfer between particles, hence suitable for fast and flash pyrolysis.

(b) Fixed bed: This is the conventional and the oldest type of pyrolysis reactor that operates in batches. It is suitable for solid char production because of the longer residence time and slow heating rate in the pyrolysis region. Usually, in this reactor settings, the biomass is heated by a conventional firing [6]. The biomass is fed through a simple vertical shaft. The condensable gases are collected with a carrier gas and the char at the end. The limitations of the rector system include; char removal difficulty, low gas velocity, but it has a simple design, and good gas cleaning and cooling system [31].

(c) Bubbling fluidized bed: This reactor comprises of a bubbling bed with inert bed solids fluidized by an inert gas. Sand properties such as high heat transfer to biomass solids and uniform temperature control make it a suitable inert bed in the reactor system. The required heat for the process is provided by burning either the solid char in a separate chamber and conveying the heat to the bed solids or burning some part of the product gas in the bed system. In the reactor system, the residence time of the solids is substantially higher than the residence time of the gas; as a result, it yields about 70–75% liquid [6]. The system is easy to scale up, simple to construct, and good heat transfer [6, 31].

(d) Circulating fluidized bed: This reactor system works in the same manner as the bubbling fluidized bed. The difference is that the bed in circulating fluidized is more extended, and the solids are constantly recycled throughout the external loop. It uses a high amount of sand, unlike the bubbling fluidized bed system. The bed provides a uniform and good temperature control, and the char is separated easily [6, 31].

(e) Rotating cone: Biomass feedstock is fed to the reactor system from the bottom of a rotating cone, and the centrifugal force pushes the biomass particles in a spiral pattern against the wall. The condensable gas product passes through a tube and the solid product through the upper rim of the cone. Usually, the solid passes through a fluidized bed where it is burned and used as a combustion fuel to heat the cone and support the process. The reactor has special characters such as a very short solid residence time, absence of carrier gas, and short gas-phase residence time [6].

(f) Rotating disk: The rotating disk reactor consist of disks and a solid bed. At high pressure, the biomass slides on the rotating hot disk by centrifugal force. The heat transfer softens the particles and causes vaporization to occur. It is a complex process, but carrier gas is not needed [31].

(g) Vacuum: This type of reactor consists of a number of mounded circular plates. The circular plates are heated, and the biomass is fed to the reactor from the upper plate and moves down to drop unto the lower plates by the help of scrapers in sequence. As the fuel material drops down to the plates, it undergoes drying and pyrolysis process. In the end, only solid char remains on the lowest plate. As a result of the systems, low residence time and heating rate, the vapor in the pyrolysis zone is short with a low liquid yield of 35–50%. In this reactor process, a high char yield is obtained [6]. The reactor design is complex, has limited heat and mass transfer between particles, but no carrier gas is required [6, 31].

h) Microwave: Here, the mechanism of the biomass decomposition is by microwave radiation. Microwave pyrolysis is among the recent technologies for biomass conversion. The system uses electricity as an energy source. In the microwave oven chamber, the process of drying and pyrolysis is carried out simultaneously. The heat is resulted by the microwave, which lies between infrared and radio frequencies. The heat transfer between particles is efficient as a result of the uniform microwave heat between the molecules and atoms [10, 31].

(i) Auger: This reactor uses the advanced screw assembly system. They are tubular in form. The solid biomass feedstock is feed from the top and transported through a rotating screw. The hot sands are driven through an anoxic reactor vessel where the heat required for the process is transported along the tubular wall of the reactor. The efficiency of mixing is high, and the reactor enhances heat and mass transfer. The screw in the reactor mixes the feed biomass and controls the residence time [31, 40].

6.3 *Factors Affecting Pyrolysis Products*

Different properties of biomass feedstock affect the pyrolysis reactor system and the final by-products. The factors that influence the yield of the pyrolysis products include: composition and properties of the biomass feedstock, the particle size of the feedstock, operating factors such as temperature, heating rate, residence time, and type of pyrolysis reactor. Some biomasses require preprocessing processes such as drying, particle size reduction, grinding, and moisture content analysis.

(a) Composition of biomass: The biomass composition greatly influences the yield of the pyrolysis by-products. These include cellulose, hemicellulose, and lignin content, and the hydrogen-to-carbon ratio. Each of these biomass compositions has its preferred decomposition properties [6]. Cellulose and hemicellulose decompose at moderate temperatures, thus promotes the formation of liquid and gases. The lignin content contributes to the solid yield because it degrades slowly [6, 41].

(b) Characteristics of biomass: Among the properties, it include moisture and ash content, fixed carbon, and volatile matter. The pyrolysis process significantly depends on the feedstock moisture content. Depending on the reactor design and operating parameters, moisture content of the biomass has a significant effect on the product yield [38]. It should range around 10% because when it is high, more levels of water are produced in the process which yields unfavorable products [6]. High moisture affects pyrolysis reactor systems. The reactors have certain criteria for feedstock ranging from smaller to larger particles depending on the reactor settings. Moisture content has a serious effect on the properties of the by-product that include calorific value, stability, corrosiveness, and pH value [27]. A higher amount of fixed carbon contributes to the good yield of the solid material. The higher the volatile materials present in the biomass, the better it can produce higher bio-oil and syngas yields [41].

(c) Particle size: Particle size has a great influence on the solid and liquid yields. Larger particles result in a larger proportion of char formation [10].

(d) Temperature: The pyrolysis temperature affects the yield and the composition of the by-product. Changing the final temperature changes the relative yields of biofuel products. The char yield decreases as the temperature was increased [24]. Low temperature favors the solid yield while a higher temperature promotes yields of the liquid and gaseous products [6, 42].

(e) Residence time: The yield and heat energy content as well as the moisture content of the pyrolysis solid product are affected by the residence time [24]. Longer residence time results in secondary char formation. The short residence time of seconds is required for liquid and gaseous product production [6]. The effects of temperature and residence time are related to the product yields [10]. However, longer residence time at lower to moderate temperature promotes solid yield, while short residence time and higher temperatures will significantly promote the products of liquid and gaseous yield [24].

(f) Heating rate: The heating rate affects the degree of biomass decomposition in the pyrolysis process. Higher char yields are obtained as a result of the slow heating rates that lead to the formation of secondary char when the primary char and the volatiles react. Rapid heating rates yields the formation of higher volatiles and more reactive solid char than those produced by a slow heating rate. A higher heating rate favors secondary cracking to effectively strip off the vapors for more liquid and gas yield [6, 10, 38].

(g) Type of reactor system: The type of the reactor system influences the pyrolysis product distribution [10]. Different reactor systems have specific characteristics for the biofuel yielding capacity, and they all have advantages and limitations that affect the by-product composition [41].

6.4 Pyrolysis Products and Their Applications

Pyrolysis is an efficient thermochemical method that receives attention these days because of its ability to produce good quality biofuels to compete or replace the use of non-renewable fuels. It involves the decomposition of biomass or the breakdown of large complex molecules into usable smaller molecules in the form of biofuels. The relative amounts and composition of these biofuel products depend on different factors that have been discussed above. These products are classified into three principal types as listed below.

(a) Solid: This is mostly in the form of char or solid carbon, i.e., the biochar and charcoal. The char primarily contains approximately 85% carbon along with oxygen and hydrogen content, and a very little inorganic ash [6]. Biochar is the black tarry solid amorphous carbonaceous material obtained during the pyrolysis process of biomass [10] and has an energy content of almost 32 MJ/kg [6]. Biochar when applied to soil has numerous positive importance through; soil enhancement improves soil quality, prevents soil erosion, water contamination, and environmental pollution, and effective means to mitigate global climate change by carbon sequestration. It is also a highly absorbent material that increases the soil's ability to preserve chemicals, water, and nutrients [43, 44]. Charcoal is a black solid material containing mainly carbon. It is a potential energy product that has been used for decades as a domestic and industrial fuel, both as a fuel and a raw material for various products [45]. Good charcoal has an ash content of 0.5–5% and a calorific or energy content of 28–33 MJ/kg. Several positive impacts of charcoal on soil similar to biochar have been pointed out by Laird [43]. They include increase the ability of the soil to adsorb plant nutrients and chemicals by easing leaching, release plant nutrients to the growing plants, increasing drainage, aeration, and root penetration, and lastly, lower the bulk density of high clay soils and increase the ability of sandy soils to retain water and nutrients [43].

(b) Liquid: The liquid obtained at the end of the pyrolysis process is a black tarry fluid containing up to 20% water known as the tar, bio-oil, or bio-crude. Bio-oil is an emulsion with a smoke-like odor [6, 46] containing a mixture of complex hydrocarbons with a higher proportion of oxygen and water content [6]. The calorific value of bio-oil is higher than that of methanol and in some cases slightly lower than that of diesel and other light fuel oils [6, 46]. For electricity generation purposes, bio-oil burns like a petroleum fuel [47], and sometimes used directly as boiler fuels and furnace or processed to produce fuels and chemical products [48]. Bio-oil's storage and transportation are simple [47]. The use of bio-oil as a fuel has distinct eco-friendly advantages over fossil fuels. The advantages include: Production of CO_2/GHG is neutral also no emissions of SO_x and NO_x as plant biomass has insignificant amounts of sulfur and nitrogen. A study in a gas turbine reveals the advantage of bio-oil fuels with diesel fuel. It generates more than 50% lower NO_x emissions when relating to diesel oil [65].

(c) Gas: This is the pyrolysis gas, in other words, biogas. It contains incondensable gases in the form of carbon monoxide (CO), hydrogen (H_2), carbon dioxide (CO_2), methane (CH_4), and light hydrocarbons (ethyne, ethane, ethane, and benzene) [6]. The composition of the producer gas or syngas varies depending on the biomass type and composition, and the pyrolysis operational process factors [10]. Biogas is used for numerous applications in industries, such as fuel for combustion, gas turbines, and pyrolysis reactors [48].

6.5 Advantages of Biomass Pyrolysis

Biomass pyrolysis has advantages over other thermochemical processes. The major advantage of the process is that it utilizes the three major biofuels (solid, liquid, and gas). The process can be accomplished at a remote location and small-scale areas where the biomass energy density would be enhanced, and also the cost of handling and transportation will be minimized. The process is adjustable and engaging in converting biomass into energy products. The fuels from the pyrolysis process are effectively used for heating, power, and chemical production [6]. Numerous biomass types are used in pyrolysis processes for clean energy as reviewed by researchers [41, 49–51]. Biomass pyrolysis has been the research of interest as a result of its high efficiency and superior environmental and sustainability performance characteristics [52].

7 Coconut Shell Potential for Power Generation

The potential of coconut shells for use as a biofuel has not been known by many small-scale industries. It is an eco-friendly, available with low cost and sustainable raw material for the production of biofuel. Almost over 40% of the coconut fruit, total energy is from the coconut shells [17]. The shells from coconut are well known as organic fertilizer by farmers as a result of its capacity to conserve moisture, and reduction in nutrient loss to the soil. The coconut shell can be fully used as a source of biofuel apart from fertilizer [53]. However, the advantages of using coconut shells biomass into biofuels include renewability benefits, carbon-neutral energy, environmental benefits, no carbon dioxide emission to the atmosphere, and socio-political benefits through employment opportunities [53].

Energy crops for the production of biofuels are discovered based on the environmental conditions and their characteristics [54]. Coconut wastes have a high potential for power generation based on the properties that it possess over other biomass as given in Table 2. Coconut shell is made up of lignin and cellulose and is used as fuel [22]. According to different studies on coconut shells, the biomass presents good quality parameters for the thermochemical conversion. The energy content available in the coconut shell is approximately 20.8 MJ/kg [21]. This is higher than all the

Table 2 Comparison of coconut shell properties with other biomass

Properties	Coconut shell	Coconut husk	Palm kernel shell	Rice husk	Oil palm frond	Wood
References	[21, 55–58]	[56, 57]	[59, 60]	[59, 61]	[58, 62]	[58, 63]
Moisture content (%)	7.82	9.96	6.11	9.80	6.15	4.25
Fixed carbon (%)	21.60	15.21	14.87	59.97	16.43	10.61
Volatile matter (%)	81.67	72.60	81.03	18.03	80.55	88.07
Ash content (%)	0.60	0.23	4.10	22.00	3.02	1.32
Carbon content (%)	47.60	48.95	49.65	38.74	42.60	43.54
Hydrogen content (%)	6.40	5.40	6.13	5.83	5.71	3.59
Nitrogen content (%)	0.10	0.40	0.41	0.55	0.42	1.00
Sulfur content (%)	0.02	–	0.48	0.06	0.29	0.16
Oxygen content (%)	45.20	43.10	43.33	54.82	51.00	51.70
Calorific value (MJ/kg)	20.80	–	20.40	14.79	17.00	17.53
Cellulose (%)	54.00	30.00	6.92	35.23	41.88	45.00
Lignin (%)	65.00	42.00	53.85	12.92	20.65	25.00
Hemicellulose (%)	27.77	27.81	26.16	24.39	33.61	25.00

biomass presented including the coconut husk. The moisture content of 7.82% [54] is within the permissible standard value [26]. It has a fixed carbon content of 21.6% [55]. The higher fixed carbon in coconut shells leads to the production of a high-quality solid residue. This solid biofuel can be used in the production of activated carbon to treat wastewater or for other applications [6].

A volatile matter of 81.67% [58] is high enough for pyrolysis oil and gas production. Ash content of 0.6% [55] makes the biomass more attractive for thermochemical conversion than other biomass. The elemental analysis of the coconut shells shows that it has 47.6% carbon content, 6.4% hydrogen content, 0.1% nitrogen, 0.02% sulfur, and 45.2% oxygen content, respectively [55]. It has 65% lignin and 54% cellulose contents [64], and hemicellulose of 27.77% [57], respectively. The low amount of moisture and ash content, high amount of lignin and volatile matter content, and the availability and cheap cost of coconut shell makes it more suitable for the pyrolysis process.

8   Fuels from Pyrolysis of Coconut Shells

The fuels from coconut shells include solid, liquid, and gaseous biofuel. Emerging coconut shells products are used as functional foods, feeds, pharmaceuticals, beauty, wellness, agricultural/environmental, and industrial markets [17].

8.1   Solid Biofuel

The solid pyrolysis by-product from coconut shells is known as the biochar and charcoal. They are highly porous, very reactive, and contain volatiles, ash, and fixed carbon. They serve in domestic and industrial usage such as chemical, pharmaceutical, and metallurgical purposes. Apart from energy purposes, biochar and charcoal are used for agriculture, fertilizer production, and pesticide production [6]. The biochar and charcoal products produced from coconut shell biomass are emerging eco-friendly products that do not contribute to global warming because the CO_2 balanced is always positive [14]. Biochars from coconut shells show excellent properties in carbon sequestration, soil improvement, bio-composites, bioremediation, and bioenergy [65]. It also adds nutrients to the soil and enhances the water-holding capacity. As a result, it reduces the need for frequent irrigation to crops and the use of fertilizers [66]. On the other side, coconut shell charcoal is a valuable resource used as industrial fuel, domestic fuel, for blacksmith and goldsmith operations as a result of its certain properties. Activated carbon is a solid biofuel obtained from the activation of the coconut shell charcoal with chemicals. It is mainly used for purification and filtration, cosmetics, and medicinal applications [31].

8.2   Liquid Biofuel

The liquid biofuel from coconut shells is the black oil obtained during the pyrolysis process. Krishna [31] analyzed the fuel-related properties (specific energy, cetane number, solidification point, viscosity, saponification value, and iodine value) of the coconut shell bio-oil, closely connected for its use as a diesel substitute. The value of the bio-oils specific energy (38.4 MJ/kg) is slightly lower than the diesel fuel (46 MJ/kg). For ignition and ability to atomize in the injector system, bio-oils present a higher cetane number (60), and viscosity than the petro-diesel. The viscosity of the bio-oil can be reduced by heating to avoid poor volatilization and spray pattern. Its highest saponification value makes the bio-oil good for ignition more quickly than other plant oils [31]. The bio-oil from coconut shells are used in petroleum refinery as feedstocks or in producing petroleum grade refined fuels by catalytic upgrading [67] to be used as petroleum to generate electricity.

8.3 Gaseous Biofuel

The coconut shell biogas is used for process heat, feed drying, and also as an inert carrier gas in the pyrolysis process. It can be used for the heating system of reactors to reduce the costs of production, or in boilers and furnaces [67].

9 Untapped Opportunities from the Utilization of Coconut Shells

With the growing global demand for coconut milk, oil, water, Nigeria is losing potentially over $1 billion yearly from the untapped coconut farming industry. There is a significant expectation regarding coconut farming in Nigeria and many other tropical countries, and nevertheless, coconut farming has remained a grossly untapped resource. As a result of the country's coconut production, it can participate in the world market of coconut production. Coconut farming accounts for over 10,000 hectares of land in the country [68]. Coconut shells serve as a raw material in the pyrolysis process for charcoal, biochar, activated carbon, bio-oil, chemicals, biogas, and adsorbent material. Despite the potentials of coconut shells, in terms of sustainable clean energy and geotextiles; coconut utilization is left behind, only 8% of the shell are used for charcoal and activated carbon production [17]. The development of a small-scale coconut processing industry near the production site will maximize the utilization of coconut bio-wastes for power generation and create employment opportunities among the nearby population and availability of high-value raw materials [69]. Small-scale power generation in the rural areas can stimulate the development of small- and medium-scale industries that will add value to the locally available raw materials, by providing additional employment opportunities and income generation.

In many developing tropical countries, more especially the low-income communities require a sustainable way of living. The coconut shell biomass management will provide an opportunity at a larger scale. It will be a source of huge potential for income generation (reduction of unemployment) and the enhancement of the sustainable development of the local community. This could help to transfer employment from the public to the private sector, which will reduce reliance on government employment for many countries. The energy sector in the country is very poor. It leads to unemployment and the closure of many manufacturing companies and industries. Therefore, if adequate energy resource is a harness, there will be more employment opportunities revenue generation and poverty alleviation. It will also provide opportunities for industry and environmental research. These can be achieved by organizing programs to increase public awareness for coconut shell waste conversion techniques (renewable energy technology program), and the development of a sound regulatory framework for the process. The government, stakeholders, private sector, and individual participation in the program should be ensured.

10 Summary

Biomass is the major renewable energy sources that supply heat and power to a large population in some developing and developed countries. Biomass energy is economical and sustainable energy, which has advantages over the use of fossil fuel. Among available technologies, the thermochemical conversion is one of the most variables with regards to the flexibility of the feedstock and the final product. With regards to the economic benefit concerning the availability of the feedstock material and sustainable development, today, the waste conversion technology is becoming a key topic of research. The chapter highlights the pyrolysis process involved in the conversion of coconut shells to future biofuels. Nigeria as a tropical country has coconut shell biomass in abundance to be used as an energy resource. The shells can be converted to energy using torrefaction to increase the biomass energy of the shells, pyrolysis to biochar, charcoal, bio-oil or biogas, gasification to synthetic gas, and for direct combustion in boilers. The major advantage of the pyrolysis process is that it utilizes the three major biofuels of solid, liquid, and gaseous products. The yields of these biofuels products depend on the reactor configuration and operational conditions. The factors that influence the yield of the pyrolysis products include; composition and characteristics of the biomass, particle size of the feedstock, type of the pyrolysis reactor, operating factors such as temperature, residence time, and heating rate. The ablative, fixed bed, bubbling and circulating fluidized bed, vacuum, rotating disk, rotating cone, plasma, microwave, and auger reactors are some of the pyrolysis conversion systems. The types of pyrolysis are slow, fast, and flash pyrolysis methods, and based on the medium, pyrolysis is further divided into hydrous-pyrolysis and hydro-pyrolysis. The potential of coconut shells for use as a biofuel has not been known by many small-scale industries. It is an eco-friendly, sustainable raw material for the production of biofuel and available at a low cost. The biochar and charcoal are the major solid fuels obtained from the pyrolysis of coconut shells. Biochars are used for carbon sequestration, soil amendment, bioremediation, bio-composites, and bioenergy. Activated carbon from coconut charcoal is used for purification and filtration, chemicals, cosmetics, and medicinal applications. The bio-oil can substitute the use of diesel and petroleum for electricity while the biogas can be used for heating systems, in boilers and furnaces.

Many research on coconut shells as fuel has been carried out by several researchers as mentioned earlier. Several opportunities have not been tapped from the processing of coconut biomass in Nigeria. This could be due to lower efficiency in biomass technologies, lack of funding, high project costs, and lack of conversion technologies awareness. Nigeria is blessed with coconut trees that can be used for economic, industrial, and social development. The coconut production and its biomass availability utilization in Nigeria will increase the country's economy by providing employment through which the quality of life of the people around the production site will be improved. Compared to other energy sources, a coconut shell can directly supply high-value products such as solid fuels, liquid fuels, chemicals, and gaseous materials. Concerning global climate change and poverty alleviation in some regions, the amount of energy available from the coconut shell is large enough to merit further studies on these energy sources. This will enhance the establishment of an investment for coconut processing, agricultural activities, and businesses which will engage the small-scale coconut farmers and stakeholders in the utilization of the coconut shells. The growth and availability of coconut palm in Nigeria have shown a drive for economic diversification through greater emphasis on coconut production. Agricultural development on coconut production can be strengthened by its biomass proper utilization as a feedstock for biofuel production. This will enhance the proper disposal of coconut palm wastes that serve as a disease vector, reduce the crisis of energy demand, and finally create employment opportunities in the country. This will be achieved by focusing more on research and development and bridging the gap between academia and industry to make this a reality in the country.

References

1. Khuenkaeo N, Tippayawong N (2019) Production and characterization of bio-oil and biochar from ablative pyrolysis of lignocellulosic biomass residues. Chem Eng Commun
2. Srinivasan V (2013) Catalytic pyrolysis of thermally pre-treated biomass for aromatic production. M.Sc, Auburn University, Auburn, Alabama
3. Demirbas M, Balat M (2006) Recent advances on the production and utilization trends of bio-fuels: a global perspective. Energy Convers Manag 47(15–16):2371–2381
4. Harvey F (2019) Greenhouse gas emission. *The guardian.* Available: https://www.the guardian.com/environment/2019/nov/26/united-nations-global-effort-cut-emissions-stop-cli mate-chaos-2030
5. Balat M, Ayar G (2005) Biomass energy in the world, use of biomass and potential trends. Energy Sources 27(10):931–940
6. Basu P (2010) Biomass gasification and pyrolysis: practical design and theory. Elsevier, Amsterdam. (The Boulevard, Langford Lane Kidlington, Oxford, OX5 1 GB, UK: Elsevier Inc. All rights reserved)
7. Yaman S (2004) Pyrolysis of biomass to produce fuels and chemical feedstocks. Energy Convers Manage 45(5):651–671
8. Ahmad RK, Sulaiman SA, Inayat M, Umar HA (2020) The effects of temperature, residence time and particle size on a charcoal produced from coconut shell. IOP Conf Series Mater Sci Eng 863:012005
9. Pestaño LDB, Jose WI (2016) Production of solid fuel by torrefaction using coconut leaves as renewable biomass. Int J Renew Energy Dev Torrefaction Biomass Coconut Leaves Renew Energy 5(3):11
10. Zaman CZ et al (2017) Pyrolysis: a sustainable way to generate energy from waste. Open Access Peer-Reviewed Chapter
11. Posibi T (2020) Biomass energy in nigeria: an overview. Bio Energy Consult. Available: https://www.bioenergyconsult.com/biomass-energy-in-nigeria/
12. Lee SY et al (2019) Waste to bioenergy: a review on the recent conversion technologies. BMC Energy 1(4)
13. Demirbaş A (2001) Biomass resource facilities and biomass conversion processing for fuels and chemicals. Energy Conv Manage 42(11):1357–1378
14. Capareda SC (2014) Introduction to biomass energy conversions. CRC Press Taylor and Francis Group, USA
15. Ragauskas AJ et al (2006) The path forward for biofuels and biomaterials. Sci Rev 311(5760):484–489

16. Sadig H, Sulaiman SA, Moni MNZ, Anbealagan LD (2017) Characterization of date palm frond as a fuel for thermal conversion processes. In: MATEC web of conferences, vol 131, p 01002. EDP Sciences

17. The whole nut: a true coconut story, impact through innovation. Coconut knowledge center 2018, Available: http://coconutknowledgecenter.com/

18. Lewin J (2020) The health benefits of coconut milk. Good food. Available: https://www.bbc goodfood.com/howto/guide/ingredient-focus-coconut-milk

19. Foraminifera (2018) Coconut chips production in Nigeria: the feasibility report. Foraminifera Market Research Limited. (Online). Available: https://foramfera.com/coconut

20. Köhler FE (2020) Cocos nucifera. In: Wikimedia commons

21. Zafar S (2019) Energy potential of coconut biomass. Bio Energy consult. Available: https://www.bioenergyconsult.com/coconut-biomass

22. Raghavan K (2010) Biofuels from coconuts. Energypedia, Utilization of solid Biofuels

23. Unsplash, Photos for everyone: coconut, ed. Unsplash, the internet's source of freely-usable images. https://unsplash.com/s/photos/coconut

24. Ahmad RK, Sulaiman SA, Yusuf SB, Dol SS, Umar HA (2020) The influence of pyrolysis process conditions on the quality of coconut shells charcoal. Platform A J Eng 4(1):73–81

25. How much does a coconut weigh? Reference.com. Available: https://www.reference.com/world-view/much-coconut-weigh-5faa7de1113733e7 (2019)

26. Krzysztof D, Krzysztof M, Taras H, Barbara D (2018) Impact of grinding coconut shell and agglomeration pressure on quality parameters of briquette. Eng Rural Dev Jelgava 23:1884–1889

27. Yun TZ (2019) Agriculture: a coconut revival. The Edge Markets (Magazine online). Available: https://www.theedgemarkets.com/article/agriculture-coconut-revival

28. Anton I, Latifah US, Meity DIP (2017) Effect of torrefaction process on the coconut shell energy content for solid fuel, 2017, vol 1826, pp 1–7. (online)

29. Yerima I, Grema MZ (2018) The potential of coconut shell as biofuel. J Middle East North Afr Sci 4(5):11–15

30. Sulaiman SA, Roslan R, Inayat M, Naz MY (2018) Effect of blending ratio and catalyst loading on co-gasification of wood chips and coconut waste. J Energy Inst 91(5):779–785

31. Kundu K, Chatterjee A, Bhattacharyya T, Roy M, Kaur A (2018) Thermochemical conversion of biomass to bioenergy: a review. In: Prospects of alternative transportation fuels Singapore. Springer, Singapore, pp 235–268

32. Matali S, Rahman NA, Idris SS, Yaacob N, Alias AB (2016) Lignocellulosic biomass solid fuel properties enhancement via torrefaction. Proc Eng 148:671–678

33. Batidzirai B, Mignot APR, Schakel WB, Junginger HM, Faaij APC (2013) Biomass torrefaction technology: techno-economic status and future prospects. Energy 62(C):196–214

34. Tumuluru JS, Sokhansanj S, Hess JR, Wright CT, Boardman RD (2011) A review on biomass torrefaction process and product properties for energy applications. Indus Biotechnol 7(5):384–401

35. Zafar S (2019) Thermal conversion of biomass. BioEnergy Consult. Available: https://www.bioenergyconsult.com/thermochemical-conversion-technologies/

36. Chan YH et al (2019) An overview of biomass thermochemical conversion technologies in Malaysia. Sci Total Environ 680:105–123

37. Dickerson T, Soria J (2013) Catalytic fast pyrolysis: a review. Energies 6(1):514–538

38. Antal MJ, Grønli M (2003) The art, science, and technology of charcoal production. J Indus Eng Chem Res 42(8):1619–1640

39. Bridgwater A (1999) Principles and practice of biomass fast pyrolysis processes for liquids. J Anal Appl Pyrolysis 51:3

40. Dhyani V, Bhaskar T (2018) A comprehensive review on the pyrolysis of lignocellulosic biomass. Renew Energy 29:695–716

41. Jahirul MI, Rasul MG, Chowdhury AA, Ashwath N (2012) Biofuels production through biomass pyrolysis-a technological review. Energies 5(12):4952–5001

42. Demirbas A (2007) Effect of Temperature on Pyrolysis Products from Biomass. Energy Sources Part A 29(4):329–336
43. Laird DA (2008) The charcoal vision: a Win–Win–Win scenario for simultaneously producing bioenergy, permanently sequestering carbon, while improving soil and water quality. Agron J 100(1):178–181
44. Aziz NS, Shariff A, Abdullah N, Noor NM (2018) Characteristics of coconut frond as a potential feedstock for biochar via slow pyrolysis. Malays J Fundam Appl Sci 14(4):408–413
45. Oyedun OA, Lam KL, Hui CW (2012) Charcoal production via multi stage pyrolysis. Chin J Chem Eng 20:455–460
46. Fardhyanti DS, Damayanti A (2017) Analysis of bio-oil produced by pyrolysis of coconut shell. Int J Chem Mol Eng 11(9):651–654
47. Demirbas A, Arin G (2002) An overview of biomass pyrolysis. Energy Sources 24(5):471–482
48. Guedes RE, Luna AS, Torres AR (2018) Operating parameters for bio-oil production in biomass pyrolysis: a review. J Anal Appl Pyrol 129:134–149
49. CZ et al (2018) Pyrolysis of torrefied biomass. Trends Biotechnol 36(12):1287–1296
50. Rangabhashiyam S, Balasubramanian P (2019) The potential of lignocellulosic biomass precursors for biochar production: performance, mechanism and wastewater application- a review. Ind Crops Prod 128:405–423
51. Clifford CB (2018) Biomass pyrolysis. In: Alternative fuels from biomass sources
52. Zafar S (2019/2020) Overview of biomass pyrolysis process. BioEnergy Conslt. Available: https://www.bioenergyconsult.com/biomass-pyrolysis/
53. Archana A, Singh MV, Chozhavendhan S, Gnanavel G, Jeevitha S, Pandian AM (2020) Coconut shell as a promising resource for future biofuel production. In: Biomass valorization to bioenergy. energy, environment, and sustainability. Springer, Singapore, pp 31–42
54. Ting TL, Jaya RP, Hassan NA, Yaacob H, Jayanti DS, Ariffin MAM (2016) A review of chemical and physical properties of coconut shell in asphalt mixture. Jurnal Teknologi 78(4):85–89
55. Dongardive SN, Mohod AG, Khandetod YP (2019) Slow pyrolysis of coconut shell to produce crude oil. Int J Innov Eng Technol 12(3):94–97
56. Wang Q, Sarkar J (2018) Pyrolysis behaviors of waste coconut shell and husk biomasses. Int J Energy Prod Manage 3(1):34–43
57. Gonçalves FA, Ruiz HA, dos Santos ES, Teixeira JA, de Macedo GR (2019) Valorization, comparison and characterization of coconuts waste and cactus in a biorefinery context using $NaClO_2$–$C_2H_4O_2$ and sequential $NaClO_2$–$C_2H_4O_2$/autohydrolysis pretreatment. Waste Biomass Valorization 10(8):2249–2262
58. Inayat M, Sulaiman SA, Naz MY (2018) Thermochemical characterization of oil palmfronds, coconut shells, and wood as a fuel for heat and power generation. MATEC Web of Conf 225:1–6
59. Abdullah SS, Yusup S (2010) Method for screening of Malaysian biomass based on aggregated matrix for hydrogen production through gasification. J Appl Sci 10(24):3301–3306
60. Edmund CO, Christopher SM, Pascal DK (2014) Characterization of palm kernel shell for material reinforcement and water treatment. J Chem Eng Mater Sci 5(1):1–6
61. Guofeng W, Ping Q, Enhui S, Zhizhou C, Yueding X, Hongying H (2015) Physical, chemical, and rheological properties of rice husks treated by composting process. Bioresource 10(1):227–239
62. Lai LW et al (2016) Study on composition, structural and property changes of oil palm frond biomass under different pretreatments. Cellul Chem Technol 50:951–959
63. Sjostrom E (1993) Wood chemistry: fundamentals and applications. Gulf Professional Publishing
64. Wang Q, Sarkar J (2018) Pyrolysis behaviors of waste coconut shell and husk biomasses. Int J Energy Prod Manage 3(9):34–43
65. Castilla-Caballero D et al (2020) Experimental data on the production and characterization of biochars derived from coconut-shell wastes obtained from the Colombian Pacific Coast at low temperature pyrolysis. Data in Brief 28:104855
66. Eliana N, Coconut shell biochar (Online). Available: https://www.lifegreencharcoal.com/coconut-shell-biochar/

67. Joardder M, Islam MR, Beg MRA (2011) Pyrolysis of coconut shell for bio-oil. In: Proceedings of the 9th international conference on mechanical engineering 2011, no ICME11-TH-025, pp 1–6
68. Edom S (2020) How to start a lucrative coconut farming business in Nigeria: the complete guide. StartupTipsDaily. Available: https://startuptipsdaily.com/how-to-start-coconut-farming-in-nig eria-africa/
69. Basrah EA (1998) Technological developments and achievements of the coconut sector in Indonesia. A case model on agro-based industrialization. In: Paper presented at the XXXV COCOTECH Meeting. Denpasar, Bali, Institute for Research and Development of Agro-based Industry (IRDABI) Agency for Industrial and Trade Research and Development Ministry of Industry and Trade Bogor, Indonesia

Performance of Biodiesel with Diethyl Ether in DI Diesel Engine

Hailu Abebe Debella and Samson Mekbib Atnaw

Abstract The increasing industrialization and motorization of the world are leading to a steep rise in the demand for petroleum products. The petroleum-based fuels are stored fossil fuels in the earth. With limited reserves of these fossil fuels, it is feared that they are not going to last long. Biodiesel is an ecofriendly, alternative diesel fuel prepared from domestic renewable resources which makes it an ideal alternative as current and future replacement of depleting fossil-based fuels. This chapter attempts to introduce biodiesel production of different seeds and their characteristics in performance, combustion and emission as a comprehensive review with and without the addition of diethyl ether in particular and *Prosopis juliflora* oil characterization in general with the future work plan, which is blended with diesel and diethyl ether in its production and properties when used in internal combustion (IC) engine. Preliminary result showed that juliflora fruit has better biodiesel yield with 80% yield of juliflora biodiesel which is free from glycerin. The performance characteristics compared with conventional (diesel) fuel and their environmental pollution characteristics are also discussed. Various aspects of non-edible feedstock, such as biology, distribution and chemistry, the biodiesel's physicochemical properties and its effect on engine performance and emission, are reviewed based on published articles. From the review, fuel properties are found to considerably vary depending on feedstocks. Analysis of the performance results revealed that most of the biodiesel generally give higher brake thermal efficiency and lower brake-specific fuel consumption. Emission results showed that in most cases, NO_x emission is increased, and HC, CO and PM emissions are decreased. These will help to reduce the environmental problems caused by the use of fossil fuels and help to reduce the global scarcity problem of conventional energy sources through the use of biodiesel as an alternative fuel.

H. A. Debella · S. M. Atnaw (✉)
College of Electrical and Mechanical Engineering, Addis Ababa Science and Technology University, Kaliti Sub City, B.O. Box 16417, Addis Ababa, Ethiopia
e-mail: samson.mekbib@aastu.edu.et

H. A. Debella
e-mail: hailuautoeng@gmail.com

© The Author(s), under exclusive license to Springer Nature Singapore Pte Ltd. 2021   177
S. A. Sulaiman (ed.), *Clean Energy Opportunities in Tropical Countries*,
Green Energy and Technology, https://doi.org/10.1007/978-981-15-9140-2_9

1   Introduction

The depletion of the natural reserves of fossil fuels and the adverse effect of their use on the environment in terms of harmful emissions make the use of alternative sources of renewable fuels inside diesel engines highly relevant [1–8]. In addition to its environmental advantage, the use of alternative and locally available biodiesel oils will reduce the dependency of developing countries like Ethiopia on imported fuels, enabling them to save spending their hard earned foreign currency for the import of petroleum-based fuels [9]. The possibility of using biodiesel fuel in diesel engines without any significant modifications makes it an ideal alternative and highly promising [10, 11]. As in the case of most organic-based biofuels, biodiesel is oxygenated by nature which enables it to undergo combustion at lower air fuel ratio (AFR) resulting in lower environmental emissions and ensuring better combustion efficiency at the same time. However, some studies [12] showed that variations in the properties of biodiesel fuel such as density, calorific value, viscosity and cleanliness and/or purity cause a challenge in terms of reduced engine performance, reduce lifetime of engine parts and reduced power output and overall thermal efficiency. The physical properties of biodiesel such as its higher viscosity compared to that of diesel are known to significantly increase the droplet size of the atomized fuel injected into the combustion chamber affecting the combustion efficiency and result in a significant drop in power output. Therefore, studies on engine modifications aimed at overcoming these disadvantages need to be given proper attention. In addition to engine modifications, the use of biodiesel and diesel blends in different proportions is commonly used in order to ensure biodiesel blends having similar physical and chemical characteristics with that of diesel. Furthermore, the use of various additives that will be mixed to biodiesel with optimal percentages has been studied with the objective of improving the properties and combustion performance of biodiesel fuels inside diesel engines.

1.1   Use of Diethyl Ether

Diethyl ether (DEE) is an alcohol group renewable fuels which can be produced from biomass, and it is easily miscible with biodiesel and diesel fuel [13]. Alcohols like methanol, ethanol, butanol and DEE are known to have low viscosity, low cetane number and higher heating value. Therefore, mixing of these additives to biodiesel has an advantage in terms of reducing its high viscosity, increasing its energy density and hence increasing the performance of biodiesel used inside a diesel engine by giving it a fuel characteristic more comparable to that of petrol-based diesel. Shown in Table 1 is the comparison of alternative fuels to be used in an engine and potential alcohol-based additives with respect to their physical and chemical properties such as heating value, density, viscosity, flash point and cetane number. As can be seen from the table, biodiesel and vegetable oil have significantly higher density, viscosity and

Table 1 Properties of various fuels suitable for diesel engines and potential alcohol-based additives

Fuel	Lower hearing value	Density @ 20 °C (kg/m^3)	Viscosity @ 40 °C (MPa s)	Flash Point (°C)	Cetane number	References
Diesel	44.8	815	2.95	70	52	[16, 17]
Vegetable Oil	40.4	916	3.42	274	37	[16, 17]
Biodiesel	40.5	855	4.57	126	52	[16, 17]
DEE	33.9	714	0.22	-45	125	[16, 17]
Butanol	33.1	808	2.63	35	25	[16, 17]
Ethanol	28.6	790	1.1	13	6	[16, 17]
Methanol	19.8	792	0.59	11	< 5	[16, 17]

flash point and lower heating values as compared to that of diesel. Whereas, though the alcohol-based additives such as DEE, butanol, ethanol and methanol have modest heating values, they are shown to have lower density, viscosity and flash point even as compared to diesel. Therefore, it can be seen that the mixing of a lower percentage of these alcohol-based additives to biodiesel will give it a more desirable characteristic that are comparable with that of diesel. From Table 1, it can be seen that among the alcohol-based additives DEE has the lowest viscosity, density and flash point temperature and the highest cetane number 125, which makes it an ideal and very promising candidate as additive to biodiesel which could give a much better result even at very small mixing percentage.

Based on its ultimate analysis, the chemical composition of DEE is given by the empirical formula $C_4H_{10}O$. Making use of this chemical formula, the percentage of oxygen by mass of DEE is 21.6%. Such higher oxygen content of DEE also makes it more suitable for efficient burning of the DEE blend biodiesel at lower AFR. However, there are only limited number of studies on blending of DEE with various types of biodiesels: waste plastic oil [14], ethanol biodiesel–diesel (EBD) and methanol biodiesel–diesel (MBD) [15], diesel [16], tire-derived fuel-derived biodiesel [17], karanja methyl ester (KME) [18] and waste plastic pyrolysis oil [19]. Most of these studies conducted experimental investigation of effect of DEE blend on the combustion performance and emission characteristics of diesel engine fueled with various biodiesel fuels. The work of Devaraj et al. [19] reported that the use of up to 10% DEE as an additive to waste plastic-fueled diesel engine has resulted in decreased NO_x emission with high engine load condition and increased the thermal efficiency by 1% from 28 to 29%. Darik et al. [18] studied exploiting biodiesel–biogas with the addition of up to 6% DEE. Compared to the dual fuel operation at full-load condition, the study showed NO_x emissions increased by 12.7%, while HC, CO and particulate matter (PM) decreased by 10.6%, 12.2% and 5.7%, respectively. In addition, the injection DEE is reported to reduce the brake-specific fuel consumption (BSFC) by 5% and increased the thermal efficiency by 2.3%. Tudu et al. [16] showed different proportion of DEE (up to 50%) addition on diesel fuel confirmed not worthily change the thermal efficiency of engine. Furthermore, NO_x emissions increased while carbon monoxide (CO) and hydrocarbons (HC) decreased. Patil et al. [13] scrutinized the addition of different proportions of DEE into diesel fuel from 2 to 25% by volume. The best-case scenario proportion for optimum engine performance was 15%, and in addition it also showed that the NO_x emission and particulate matter (PM) had reduced the mutual benefit between them due to the addition of DEE. The DEE was also used up to 4% to a mixture of tire-derived fuel which was 40% and diesel fuel 60%, and similarly Tudu et al. [17] showed as compared to diesel engine operation when the engine was at full-load condition, the NO_x emission decreased by 25% and BSFC decreased by 6%. Kaimal et al. [14] showed blending of DEE on waste plastic oil with up to 15% proportions increased the thermal efficiency and in addition NO_x and soot emission were significantly reduced. Venu et al. [15] investigated blending of up to 10% DEE into diesel–biodiesel–methanol and diesel–biodiesel– ethanol brought an opposite result on them. When adding DEE to diesel, biodiesel– methanol blend increased smoke emissions and decreased cylinder pressure, BSFC and combustion duration. In other way, the addition of DEE to diesel–biodiesel– ethanol blend decreased NO_x and smoke emissions, whereas the BSFC, combustion duration and cylinder pressure increased.

1.2   Juliflora Tree

Prosopis juliflora grows in a hot climate, and its group is from the Fabaceae family. It is known for its special characteristics that it does not need any water from the soil for survival as it is able to obtain water through moisture from the atmosphere. Juliflora is widely available in Asia and the Caribbean as well as the hot climate regions of Ethiopia such as Diredawa, Afar (awash and Metehara, etc.) and lowland areas of the Somali region. Utilizing this fruit for a biodiesel has economic benefits and keeps a better biodiversity of the land since the flora affects other plantations and agricultural activities. It is commonly used as a source for wood, forage and as a green cover for environmental protection. Significant chemical and physical properties of a biodiesel from juliflora seed oil are found to be similar as compared to other vegetable biodiesel oil (Fig. 1).

1.3   Preliminary Study Results for Juliflora Biodiesel

The preliminary characterization and study of the properties and yield of juliflora biodiesel oil were carried out by Asokan [21]. The result of the study showed that the fuel properties of juliflora oil was found to be comparable with other vegetable oils as shown in Table 2. This shoes that it is possible to use juliflora biodiesel as a fuel for diesel engines with minor or no modifications needed [21]. However, it is recommended to use additives that will improve the characteristics of the biodiesel

Fig. 1 Juliflora tree with its fruits [20]

Table 2 Comparison of fuel properties of juliflora oil with other vegetable oil and diesel [21]

Property	Density @40 °C, (kg/m³)	Kinematic viscosity @ 40 °C, in (mm²/sec)	Flash point (°C)	Gross calorific value (MJ/kg)	Cetane number
Juliflora	930	38.84	181–376	37.6	49
Jatropha	901–940	24.5–52.7	180–280	37.2–43	33.7–52
Karanja	870–928	27.8–56	198–263	36–42.1	45–67
Mahua	891–960	24.6–37.6	212–260	36.8–43.0	43.5
Cotton seed	911–921	32.8–36.0	210–243	40.1–40.8	4.2–59.5
Linseed	865–950	16.2–36.6	108–242	37.5–42.2	28–35
Diesel	-	2.0–2.7	45 min	-	45 min

in order to ensure acceptable combustion efficiency and lower emissions with the smallest power drop while using the biogas inside a diesel engine with no modification. This way long-term operation problems of biodiesels inside diesel engines without optimum mix of additives such as engine choking, clogging of the fuel injector, gum formation and piston sticking could be avoided.

The amount of biodiesel yield in general terms can be obtained from the ratio of percentage yield methyl ester produced with respect to oil used for reaction, and this can be expressed as follows:

$$\% \ \text{yield} = \frac{m(\text{met})}{m(\text{oil})} 100 \tag{9.1}$$

It is worthwhile to note that 80% yield of juliflora biodiesel obtained is free from glycerin.

2 Biodiesel Production

The most common sources of biodiesel production include animal fats, vegetable oil and waste cooking by lipid transesterification. It is known that much earlier that the production of the diesel engine in 1893, the inventors were using vegetable oil as engine fuel instead. The common way of producing biodiesel involves transesterification of non-edible oil with methanol in the presence of either base or strong acid catalysts. The transesterification reaction is known to be heavily affected by various operating parameters. The major variables affecting the process are being acid composition and the free fatty acid content of the oil. Others include reaction temperature, ratio of alcohol to vegetable oil, catalyst, mixing intensity and purity of reactants. This chapter focuses on the major factors affecting the process and the main fuel properties of biodiesel as well as the production process.

2.1 Experimental Procedure

2.1.1 Biodiesel Oil Extraction

The seeds for producing biodiesel can be collected and usually cleaned by water. Then the seeds that can be sun-dried for two weeks and separated from the fruit manually. The dried seeds were placed in an oven and baked for 24 h at a temperature of 500 °C. After drying, the seeds are crushed mechanically into small pieces. Soxhlet apparatus is often used for the extraction of oil. The prepared sample placed inside a thimble is made from thick cotton, which is loaded into the main chamber of the Soxhlet extractor [22] which also has a condenser, where hexane was used as solvent in the extraction. The mechanism of operation is the rise of vapor inside the distillation arm and then heading to thimble housing chamber. The chamber was filled with warm solvent that contains crushed pieces of seeds. When the Soxhlet chamber was almost full, it was automatically emptied by a siphon side arm, with the solvent running back down to the distillation flask. After many cycles, the desired compound was concentrated in the distillation flask. The seed oil obtained had a higher viscosity and lower volatility. Then the oil is converted to biodiesel where its performances and properties are the same as that of diesel [23]. Various studies reported that transesterification is more authentic due to its short reaction time, low cost, high yield, low temperature and pressure and direct conversion process [24, 25].

2.1.2 Transesterification

Transesterification involves exchanging the organic group R of an ester with that of an alcohol. Catalysis of the reactions is often carried out with acidic or base catalysts.

Oil is preheated to 100 °C to remove the moisture in order to avoid the negative effect of moisture effect on the yield of methyl esters [21].

The next step after transesterification is alkali transesterification. Factors affecting this process include vegetable oil/methanol molar ratio, catalyst concentration, nature of catalysts and temperature [23]. Reaction time is also known to play an important role in obtaining biodiesel with the required purity. The abovementioned parameters were investigated for optimizing the biodiesel process. The transesterification reactions were conducted with various percentages of methanol of 15, 18, 21 and 24%. The catalyst concentrations were varied from 3, 3.5, 4, 4.5 and 5%, while the temperature was varied from 45 to 70 °C at an interval of 5 °C. The optimum range of methanol and catalyst concentrations and operating parameters were reported to be 15% for methanol and 4% (w/v) of NaOH with 1 L of juliflora oil heated to 60 °C in 60 min time and allowed to settle down for 24 h in a separating flask.

3 Properties of Biodiesel Fuel

Biodiesel fuel properties are viscosity, density, flash point, cloud point, fire point, cetane number, pour point, calorific values, etc. Biodiesel is one of the main renewable fuel sources derived from organic feedstock. A number of oil feedstock's can be extracted to biodiesel fuel. Some of the biofuel sources include vegetable oil, animal fat and waste cooking oil while rapeseed and soybean oils being the most widely used as a source of biodiesel. Soybean oil alone covered nearly 65% of the US production in 2013. The use of waste cooking oil for biodiesel extraction is having the additional advantage of avoiding the need for other ways of disposal, and it was found to result in a reduction of up to 82% GHG emission when compared with diesel [26, 27]. The selection of feedstock depends on availability, production cost, agricultural strategies and environmental impact. Preferences of feedstock differ in different countries, based on regional production cost, environmental impact and agricultural strategies. According to these preferences, feedstock is categorized into the following generations of vegetable oils.

3.1 First-Generation Feedstock

The first-generation feedstock for biodiesel production was edible vegetable oils. However, as the use of edible oils would jeopardize food supplies and biodiversity, most developing countries like India strictly prohibited the utilization of edible oil for biodiesel production in their food policies; this involved, for example, corn, nuts, soya bean, palm oil and coconut oil [27, 28].

3.2 Second-Generation Feedstock

The focus of second-generation biodiesel at first stage was the use of non-vegetable sources such as jatropha curcass oils, etc., and agricultural residues. This has been a successful example of sustainable commercial biodiesel production from residues and is currently practiced in developed countries such as the USA and Europe [27]. The efficient use of ample resources and being non-edible sources makes feedstock like algae to be a viable candidate for biodiesel production. However, considering the higher cost of extraction calls for studies on effective ways for commercialization [27]. The complexity and cost of extraction of biodiesel fuel could be countered with proper feedstock selection and better technology. The other factor in favor of biodiesels is its higher potential of reducing WTW GHG emissions up to 50% especially for application not on grasslands or deforestations.

3.3 Third-Generation Feedstock's

The third-generation fuels are synthetic fuels produced from gasification and Fischer–Tropsch (FT) process of any hydrocarbons or coal feedstock. The synthesis gases are CO and H2, and FT synthesis is gas to liquid (GTL), coal to liquid (CTL) and biomass to liquid (BML) [27].

4 Characteristics of Biodiesel

4.1 Fuel Emissions Perspectives

The U.S. Energy and Power Authority (EPA) published biodiesel emissions data for heavy-duty engines. According to this date for NO_x, CO, PM and hydrocarbons (HC), the NO_x, CO, PM concentration is shown to decrease with the increase in the volume percentage biodiesel. Hence, it can be observed that increase in biodiesel volume percentage is inversely proportional to the concentrations of CO, PM and HC and directly proportional to that of NO_x. This, in turn, signifies that higher amount of biodiesel concentration results in higher combustion efficiency but at the same time leads to higher NO_x emission, i.e., when biodiesel percentage by volume increases the oxides of nitrogen will increase too. This can be explained in terms of the higher natural oxygen content of biodiesel, where more oxygen was available for burning, and for reducing the emissions in the exhaust. It also helped to promote a stable and complete combustion by delivering oxygen to the pyrolysis zone. In addition, studies showed that due to the strong chemical bonding between the oxygen and carbon atoms in the fuel, it prevented availability of the carbon for soot formation and hence resulted in lower soot production. The strong oxygen and carbon bond remained intact, preventing it from becoming available for soot production [27].

Various studies tried to investigate the reason behind the increase in NO_x emissions in the case of biodiesel fuels. One plausible explanation is the increased flame temperature with biodiesel, caused by a reduction in the heat dissipation through radiation as a consequence of the lower amount of soot emitted. However, there was no evidence for significant difference in the calculated stoichiometric adiabatic flame temperature between biodiesel and diesel. But studies showed that the reason for the higher flame temperature in the case of biodiesel combustion is actually caused by the effect of lower soot emissions reducing the radiation heat loss by PM. The radiative cooling caused by PM is reported to reduce the flame temperature by -248.15 to -223.15 °C, corresponding to an estimated NO_x reduction of 12–25%. Therefore, it can be concluded that the combustion of biodiesel could result in elevated in-cylinder temperature, as a result of the lower PM emissions and the higher NO_x emissions [27].

4.2 Effects of Biodiesel Addition on Engine Performance

The study of Erkan et al. [29] reported that the most important biodiesel characteristics that highly influenced the specific fuel consumption of a biodiesel fuel run diesel engine included the heating value, density, viscosity and surface tension. The quantification of fuel injected by the injection pump of a compression ignition (CI) engine was volumetrically controlled, and as a result an increase in the density of the biodiesel directly resulted in injection of more fuel by mass. On the other hand, higher viscosity of biodiesel has been known to cause higher line pressure and even injection of higher volume of fuel due to decreased delivery losses. Hence, it can be concluded that the higher the density and viscosity of the biodiesel fuels would cause injection of higher amount biodiesel both in terms of volume and mass as compared to that of diesel. In addition, the reduction in effective atomization and combustion of the highly viscous and resulting higher surface tension biodiesel fuel would result in further increase in the fuel consumption.

The calorific or heating value of diesel and biodiesel is not similar. Studies revealed that due to the high oxygen and viscosity complexion of biodiesel, the heating value of biodiesel would be lower than that of petrodiesel; this led to increase in specific fuel consumption and decrease in brake power. The effects of biodiesel additions on the specific fuel consumption and the break thermal efficiency on the engine load are shown in Table 3. The 5% biodiesel mix has a very similar effect on specific fuel consumption since the biodiesel percentage share by volume is less but at low load the brake-specific fuel consumption is higher as compared to D100. For the case of 10% biodiesel addition, considerable increase is in specific fuel consumption at any loads of the experiment. This is because of the lower heating value of biodiesel mixture than petrodiesel. Therefore, biodiesel gets higher specific fuel consumption which is required to supply more fuel in order to have the same power.

Table 3 Effects of biodiesel addition on BSFC and thermal efficiency [29]

Engine load %	D100		B10		B5	
	BSFC	BTE	BSFC	BTE	BSFC	BTE
25	435	18	470	17	450	18
50	320	26	335	25	325	27
75	300	17	335	16	300	17
100	335	19	347	18	336	18

The break thermal efficiency is more related to lower heating value and specific fuel consumption. Studies showed that the brake thermal efficiency upsurged when the engine load increased. The brake thermal efficiency of B5 had similar trend with that of brake-specific fuel consumption. The result obtained showed that the B5 brake thermal efficiency values were nearly same with that of D100 fuel. It is understood that the reason for better combustion efficiency for B5 was due to the better oxygen content of the biodiesel waged for the lower heating value of B5. But for the case of B10, the brake thermal efficiency was reduced at all loads. The reductions of brake thermal efficiency are evidenced from the lower heating value and the higher brake-specific fuel consumption of the biodiesel. Furthermore, due to the higher viscosity nature of B10, the thermal efficiency and specific fuel consumption adversely affected from combustion deterioration.

4.3   Effect of Biodiesel Addition on the Combustion Parameters

One way to evaluate the performance and emission level of engines is the combustion parameters such as in-cylinder gas pressure, heat release rate, ignition delay, duration of premixed and diffusion phases. These combustion parameters make comparison of performance of different types of fuels with similar operating conditions. The most significant operating conditions that affect the parameters in-cylinder gas pressure and heat release profiles include air–fuel mixing ratio, the amount of fuel and heating value of fuel. Whereas the effect of blend ratio on these parameters was reported to be minimal at full-load operation [29]. At full load, it was further observed that the in-cylinder pressure (CP_{max}) and the maximum heat release rate (HRR_{max}) decrease with an increase in blend ratio. The decrease in CP_{max} and HRR_{max} with increase in blend ratio could be explained in terms of the lower amount of the fuel burned at premixed phase, poor atomization and inadequate air–fuel mixing of biodiesel blends. $LOHRR_{max}$ was generally shifted to BTDC with the increase in the biodiesel ratio. Also, $LOCP_{max}$ was slightly shifted to ATDC at high loads while it was almost stable at low loads. This situation at high loads may be the result of deterioration of atomization and combustion with the increase of biodiesel fuel in the blend [29].

The high cetane number of the fuel due to its higher oxygen content is known to decrease ignition delay. In addition, the combined effect of the lower heating value of the biodiesel fuel that results in faster warming and the higher volatility of the biodiesel causes a further reduction of the ignition delay. The ignition delays for static injection advance (ID-S) decreased gradually with addition of the biodiesel in the blend. However, the significant variations on ignition delays for the dynamic injection advance (ID-D) were not observed. Slight variations on PCD at low and partial loads were reported with increase in their values as the amount of biodiesel fuel at full load. DCD and combustion durations (CD) generally increased with the biodiesel addition. The variations in these durations were significantly influenced by the spray characteristics which again depend on characteristics of the biodiesel such as high viscosity and surface tension. In general, higher density, viscosity and surface tension characteristics of biodiesel decrease the fuel evaporation rate and reduce the effectiveness of the air–fuel mixing ratio as a result of bigger diameter droplets of the atomized fuel. As a result, the combustion phase further shifts toward the expansion period [29].

5 Effect of Different Biodiesel Oilseeds

Different researchers made a study on biodiesel fuel on different oil seeds from vegetables and non-vegetable feed stocks to obtain fuels with different performance, combustion and emission characteristics. With respect to this, Athul et al. [30] tested karanja oil methyl ester (KOME) biodiesel in transportation engine to investigate its performance, emissions and combustion characteristics. They showed that karanja biodiesel blends B5-B100 had attained maximum torque of 10% and 20% KOME blends. These also indicate that up to 20% there was no need for engine modification. And from the study, KOME obtained higher torque than pure diesel with lower CO, HC and CO_2 but at higher blends higher NO_x and lower cylinder pressure are attained.

Kalam et al. [31] tried to compare the engine performance and combustion characteristics of grape seed biodiesel and waste cooking oil blends in a diesel engine and showed that B5 grape seed biodiesel blend performed better than both B10 grape seed blend and all waste cooking biodiesel blends. In addition, the performance of B5 grape seed biodiesel was found to be comparable with that of ultra-low sulfur diesel. GS-B5 exhibits a slightly better performance due to its less ignition delay when compared to all other blends (GS-B10, WC-B5 and WC-B10). Further study on the performance emission and tribological behavior of GS-B5 blends is recommended to further investigate the sustainability of the fuel.

Mofijur et al. [32] showed the comparative evaluation of performance and emission characteristics of Moringa olifera and palm oil-based biodiesel in a diesel engine. The performance of these fuels was conducted an experiment on a multi-cylinder diesel engine at various speeds and under the full-load condition while emissions were assessed under the full-load and half-load condition. The physiochemical properties of Moringa olifera methyl ester (MOME) were determined, and the property of 5 and 10% (by volume) blends (MB5 and MB10) was compared with those of palm oil blends (PB5 and PB10) and diesel fuel (B0) performances of these fuels were assessed.

The results indicated that BP of B10 of both 4.2% lower than the average B0 and BSFC was slightly higher (5.13%) than the average B0. These results were attributed to the higher viscosity and density and the lower energy content of these biodiesel blends. The PB5, MB5, PB10 and MB10 blends reduced the average CO and HC emissions but the PB5 and MB5 blends slightly increased the NO_x emissions of diesel fuel. Both B5, B10, M5 and M10 CO_2 level are slightly increased. The results were attributed to the higher oxygen content and cetane numbers of the biodiesel blended fuels. The P10 and M10 of NO_x were found to be similar with B0. B5, B10 and M5, M10. Results were comparable, and with these blends it was confirmed that there was no need of engine modification. In addition, the study showed that Moringa olifera oil was a potential feedstock for biodiesel production, and the performances of the MB5 and MB10 biodiesel blends were comparable with those of PB5 and PB10 biodiesel blends and diesel fuel.

Mofijur et al. [33] used jatropha curcas methyl ester (JCME) blending with diesel to evaluate performance and emission characteristics. The physiochemical (T, BSFC, BP) properties of jatropha biodiesel and its blends with diesel followed by engine performance and emission characteristics of B10, B20 and B0 were studied. The result from the study indicated that calorific value decreased as the amount of jatropha biodiesel increased in the blends, while density was found to increase as the percentage of jatropha biodiesel increased in the blends. An average reduction in brake power for B10 and B20 by 4.67% and 8.86%, respectively, was observed as compared to B0. The performance and emission perspectives indicated that higher BSFC and decreased torque but the viscosities of B10 and B20 are closer to diesel. Similarly, the flash point was considered to be safe to store and to be used in diesel engines since jatropha biodiesel had higher flash point than diesel fuel; hence, the B10 and B20 can be used in diesel engine without any modification. For the case of emission compared to B0, a reduction in hydrocarbon (HC) emission of 3.84% and 10.25% and carbon monoxide (CO) emission of 16% and 25% was reported using B10 and B20. However, the blends give higher nitrogen oxides (NO_x) emission of 3% and 6% using B10 and B20. In general, jatropha crop can be considered as a promising source of biodiesel production.

Chandra et al. [34] used Pithecellobium dulceseed seed oil (PDSO) methyl ester blending with diesel engine as B20, B40, B60 and B80 by vol.% and compared with diesel engine for biodiesel production process optimization. From this two-step transesterification consisting of acid and base phases, the PDSOME and its blends showed a drop in-cylinder pressure, heat release rate (HRR) and exhaust gas temperature (EGT).

Additional results obtained include increase in the brake-specific fuel consumption (BSFC) and resulting reduction in brake thermal efficiency (BTE) in the case of using PDSOME and its blends. The exhaust emissions of CO, HC and NO_x were decreased while CO_2 and smoke intensity were increased with increase in PDSOME blending percentage in comparison with diesel fuel but B5 and B10 properties and performances were nearly same with B100. In general, at full load as the biodiesel blend increases it leads to a decrease in the concentration of HC and CO while the NO_x emission, BSFC, BTE, EGT and HRR increase slightly.

Wai et al. [35] used the free fatty acid Colophyllum inophyllumoil (CIO) for optimum production of the fuel and performance in compression ignition on engine and samples of CIOB10, CIOB20, CIOB30 and CIOB50 were tested on single-cylinder DI engine with dynamometer and gas analyzer device. The CIO biodiesel was produced using a laboratory scale reactor via degummed, acid esterification, neutralization and then undergoing base transesterification process. A 98.92% ester yield was obtained by using 1 wt. % catalyst concentration, with 9:1 ratio of methanol to oil at 50 °C for 60 min.

The present results obtained showed that degummed and neutralization process improved the fuel properties of the CIO biodiesel. From these experimental activities, the performance and emission of 10% CIO biodiesel blends (CIOB10) were found to give a satisfactory result in diesel engines as the brake thermal efficiency found to increase 2.3% and fuel consumption decreases 3% compared to diesel. Compared to diesel, the CIOB10 emitted CO and smoke opacity level were reduced but NO_x emission was slightly increased. CIOB5 performance and emission were reported to be nearly similar with that of D100 and with respect to D100 CIOB10 had better BTE and BSFC and reduced emission except NO_x but further studies on tribology and fuel economy need to be carried out before the CIO biodiesel–diesel blends could be utilized widely in the future. In short, CIO biodiesel was found to be a good candidate as an alternative fuel in the future.

Erkan et al. [29] recently studied the performance, emission, combustion and injection characteristics for CI engine running with diesel fuel blended with canola oil–hazelnut soap stock biodiesels. The experiment conducted with 5% (B5) and 10% (B10) biodiesel blends showed an increase in biodiesel addition resulted in reduced injection and ignition delays and lower HRR_{max} while the injection and combustion durations were increased. In addition, it was determined that the oxygen content of B5 enhanced the combustion process, and this resulted in increased NO_x emission and decreased in total hydrocarbons (THC), CO and smoke emissions. However, B10 fuel resulted in a less efficient combustion which may be due to its increased values of density, viscosity and surface tension. As a result, values of THC, CO and smoke emissions were reported to increase while reduction in NO_x emission level and with no significant difference are in terms of CO_2 emissions as compared to diesel. Similar to other studies, the injection and ignition delays, and HRR_{max} decreased with increase in blend ratio. The use of lower blend ratios (5–10% in vol.) for the mixture of the canola oil–hazelnut soap stock biodiesels in diesel engines without any modification was demonstrated to be safe while for the case of higher blend ratios further investigations need to be done.

Meisam et al. [36] studied and compared the combustion, performance and emission of a diesel engine running with diesel, biodiesel and ethanol in the blended and fumigation modes of ternary fuel. The test was done at five loading conditions for a speed of 1800 rpm with a fixed mixture percentage of D80B5E15. In addition to this to provide the same fuel composition, the three fueling modes for comparing the effects of the blends like blended mode 80%D, 5%B, 15%E and with fumigation mode BE with diesel in the first phase and DB mixed with diesel for second phase are injected into the intake manifold. Combined fumigation seven blended mode (F + B) samples are used. According to the average of five loads, compared to the pure diesel operation, the blended mode increases in peak heat release rate (HRR), ignition delay (ID), BSFC, CO, HC and NO_2; drops in duration of combustion (DOC), CO_2, NO_x, PM, total number concentration (TNC) and geometric mean diameter (GMD); and similar peak in-cylinder pressure and BTE. While the fumigation mode rose to higher peak HRR, DOC, BSFC, CO, HC, NO_2 and TNC; lower ID, BTE and NO_x; and similar peak in-cylinder pressure, CO_2, PM and GMD in comparison with the diesel mode. In addition, the F + B mode has the effects between those of the fumigation and blended modes. Better engine performance and emission levels (except for NO_x) were obtained corresponding the blended mode. However, the lowest engine performance and emissions were obtained for the case of the fumigation mode.

Meisam et al. [37] also investigated combustion performance and emission level of a CI engine running with diesel–biodiesel–alcohol blends having the same oxygen concentration and experiment were conducted on five engine loads and at an engine speed of 1800 rpm seven fuels including diesel used like samples of diesel (D), waste cooking oil biodiesel (B), methanol (M), ethanol (E), 2-propanol (Pr), n-butanol (Bu) and n-pentanol (Pe)) were used to produce six blended fuels, labeled as DB, DBM, DBE, DBPr, DBBu and DBPe. Each blended fuel has the same oxygen content of 5.0% and very close carbon and hydrogen contents and LHV. According to the average results of five loads, the blended fuels in general were found to cause: (a) increases in peak HRR (except DB), ignition delay (except DB), COV_{IMEP} (except DBM), COV_{Max} (dP/dq) (except DB and DBM) and BSFC; (b) slight decrease in duration of combustion (except DB) and (c) similar peak in-cylinder pressure and BTE (except DBM and DBBu) compared to diesel fuel. Moreover, all the blended fuels lead to reductions in CO_2 (except DB), CO, HC, NO_x (except DB), PM, total number concentration (except DBPr) and geometric mean diameter, compared to diesel fuel. Overall, DBM showed the highest BTE, the lowest BSFC and the lowest CO_2, CO, HC, PM, NO_x (after DBPr), COV_{IMEP} and COV_{Max} (dP/dq) (after DB), while DB had the lowest influence, among all the tested blended fuels. All the blended fuels had almost the same C/H/O compositions and lower heating values. Based on the average of five engine loads, it was observed that all the tested blended fuels using alcohols had almost similar trends on the engine combustion, performance and emissions. The use of biodiesel blend (DB) had opposite trends in some parameters PM is higher, HRR lower, relatively higher ID and higher NO_x.

Palm oil-based additive biodiesel was used in the work of How et al. [38] inside a direct injection compression ignition engine in order to investigate performance and exhaust gas emission. The study investigated the ideal biodiesel, diesel and additive mixture that produce the optimum engine emission and performance under ASTM standards using emission analyzer and engine dynamometer. For the experiment, the mixture or sample ratios used were 10%, 20% and 30% of biodiesel with and without the additive (palm oil-based additives 0.8% and 2.0%ml). The results of the experiment showed PB10 with 0.8 ml additives produced the highest braking power and lowest fuel consumption as compared to the diesel and the rest of biodiesel blends. The presence of biodiesel and additives was found to not only improve the engine performance, but also leads to the reduction of carbon emission. Diesel, biodiesel and additives demonstrated low smoke emission with a complete combustion with a slight increase observed in the NO_x and CO_2 emissions. In conclusion, PB10 was seen as the most ideal blend for diesel engine in terms of providing the most optimum engine emission and performance.

Kumar et al. [39] used WCO oil, jatropha and karanja oil-derived biodiesel to investigate comparative evaluation of diesel engine performance, combustion and emission characteristics with trace metals in particulates on single-cylinder diesel genset engine. Investigations were carried out at constant engine speed of 1500 rpm and six engine loads ranging from 0 to 100% were used, for particulate matters quartz filter paper flow dilution tunnel at 50 and 100% engine loads, for trace metal analyzing using inductively coupled plasma optical emission spectroscopy (ICP-OES). Samples of karanja, jatropha and WCO were used in experiment conducted and compared with diesel.

The result showed WCO biodiesel to have a slightly increased heat release rate (HRR) compared to baseline mineral diesel, while it was slightly lower for karanja and jatropha biodiesels. The concentrations of hydrocarbons (HC) and oxides of nitrogen (NO_x) emission levels were observed to reduce while comparative increase in carbon monoxide (CO) emission was observed as compared to diesel. In addition, WCO biodiesel exhibited comparable smoke opacity with baseline mineral diesel except at operating conditions corresponding to full load. Tests conducted on the PM showed minimum amount of Ca, Cu, Fe, K, Mg, Na, Zn concentrations while significant concentration of Al is observed, while Ba, Cd, Cr, Mn and Mo showed relatively lower concentrations. It needs further studies to evaluate the toxicity effects of adsorbed hydrocarbons.

6 Effect of Diethyl Ether on Biodiesel Fuel

The use of biodiesel and blends of the same are nowadays being mandated by regulation in some countries [40]. It is also well established that fuels with lower blending ratios do not generally require modification of engines as the physical and chemical properties of such low blend ratio fuels are closely similar with that of diesel. However, some researchers focused on the effect of the slight differences in the characteristics of low blending ratio biodiesels and that of diesel.

Ibrahim et al. [40] recently reviewed on using diethyl ether in a diesel engine operated with diesel–biodiesel fuel blend to enhance the performance of a compression–ignition engine by using a single-cylinder direct injection, four-stroke engines with hydraulic dynamometer used to conduct all tests. Four fuels were examined for the experiment with 100% diesel, 70% diesel +30% biodiesel (% by volume) D70B30, 70% diesel +25% biodiesel +5% DEE (% by volume) D70B25DEE5, 70% diesel +20% biodiesel +10% DEE (% by volume) D70B20DEE10. From the study reported with diesel–biodiesel blend increased the minimum brake-specific fuel consumption (BSFC) and decreased the maximum thermal efficiency by 8.1% and 6.8%, respectively, compared to diesel fuel and the optimum blending proportion was 5% because blending the diesel–biodiesel mixture with 5% DEE led to a significant improvement in engine performance as the BSFC decreased and thermal efficiency increased at most engine loads compared to diesel. However, increasing the proportion of DEE to 10% decreased thermal efficiency.

Kapura et.al [41] used a tire-derived fuel–diesel blend to investigate the effect of diethyl ether in DI diesel engine with sample parameters of diethyl ether on 60% diesel with 40% pyrolyzed tire oil with 1–4% (X1-X4)-diethyl ether and the emission and BSFC compared. They found that the 4% DEE addition to 40LFPO (X4) gave better performance and lower emissions than those of X1, X2 and X3 at full load. The brake-specific fuel consumption (BSFC) of X4 was lower by about 6% than that of diesel operation at full load. The NO_x emission of X4 was lower by about 25% than that of diesel operation at full load.

Sivalakshmi et al. [42] identified the effect of biodiesel and its blends with diethyl ether on the combustion, performance and emissions from a diesel engine. Experiments have been conducted in a single-cylinder four-stroke naturally aspirated, direct injection diesel engine with eddy current dynamometer. Sample parameters of neem oil with two-step esterification (H2SO4, KOH) and D100, BD5, BD10 and BD15 were used. Enhancement of the useful physical and chemical properties of biodiesel was possible through the addition of DEE. The highest values of C_{pmax} and HRR_{max} were reported corresponding to BD5. Addition of DEE with BD5 resulted in improved brake thermal efficiency and brake-specific fuel consumption (BSFC) while hydrocarbon (HC) emission remained high for all blends of DEE and biodiesel compared to that of the biodiesels. From these results, it could be concluded that addition of DEE up to 5% (by vol.) is highly promising for use inside a CI engine with no modification.

Dimitrios et al. [43] investigated the combustion performance and emission levels of a high-speed direct injection diesel engine operated with biodiesel and DEE blend. In this study, investigation on the efficient use of DEE as additive to normal diesel fuel inside a high-speed, direct injection CI engine was carried out keeping a fixed engine speed of 2000 rpm at three different loading conditions. Measurement of smoke and emission levels was carried out while the break-specific fuel consumption (BSFC) and the engine thermal efficiency were determined based on measure values of the fuel volumetric flow rate, as well as its density and calorific value. It was observed from the experimental results that the use of DEE biodiesel blends resulted in reduced NO_x and CO emissions levels with higher emission reductions obtained for higher DEE additions while the reverse is true for HC emissions since HC emissions increase with the higher percentage of DEE. As of the performance of the engine higher specific fuel consumption was reported corresponding to higher percentage of DEE as compared to the pure diesel fuel while the brake thermal efficiency remains nearly the same and a slight increase in the exhaust gas temperature was reported. In general, it was observed that addition of bio-based DEE with diesel is found to give a significant improvement in the performance of diesel engines.

Swaminathan et al. [44] identified the investigated effect of DEE addition on the performance and emission characteristics of a single-cylinder diesel engine operating with fish oil biodiesel. Samples used were BFO with 1–3% DEE addition and compared with diesel fuel. The results obtained were BFO, of 2% operated with EGR resulted in the least level of all forms of emission gases. The study further reported that brake thermal efficiency at all loading conditions was found to be better for BFO and BFO blends when compared with pure diesel. The exact values of the percentage reduction in the emission levels were reported as CO-91%, CO_2-62%, NO_x-92% and C_xH_y-90% for the engine running at maximum loading using BFO with 2% additive with EGR. In the case of NO_x, an increase of 48% at maximum load was reported for BFO in comparison with pure diesel, which could be due to addition of oxygenates and use of EGR technique.

Similarly, the study of Dimitrios et al. [45] on cotton seed biodiesel with DEE additives reported reduction in ignition delay and increased ignition delay. At the start of combustion, the temperature and cylinder pressure were reduced due to the DEE additives. Higher reductions of all emissions were observed with the diesel fuel diethyl ether blends. The exhaust gas temperature was slightly higher but the brake thermal efficiency and BTE remained nearly the same while emissions were significantly reduced.

7 Summary

Biodiesel is an alternative source of energy which benefits the environment through keeping clean atmosphere and replaces the fossil fuel and helps the countries economically, socially and politically. A comprehensive coverage of a review of biodiesel production technique and process was given in this chapter. The advantage of using additives on a biodiesel to overcome its drawbacks in terms of higher viscosity, calorific value, pour point, etc., was evidenced from the results presented in the works of different researchers. Finally, the performance, emission and combustion performance of biodiesel oil in comparison with normal diesel oil and that of blending of additives were presented comparatively in order to highlight their advantage for researchers, policymakers and manufacturers.

The summary of this study can be highlighted as follows:

- The main factors that affect the combustion characteristics of biodiesel in the engine are injection timing, fuel properties, operating conditions, engine types and feedstock types. An emission and performance characteristic of a biodiesel in a diesel engine vary depending on the country of origin since different biodiesel has a different behavior.
- Utilization of biodiesel drastically reduces harmful emissions such as CO, HC, PM emissions and the CO_2 emissions from biodiesel can be ignored since it is absorbed by the greenhouse gas exchange with the crop plants but slightly increases NO_x emissions. The NO_x can be reduced by using some additives or by

exhaust gas recirculation (EGR). However, employment of biodiesel also stubbly increases fuel consumption and reduces engine power.

- The addition of DEE on biodiesel in different percentage indicates it dilutes the high viscosity biodiesel pour point, flash point, effect and in result reduces harmful emissions and improves engine performance and combustion phenomena.
- Two-third of total production cost is feedstock selection so the total percentage of biodiesel yield obtained from a given mass of feedstock matters the production cost of a biodiesel.
- Transesterification biodiesel oil production method is the most economical, convenient, simplicity and cost-effective process among other biodiesel production process.
- The fatty acid constituents of biodiesel oil influence the characteristics of biodiesel produced.
- Further studies focus on juliflora fruit biodiesel and its performance, emission and combustion characteristics with the addition of different percentage levels of DEE and performances are compared.

In general, DEE as an additive in different proportions with biodiesel has great role towards improving performance, power, combustion and emission levels in diesel engines. In addition, it being a renewable and environmentally friendly fuel source increases the viability of biodiesel as an alternative and renewable fuel.

Acknowledgements The authors would like to acknowledge the support of Addis Ababa Science and Technology University (AASTU), College of Electrical and Mechanical Engineering and Renewable Center of Excellence of the University.

References

1. Krishna KV, Sastry GRK, Krishna MM, Barma JD (2018) Investigation on performance and emission characteristics of EGR coupled semi adiabatic diesel engine fueled by DEE blended rubber seed biodiesel. Eng Sci Technol An Int J 21:122–129
2. Feroskhan M, Ismail S, Reddy MG, Teja AS (2018) Effects of charge preheating on the performance of a biogas-diesel dual fuel CI engine. Eng Sci Technol An Int J 21:330–337
3. Kurtgoz Y, Karagoz M, Deniz E (2017) Biogas engine performance estimation using ANN. Eng Sci Technol An Int J 20:1563–1570
4. Sivasubramanian H, Pochareddy YK, Dhamodaran G, Esakkimuthu GS (2017) performance, emission and combustion characteristics of a branched higher mass, C3 alcohol (isopropanol) blends fueled medium duty MPFI SI engine. Eng Sci Technol An Int J 20:528–535
5. Acharya N, Nanda P, Panda S, Acharya S (2017) Analysis of properties and estimation of optimum blending ratio of blended mahua biodiesel. Eng Sci Technol An Int J 20:511–517
6. Milano J, Ong HC, Masjuki HH, Silitonga AS, Chen WH, Kusumo F, Dharma S, Sebayang AH (2018) Optimization of biodiesel production by microwave irradiation-assisted transesterification for waste cooking oil-*Calophyllum inophyllum* oil via response surface methodology. Energy Convers Manage 158:400–415
7. Silitonga AS, Mahlia TMI, Ong HC, Riayatsyah TMI, Kusumo F, Ibrahim H, Dharma S, Gumilang D (2017) A comparative study of biodiesel production methods for *Reutealis trisperma* biodiesel. Energy Source Part a. 39(20):2006–2014

8. Damanik N, Ong HC, Chong WT, Silitonga AS (2017) Biodiesel production from *Calophyllum inophyllumpalm* mixed oil. Energy Source Part a 39(12):1283–1289
9. Ibrahim A (2016) Investigating the effect of using diethyl ether as a fuel additive on diesel engine performance and combustion. Appl Therm Eng 107:853–862
10. El-Adawy M, Ibrahim A, El-Kassaby MM (2013) An experimental evaluation of using waste cooking oil biodiesel in a diesel engine. Energy Technol 1(12):726–734
11. Ibrahim A, El-Adawy M, El-Kassaby MM (2013) The impact of changing the compression ratio on the performance of an engine fueled by biodiesel blends. Energy Technol 1(7):395–404
12. Mahmudul HM, Hagos FY, Mamat R, Abdul Adam A, Ishak WFW, Alenezi R (2017) Production, characterization and performance of biodiesel as an alternative fuel in diesel engines—A review. Renew Sustain Energy Rev 72:497–503
13. Patil KR, Thipse SS (2015) Experimental investigation of CI engine combustion, performance and emissions in DEE–kerosene–diesel blends of high DEE concentration. Energy Convers Manage 89:396–408
14. Kaimal VK, Vijayabalan P (2016) An investigation on the effects of using DEE additive in a DI diesel engine fuelled with waste plastic oil. Fuel 180:90–96
15. Venu H, Madhavan V (2017) Influence of diethyl ether (DEE) addition in ethanol biodiesel-diesel (EBD) and methanol-biodiesel-diesel (MBD) blends in a diesel engine. Fuel 189:377–390
16. Lee S, Kim TY (2017) Performance and emission characteristics of a DI diesel engine operated with diesel/DEE blended fuel. Appl Therm Eng. 121:454–461
17. Tudu K, Murugan S, Patel SK (2016) Effect of diethyl ether in a DI diesel engine run on a tyre derived fuel-diesel blend. J Energy Inst 89:525–535
18. Barik D, Murugan S (2016) Effects of diethyl ether (DEE) injection on combustion performance and emission characteristics of Karanja methyl ester (KME)–biogas fueled dual fuel diesel engine. Fuel 164:286–296
19. Devaraj J, Robinson Y, Ganapathi P (2015) Experimental investigation of performance, emission and combustion characteristics of waste plastic pyrolysis oil blended with diethyl ether used as fuel for diesel engine. Energy 85:304–309
20. Accessed from online https://commons.wikimedia.org
21. Asokan MA (2019) Performance, combustion and emission characteristics of juliflora biodiesel fueled DI diesel engine. Energy 173:883–892
22. Islam MN, Sabur A, Ahmmed R, Hoque ME (2015) Oil extraction from pine seed (*Polyalthia longifolia*) by solvent extraction method and its property analysis. Proc Eng 105:613–618
23. Silitonga AS, Mahlia TMI, Ong Hwai Chyuan, Riayatsyah TMI, Kusumo F, Husin Ibrahim A (2017) Comparative study of biodiesel production methods for Reutea-listrisperma biodiesel. Energy Sources, Part A Recovery Util Environ Eff 39(20)
24. Milano J (2018) Optimization of biodiesel production by microwave irradiation-assisted transesterification for waste cooking oil *Calophyllum inophyllum* oil via response surface methodology. Energy Convers Manag 158:400–415
25. Ashraful AM (2014) Production and comparison of fuel properties, engine performance, and emission characteristics of biodiesel from various non-edible vegetable oils. Energy Convers Manage 80:202–228
26. Ghazali W, Mamat R, Masjuki HH, Najafi G (2015) Renewable sustainable. Energy Rev 51:585–602
27. Bae C (2017) Alternative fuels for internal combustion engines. Proc Combust Inst 36:3389–3413
28. Edwards R, Hass H, Larive J, Lonza L, Maas H, Rickeard D (2014) Well-to-WheelsReportVersion4.a, JEC Well-to-Wheels Analysis, European Commission Joint Research Centre Institute for Energy and Transport, 2014 ISBN 978-92-79-33887-8
29. Erkan Ö (2015) Performance, emissions, combustion and injection characteristics of a diesel engine fueled with canola oil–hazelnut soap stock biodiesel mixture. Fuel Process Technol 129:183–191

30. Atul D (2014) Performance, emissions and combustion characteristics of Karanja biodiesel in a transportation engine. Fuel 119:70–80
31. Kalam A (2019) performance and combustion analysis of diesel engine fueled with grape seed and waste cooking biodiesel. Energy Proc 160:340–347
32. Mofijur M (2014) Comparative evaluation of performance and emission characteristics of *Moringa oleifera* and Palm oil based biodiesel in a diesel engine. Ind Crops Prod 53:78–84
33. Mofijur M (2013) Evaluation of biodiesel blending, engine performance and emissions characteristics of *Jatropha curcas* methyl ester: Malaysian perspective. Energy 55:879–887
34. Chandra Sekhar S (2018) Biodiesel production process optimization from *Pithecellobium dulceseed* oil: Performance, combustion, and emission analysis on compression ignition engine fueled with diesel/biodiesel blends. Energy Convers Manage 161:141–154
35. Hwai Chyuan Ong (2014) Optimization of biodiesel production and engine performance from high free fatty acid *Calophyllum inophyllumoil* in CI diesel engine. Energy Convers Manage 81:30–40
36. Meisam A (2019) Study of combustion, performance and emissions of a diesel engine fueled with ternary fuel in blended and fumigation modes. Fuel 235:288–300
37. Meisam A (2018) Study of combustion, performance and emissions of diesel engine fueled with diesel/biodiesel/alcohol blends having the same oxygen concentration. Energy 157:258e269
38. How CB (2019) Performance and exhaust gas emission of biodiesel fuel with palm oil based additive in direct injection compression ignition engine. Int J Autom Mech Eng 16(1):6173–6187. ISSN: 2229-8649 (Print); ISSN: 2180-1606 (Online)
39. Chetankumar P (2019) Comparative compression ignition engine performance, combustion, and emission characteristics, and trace metals in particulates from Waste cooking oil Jatropha and Karanja Oil Derived Biodiesels. Fuel 236:1366–1376
40. Ibrahim A (2018) An experimental study on using diethyl ether in a diesel engine operated with diesel-biodiesel fuel blend. Int J Eng Sci Technol 21:1024–1033
41. Kapura T (2016) Effect of diethyl ether in a DI diesel engine run on a tyre derived fuel-diesel blend. J Energy Inst 89:525–535
42. Sivalakshmi S (2013) Effect of biodiesel and its blends with diethyl ether on the combustion, performance and emissions from a diesel engine. Fuel 106:106–110
43. Dimitrios CR (2012) Characteristics of performance and emissions in high-speed direct injection diesel engine fueled with diethyl ether/diesel fuel blends. Energy 43:214–224
44. Swaminathan C (2012) Performance and exhaust emission characteristics of a CI engine fueled with biodiesel (fish oil) with DEE as additive. Biomass Bioenergy 39:168–174
45. Dimitrios CR (2014) Influence of properties of various common bio-fuels on the combustion and emission characteristics of high-speed DI (direct injection) diesel engine: Vegetable oil, bio-diesel, ethanol, n-butanol, diethyl ether. Energy 73:354–366

Geothermal Power Potential in Ethiopia

Fiseha M. Guangul and Girma T. Chala

Abstract Ethiopia is endowed with various types of renewable energy resources. The country has significant amount of geothermal energy source that can be developed to produce more than 10,000 MW electric power. The unique location of Ethiopia where The Great Rift Valley extending 1000 km from the Afar depression at Red Sea–Gulf of Aden junction north-west to Turkana depression on the south enables the country to share the abundant geothermal energy of the rift valley. Currently the total electric energy produced from geothermal source is not more than 8 MW. The disparity between the power demand and supply gap available in the country and the need to diversify the hydroelectric energy to other renewable sources enhance the development geothermal energy source in the country. High initial investment cost for geothermal energy development as opposed to hydropower energy, lack of technical competence and low tariff regime of electric power are some of the challenges faced by the country to develop geothermal energy. In this chapter, the opportunities and challenges of developing geothermal energy in Ethiopia are discussed. In addition, some recommendations to overcome the challenges are highlighted.

1 Introduction

Currently fossil fuel takes the major share in the global energy supply. Nevertheless, due to the adverse effect of fossil fuels on the environment and the limited supply, the focus of many countries has been shifted to renewable energy sources [1–6]. Ethiopia is the second most populated nation in Africa next to Nigeria with 108 million people

F. M. Guangul (✉)
Department of Mechanical Engineering, Middle East College, PB 79, PC 124, Al Rusayl, Muscat, Oman
e-mail: fiseham2002@yahoo.com

G. T. Chala
Mechanical (Well) Engineering, International College of Engineering and Management, C.P.O Seeb, P.O. Box 2511, P.C. 111, Muscat, Oman
e-mail: girma_tade@yahoo.com

© The Author(s), under exclusive license to Springer Nature Singapore Pte Ltd. 2021
S. A. Sulaiman (ed.), *Clean Energy Opportunities in Tropical Countries*,
Green Energy and Technology, https://doi.org/10.1007/978-981-15-9140-2_10

as of 2018, and one of the fastest growing economy in the world. Yet, the country is still regarded among the poorest countries in the world with annual per capita income of $953. It was projected that the country would reach to lower–middle–income status by transforming the agriculture sector into manufacturing sector by 2025 [7]. However, energy supply remains as one of the bottlenecks to achieve the targeted growth. In the Growth and Transformation Plan I (GTP I), which covered between 2011 and 2015, the government planned to increase the power supply from 2000 to 8000 MW. Although the power generation was grown by more than one fold to 4180 MW, the targeted plant was not achieved [8].

In the second Growth and Transformation Plan (GTP II), which is planned to be implemented between 2016 and 2020, The government has targeted to enhance the electric power supply to 17,208 MW from 4180 MW, of which 13,817 MW is expected from hydropower, 1224 MW from wind power, 300 MW from solar power, 577 MW from geothermal power, 509 MW from reserve fuel (gas turbine), 50 MW from wastes, 474 MW from sugar bagasse and 257 MW from biomass [8].

Currently hydropower provides 90% of the country electric power supply. Ethiopia has a potential to generate 45,000 MW electric power from hydropower; however, the country has only used not more than 10% of the resource to generate power. Though the country depends almost entirely on hydropower energy to cover its demands, this energy source is currently facing a challenge with rainfall fluctuation as a result of the world climate change resulting in shortage of water [9]. Hence, the government has intended to diversify the renewable energy source and as a result geothermal energy was set to be considered as the second priority.

The Ethiopian Rift extends from the Ethiopia–Kenya border to the Red Sea in a NNE direction for over 1000 km within Ethiopia and covers an area of 150,000 km^2. The geothermal locations of Ethiopia are shown in Figs. 1 and 2 [10, 11]. The exploration of geothermal energy in Ethiopia was started in 1969 with a regional

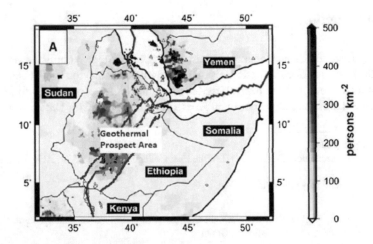

Fig. 1 Geothermal prospect areas in Ethiopia [10]

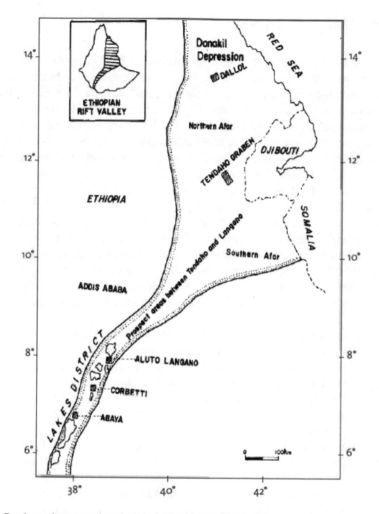

Fig. 2 Geothermal resource locations in Ethiopian Rift Valley [11]

geological–volcanological mapping and hydrothermal manifestation inventory in most of the Ethiopian Rift. Even though the inventory work is not completed in all parts of the country, especially in the highland areas, the rift valley areas are well covered, and about 120 locations are found to have independent heating and circulation system. Among these locations, above two dozen were identified to have high enthalpy resources for potential application of electric power generation. Previous studies showed that an electric power ranging between 4200 and 10,800 MW could be generated from the sites [12, 13]. The objective of this chapter is, therefore, to discuss the opportunities and challenges of developing geothermal energy in Ethiopia.

1.1 Geothermal Resource

Geothermal energy is obtained from the Earth natural heat, primarily due to natural decay of radioactive isotopes of uranium, thorium and potassium. The temperature of the earth increases with depth measured from the surface to the interior. On average, the increment on temperature reaches 20–30 °C/km. The temperature at the base of the continental crust ranges from 200 to 1000 °C, whereas it could reach 3500–4500 °C at the Earth's centre. Because of the heat gradient, the flow of heat takes place from the interior to the surface depending on the geothermal gradient and thermal conductivity of the rock. The amount of heat flow at the surface to the atmosphere ranges between 40 and 90 mW/m^2, and the total global output could reach over 4 × 10^{13} W. Currently, the total global energy consumption is about 10^{13} W, and this is a quarter of the energy that dissipates from the surface of the earth. Hence, the thermal energy of the Earth is immense; however, only a fraction can be utilized. Most geothermal exploration and use occur at the higher temperature gradient and where drilling is shallower and less costly. Currently the utilization of geothermal energy is limited to areas where heat transfer can be achieved from deep hot zone to the surface or near the surface by using heat carrier, i.e. liquid water or vapour [14, 15].

Economical geothermal energy development is achieved at a higher temperature gradient along the depth and higher temperature can be obtained in shallower depth. The shallow depth geothermal energy can be obtained when: (1) a great quality of heat is brought up by intrusion of molten magma from deep surface to the shallow surface; (2) the heat flow is high as a result of thin crust and high temperature gradient; (3) the groundwater is able to circulate to depth of several kilometres and heat transfer occurs as a result of normal temperature gradient; (4) the thermal conductivity is low due to the thermal blanket shale formation in the deep rock; and (5) the radioactive elements which may be intensified by thermal blanketing create heat in the shallow rock [16].

Geothermal resources that are used in large-scale power generation are hydrothermal, geopressured brines, hot dry rock and magma. Commonly, hydrothermal resources are used in commercial scale plants. Water heated by descending to shallow depth from few hundred to 3000 m and escaped as steam or hot water is utilized in binary power plant for electric power generation. The resource is classified into three main groups based on the temperature of the water or steam obtained: high for temperature above 200 °C, medium for temperature range between 100 and 200 °C, and low for temperature below 100 °C. The geothermal resource below 100 °C is not suitable for commercial power generation. The geothermal resource with high water content has low efficiency when compared to that of conventional steam power plant since the heat in the water is not convertible to electric energy [17].

Geopressured brains are hot pressurized waters with temperate range between 149 and 204 °C containing methane and found at 3000–6000 m below the surface of the earth. Three types of energy can be produced from geopressured brines: thermal energy produced from high temperature fluid, hydraulic energy produced from high pressure and chemical energy produced from burning of the dissolved methane gas. The geopressured brains become unusable from economic point of view once enough quantity of water is removed, and the pressure and the brine are dropped.

Dry rocks that are commonly found at depths of 4000 m and below are exceptionally hot. Two wells need to be drilled to access the hot rocks. The preexisting fractured rocks between the wells would help water to be pumped into the dry hot rock at high pressure. The water pumped is then heated and converted into steam and drawn up with the other two wells. By using a binary of flash power plant, the thermal energy of the steam is converted to electric power [17].

1.2   Types of Geothermal Power Plants

Geothermal plants operate over a temperature range of 50 and 250 °C, which is lower than the conventional fossil fuels or nuclear plants (usually at 550 °C). Geothermal power plant can be classified into three types: flash, binary and dry steam plants.

Dry steam plant

Dry steam plant is the oldest and simplest geothermal plant. The turbine operates with the steam coming directly from the production well. After using the steam and operating the turbine, the exiting steam or water is cooled and pumped back to the injection well [18].

Flash steam plant

Currently flash steam geothermal power plant is commonly used type power plant. The power plant uses water above 180 °C that flows upwards from a high-pressure reservoir well. As the water ascends to the surface, the pressure drops and the hot water partially converted into steam. The steam is sent to the turbine to drive after separating the water from the steam using steam separator. The condensed steam that exits the turbine and the unused water are pumped to the injection well as depicted in Fig. 3 [19].

Binary steam plant

Binary power plant is recently developed. In this plant, the water is used to heat another organic fluid that has lower boiling temperature. Hence, the organic fluid is converted into gas and used to drive the turbine instead of the hot water or steam. The advantage of this power plant is the possibility of electricity generation from a low water reservoir temperature of below 150 °C. The power plant uses heat exchanger to transfer the heat from the water to the secondary working fluid. The water is allowed to cool further in a condenser and injected to the well, as shown in Fig. 4 [19].

Fig. 3 Geothermal power
plant: **a** dry steam and
b flash steam [19]

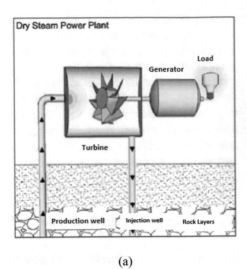

(a)

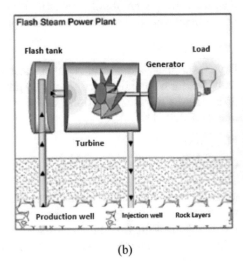

(b)

1.3   Geothermal as a Source of Electrical Energy

The global utilization of geothermal energy in the current electric power generation
is not more than 0.3% of the total electric production. The contribution of geothermal
energy for electric power generation did not also exceed 1.5%. Figures 5, 6 and 7
show the geothermal energy contribution to generate electricity and its application
in various areas [20].

Fig. 4 Binary cycle
geothermal power plant [19]

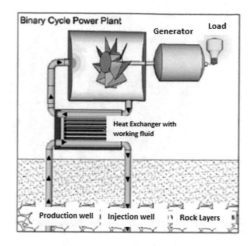

Fig. 5 World electric
production from renewable
and non-renewable sources
in 2012

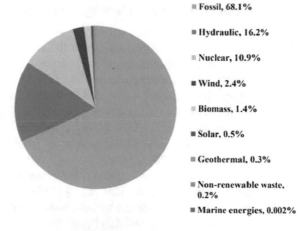

- Fossil, 68.1%
- Hydraulic, 16.2%
- Nuclear, 10.9%
- Wind, 2.4%
- Biomass, 1.4%
- Solar, 0.5%
- Geothermal, 0.3%
- Non-renewable waste, 0.2%
- Marine energies, 0.002%

Fig. 6 World electric
production from renewable
source in 2012

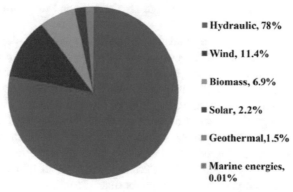

- Hydraulic, 78%
- Wind, 11.4%
- Biomass, 6.9%
- Solar, 2.2%
- Geothermal, 1.5%
- Marine energies, 0.01%

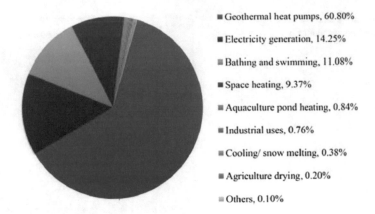

Fig. 7   Worldwide geothermal energy applications in 2014

1.4   Global Geothermal Development and Ethiopia's Position

The contribution of geothermal energy to the world's primary energy is small in proportion. In 2015, the new installed capacity of geothermal energy was about 315 MW and the total capacity reached 13.2 GW by the end of the same year. Half of the new installed capacity was added by Turkey followed by the USA, Mexico, Kenya, Japan and Germany. The countries benefited from geothermal energy are not many. For instance, countries like China, Iceland, Turkey, Hungary, Japan, USA and New Zealand comprise 70% of the total in utilizing direct geothermal energy for different applications in 2015.

The earth has an estimated stored thermal energy of 43×10^6 EJ within the continental crust in 3 km depth. This energy is higher than the world's total primary energy consumption. A significant percentage reaching 43% out of the total installed generation capacity resides in island countries. This geothermal source provides not only electric power but also heating and heat storage over a wide spectrum of conditions [21].

Iceland geothermal development is a good model and can be considered as a benchmark for Ethiopia geothermal development plan as both countries have similar geological features [9]. A report in 2013 indicated that 69.2% of the primary energy source in Iceland was obtained from geothermal energy. In 2012, about 5210 MW of electric power accounting 30% of the total demand was generated from geothermal energy. The history of using geothermal energy started late 1918 when the coldest winter hit the country, and coal price escalated due to the Second World War. During the same year, two-third of the population were affected by flue as a result of severe weather condition. By 1926, the Government of Iceland initiated to increase the usage of geothermal energy for district cooling. Geothermal energy was not applied to generate electric power until 1944. In 1944, a small turbine was used to generate electricity from a geothermal hotspot. The first commercial power plant that generated 3 MW was established in 1969 after the establishment of National Energy

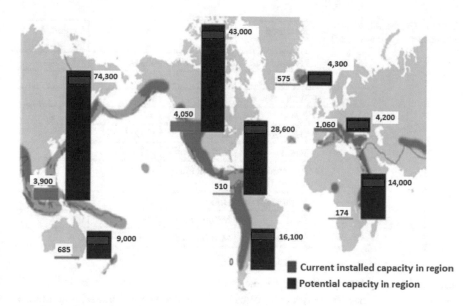

Fig. 8 Global potential and installed geothermal power in MW [22]

Authority. The country has able to generate 5210 GWh, which is 30% of the total electrical power produced, from geothermal source after the government took several steps to improve various initiatives and legal frameworks in 2012. Studies indicated that Iceland and Ethiopia have similar geological profile and Ethiopia can benefit from Iceland experience in developing geothermal energy [9].

Ethiopia is among countries that have been endowed with high potential of geothermal energy. As shown in Fig. 8, the potential of geothermal energy of the globe is much higher than the installed capacities and geothermal projects under installation. Ethiopia's geothermal energy is estimated to reach 10,000 MW, and this is higher than the available geothermal potential even to some of the continents.

1.5 Geothermal Exploration and Development in Ethiopia

Ethiopia is located in the horn of Africa that lies between 3 °N and 15 °N latitude, and 33 °E and 48 °E longitude. The country has approximately 1.14 million km^2 area [12]. Currently, the population is estimated to reach up to 110 million. The country is known as one of the least developed country (LDC) in sub-Saharan Africa. Nevertheless, during the last decade, the country has recorded more than 10% GDP growth per annum and became one of the fastest growing economies in the world [23]. In the two Growth and Transformation Plans of the country, i.e. GTP I & GTP II, the Government of Ethiopia had planned to transform the economy from agricultural lead to manufacturing lead economy. In achieving the transformation plan, power has

Table 1 Exploitable and exploited energy source [24]

Resource	Unit	Exploitable reserve	Exploited percent (%)
Hydropower	MW	45,000	<5
Solar/day	kWh/m^2	4–6	<1
Wind: Power	GW	100	<1
Speed	m/s	>7	
Geothermal	MW	10,000	<1
Wood	Million tons	1120	50
Agricultural waste	Million tons	15–20	30
Natural gas	Billion m^3	113	0
Coal	Million tons	300	0
Oil shale	Million tons	253	0

remained as one of the challenges for the government. The energy source in Ethiopia can be classified into two major categories: traditional source (biomass) and modern source (electricity and petroleum). The traditional energy source is the principal source of energy as more than 80% of the population are living in the rural area and engaged in the small-scale agricultural sector [12]. To address the energy shortage of the country in the rural areas and newly flourishing industrial areas, the government is investing huge amount of money to generate electric power from various renewable sources. Ethiopia has abundant unexploited renewable energy sources. The electric power generated from hydro and geothermal alone exceeds 55,000 MW. The potential of various energy sources and their utilization percentage are shown in Table 1. Ethiopia can generate and cover the future power from renewable sources at low cost. As shown in Table 1, from hydropower 45,000 MW and from geothermal 10,000 MW, electricity can be generated. The country can also generate huge electric power from solar and wind energies.

Currently, the total electric power generation capacity is 4180 MW. As per the GTP II, the government has planned to generate 7579 MW power from hydropower. In the last several years, the country has faced seasonal electric power supply fluctuation when the amount of rainfall fluctuated as a result of climate change. To ensure a reliable and climate resilient power supply, the Government of Ethiopia is striving to diversify the renewable energy sources to generate power from solar, geothermal, wind, biomass, etc. The government has agreed with two companies to generate 1000 MW electricity power from geothermal energy on two sites (Corbetti & Tulu Moye) [25, 26]. Currently the country produces 324 WM electric power from wind energy and by the end of 2020 this would reach 5200 MW. There is also an ambitious plan to install 5200 MW solar energy conversion plants on different sites [26]. The implementation plans for the solar energy and wind energy development do not seem achievable as only one year is left for GTP II to be concluded.

In the last four decades, substantial geothermal exploration has been carried out and promising and feasible results were obtained. Since 1970 geoscientific surveys that comprise geology, geochemistry and geophysics were carried out covering from south to north areas, which included Abaya, Corbetti, Aluto–Langano, Tulu Moye and Tendaho. From early to 1985, eight exploration drilling operations were carried out and five of them were proved to be productive. After continuing the exploration work until 1998, the pilot plant with 7.2 MW became operational. During this exploration period, three deep wells up to 2100 m and three shallow wells up to 500 m were drilled, and the temperature was found to be above 250 °C. In the preliminary operation test, from shallow wells, a steam that can produce up to 5 MW electricity was obtained, and in the deep wells, a steam that can produce up to 20 MW electricity was obtained [12, 27].

2 Opportunities of Geothermal Energy in Ethiopia

2.1 The Unique Geographical Location

The Great Rift Valley crosses Ethiopia and extends more than 1000 km in the N-S direction starting from the Afar depression at Red Sea–Gulf of Aden junction outwards to the Turkana depression. The typical rift morphology is developed in the main Ethiopian Rift covering a length of 500 km [9, 28]. The Rift Valley of Africa is known by its tremendous geothermal energy. The utilization of this resource for electricity generation and other applications are at the infant stage with the notable exception of Kenya. With the current production technologies, more than 15,000 MW electric power can be generated from East African countries geothermal resource (Fig. 9) [29, 30]. Exploration studies show that Ethiopia, Djibouti, Uganda, Kenya and Tanzania have enormous geothermal resources to generate electric power [30].

Currently Ethiopia generates 4180 MW of electricity and only 27% of the population have access to electricity grid. Hence, the geothermal resource available in the country could be one feasible source to address the power shortage of the country.

2.2 Government Policy and Trend

The Ethiopian National Energy policy which was first drafted by the Transitional Government of Ethiopia in 1996 sets two main goals. The first goal was to provide electric power at affordable price to support the agricultural led industrial development. The second goal was to ensure an electric power supply to the community and shift the traditional energy source to a sustainable, modern and renewable energy source with the aim of achieving comprehensive rural energy development [9]. With special emphasis on renewable energy, the government has introduced feed-in tariffs

Fig. 9 Potential geothermal
sites in East Africa [31]

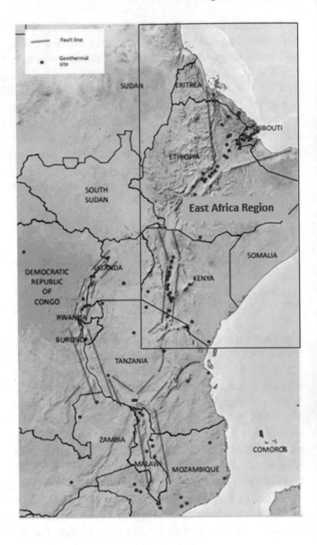

to promote investment to generate electric energy from renewable sources. The electricity feed-in tariff law would attract investors to involve in diversified renewable energy sources with financial and economic benefits as the competitiveness of some of the renewable energy technologies are not comparable with other non-renewable sources [29].

2.3 Transformation to Modern and Renewable Energy Source

Over 80% of the population in Ethiopia are residing in the rural area and use traditional energy sources such as wood, dung and agricultural residue. In sub-Saharan Africa, Ethiopia has the largest population with high level of dependence on traditional solid fuels for cooking purpose. The biomass fuels have, however, increasingly become challenging to get in abundance as the population increases and free land becomes scarce. In addition, overutilization of these resources will adversely affect the soil moisture, recycling of soil nutrients and conservation of water, soil and wildlife. Henceforth, the Government of Ethiopia has promoted diversification of the energy portfolio giving more focus on renewable energy. Although the utilization of biomass as a source of energy has an adverse effect on the environment, substitution of this resource with other renewable energy source remains slow-moving. The utilization of the renewable energy is insignificant compared to the available potential. For instance, the utilization of hydropower energy is below 10% and the utilization of geothermal energy is almost negligible [9].

Successful development of renewable energy utilization in Ethiopia is imperative for the following reasons:

1. Renewable energy is suitable for decentralized applications.
2. Ethiopia is endowed with renewable energy sources such as hydropower energy, geothermal energy, wind energy, solar energy and biomass. These sources can be used depending on their availability in a specific location.
3. It can save the hard currency spent to import the fossil fuels.
4. Utilization of renewable energy fosters environmental security [9].

2.4 Power Demand Growth

Currently the energy sector is characterized by the extraordinary demand growth which has huge disparity with the supply growth. According to the Ministry of Water and Energy, the target scenario would have a growth by 32% from 2011 to 2015. This huge demand growth rate is attributed to number of factors. Some of the factors were the fast GDP growth of the country amounting to above 10% for the last 10 years, steep population growth, expansion of the national grid to the rural areas, manufacturing boost and construction sectors [9, 32]. Considering two scenarios, i.e. CENT, where the demand is supplied only from the centralized grid and SPLIT where the demand is supplied both from the central grid and off-grid sources. Longa et al. 2018 have projected the electric generation capacity until 2050 in Ethiopia. In the projection, the amount of electric generation in 2050 will reach up to 200 TWh as shown in Fig. 10 [33].

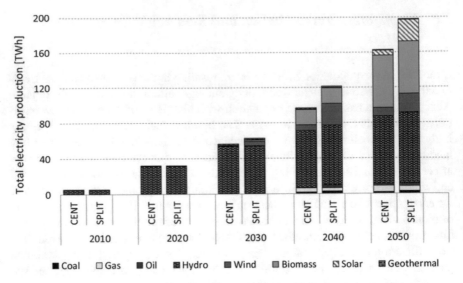

Fig. 10 Total electricity production projection in the SPLIT and CENT scenarios [33]

To address the unprecedented demand, growth in the energy sector of the country developing the geothermal energy source is one crucial option to complement other sources.

2.5 Issues with High Dependency on Hydroelectric Power

Ethiopia's electric power generation is heavily dependent on hydroelectric power. However, there is apparent threat on the heavily reliance on hydroelectric power due to a continuous variation in rainfall amount from year to year. The climate change poses a considerable uncertainty on the energy production in the country. Particularly the effect of climate change on hydropower energy sector would become more severe in the long run. To overcome the effect of climate change, Ethiopia should plan to use other renewable energy sources in the future. Geothermal energy has high capacity, longevity and stability as opposed to wind and solar energy resources, which are seasonal/intermittent and thus require storage facilities encoring additional costs [34].

3 Challenges of Using Geothermal Energy in Ethiopia

3.1 High Capital Cost

The development of geothermal energy requires high initial cost. Currently the main source of electricity in Ethiopia is hydropower. Based on the recent project costs, for instance, Gibe III and Ethiopian Grand Renaissance Dam cost around $1000 per installed kW, whereas taking in to consideration the estimated development of Corbetti and Tulu Moye geothermal power plants, the cost of electric power generation reaches $4000 per installed kW [9]. Hence, the initial investment cost of geothermal power per kW is four folds of the hydroelectric power generation. In addition, the maintenance and operation cost of hydropower plant is very little compared to geothermal power plant.

The 'levelized cost of energy' (LCOE) which is the break-even cost of generating power helps measure the competitiveness of renewable energy. This cost considers the initial investment costs, interest rates, annual operating costs and devaluation rates of power generation. The levelized cost of energy is defined as lifecycle cost of the power plant per lifecycle power generated and is given as:

$$LCOE = \text{Life cycle cost/life cycle energy} \tag{10.1}$$

Table 2 shows the annual present capital cost, capital subsidy, LCOE and the LCOE difference of other renewable electric generation sources with hydroelectric generation [34]. The LCOE value of concentrated solar power is the highest in the current context (US$ 0.189/kWh). Biomass and wind are the most expensive energy sources for electrical power generation after solar with LCOE of US$ 0.122/kWh and US$ 0.102/kWh, respectively. Hydro and geothermal sources have lower LCOE of US$ 0.051/kWh and US$ 0.080/kWh, respectively.

As the difference in capital cost and LCOE for alternative renewable sources such as solar, geothermal, wind, biomass has remained high compared to hydropower

Table 2 Required estimated capital subsidies and the difference in LCOE (US$/kW) if hydropower is substituted with other renewable energy sources [34]

Energy source	Annual present capital cost US$/kW	Capital subsidy (US$/kW)	LCOE US$/kWh	LCOE difference over hydroelectric
Wind energy	131.4	114.7	0.10	0.05
Solar energy	280.0	263.3	0.19	0.14
Hydroelectric energy	16.7	0.0	0.05	–
Geothermal energy	134.8	118.1	0.08	0.03
Biomass	137.1	120.4	0.11	0.07

source, it has become a barrier to alternative energy diversification in Ethiopia. To provide energy for communities located in remote areas, Ethiopia should have an optimal strategy to support the investment on other alternative renewable resources and provide incentives to private, household or cooperative associations to harness the required energy. Related policies such as capital subsidies should target reducing upfront capital investment costs to make alternative renewable resources competitive with hydroelectric power [34]. The capital subsidy indicated in the table for wind, solar, geothermal and biomass refers the subsidy provided by the government to make the resources competitive and attract the private investors, household or cooperative associations. To make solar energy competitive with hydroelectric energy, the Ethiopian Government must provide a capital subsidy of US$ 263/kW, followed by geothermal with US$ 118/kW, biomass with US$ 120/kW and wind with US$ 115/kW.

3.2 Lack of Technical Competence

Geothermal resources development requires a multidisciplinary professionals and institutions with different technical background and experience. The major technical competences or disciplines that are required for geothermal development include geology, geochemistry, geophysics, reservoir engineering, environmental science and geothermal engineering [30, 35]. Generally, in East Africa, studies show that the geothermal energy development is dependent mainly on the foreign institutions and personnel. Ethiopia's case is also not different. Lack of technical capacity in design, development, operation, maintenance, consultancy and provision for electrical equipment as well as appliances are some of the technical challenges of the country [36]. As geothermal energy development is at infant stage in the country lack of experience in the field will compel the country to depend on foreign companies and spend high foreign currency.

In the short run, outsourcing and contracting projects partly to experienced foreign companies could be inevitable. However, defining the requirement of core institutional and organizational capabilities and creating an action plan for developing business skills and execution capacity in domestic institutions are an ultimate solution to exploit the resource adequately [36].

3.3 Low Prevailing Tariff Regime

Broadly speaking, tariffs must be high enough to allow the utilities to cover their costs and finance new investment, but not so high as to frustrate demand and deny access to poor households that consume small amounts of electricity (subsistence consumers) [37]. Ethiopians have been enjoying low electricity tariffs for over a decade compared to their neighbouring countries in the east African region. Current

Table 3 Comparison of electric tariff between Ethiopia and other countries [39]

Countries	Price ($/kWh)	Difference (%)
Ethiopia	0.02	
Burkina Faso	0.032	60
Brazil	0.222	1010
Costa Rica	0.044	120
Swaziland	0.572	2760
Egypt	0.032	60
Madagascar	0.064	220

electricity tariffs in Ethiopia seem to be too low to guarantee continued interest of foreign investors to invest into electricity generation projects [38]. Table 3 compares the tariff of other African Countries with that of Ethiopia.

The price for electricity (per 1 kWh) in Ethiopia is 2760% lower than the electricity price in Swaziland. The existing cost of electricity, which is below the break-even, is currently hampering investment in the sector. Subsequently, the negotiation power of the Ethiopian Government with multinational companies is adversely affected [38].

The low prevailing electricity tariffs have also an effect on swelling of the country debts, threatening the credit worthiness of government-owned power companies and dragging down the national budget. The government is considering revision on the tariff structure in order to cover the electric production cost and attract the private sectors in the power market. The tariff reform is planned to be implemented step by step in three-year time instead of doing it at a time with abrupt change. The tariff increment would also consider the amount of electric consumptions and consumer type. The change in tariff for households would be lesser than that of enterprises. Block rate tariff which will have a progressive increment based on the consumption increment would be implemented. Complementary policy measures are also being considered to encourage households and enterprises to engage in energy efficiency and conservation activities [40]. In addition, incentives such as waiver of import duties, grant of tax holidays on dividends and interest incomes could also be considered to attract private investors. These measures will contribute in making tariff affordable to consumers as a result of lowering electric production cost.

4 Summary

Ethiopia is endowed with various natural resources that can enable the country to develop sustainable and renewable energy. The country can produce more than 10,000 MW electric power from geothermal energy. The Great Rift Valley of Africa crosses the country starting from the Afar depression, at Red Sea–Gulf of Aden junction north-west to Turkana desperation on the south covering about 1000 km. As the Great Rift Valley of Africa is the source of tremendous geothermal energy,

Ethiopia shares significant areas of this region and can tap the resource for the development of electric power. The low level of electrification of the country is another factor for Ethiopia to focus and develop the geothermal source to fill the gap. From the 100 million estimated population of the country, 80% of the population are living in the rural areas and use biomass sources as a fuel. However, obtaining the biomass resource in sufficient amount has increasingly become difficult due to a surge in population. This prevailing condition of the country pushes the government to develop a strategy to use renewable energy sources such as geothermal energy. The other pushing factor to develop geothermal energy source is the over-dependence on hydroelectric power plants. More than 90% of the electric power is generated from hydroelectric power which, in many instances, is affected by unsustainable rainfall due to seasonal fluctuation of the weather condition. Consequently, in the past decade, the country has experienced unsteady electric power supply. In conjunction with the urban growth and above 10% GDP growth of the country in the past 12 years, the annual power demand growth has also reached to 30%. To cope up with the power demand growth, diversifying the energy source and developing geothermal energy source would be one strategy.

Despite the many opportunities to develop geothermal energy in the country, there are still challenges hampering the development. The development of geothermal energy requires high capital cost when compared with that of the hydropower energy. For instance, the initial investment cost of Corbetti geothermal power plant was more than four folds per kilowatt of Grand Renaissance Dam. The other significant challenge in the development of geothermal energy is the lack of institution and expertise. Developing geothermal energy is a new phenomenal compared to hydropower energy in the country which has more than 80 years of history with Aba Samuel hydropower plant commenced operation in 1932 with an installed capacity of 6.6 MW. The low electric power tariff rate of the country is also another challenge to involve private sector in the development of the sector.

To overcome the aforementioned challenges, the Government of Ethiopia should have a strategy in short run and long run. In the short run, involving experienced foreign companies in the development of the resource could be of a solution. However, defining the requirement of core institutional and organizational capabilities and creating an action plan for developing business skills and execution capacity in domestic institutions are an ultimate target to exploit the resource adequately. The government has to conclude the revision of the tariff structure and implement it to solve the low tariff rate problem, which is a hindrance in generating adequate amount of income from the service to cover the initial capital cost and running cost of geothermal power plant for the private sector. Considering incentives to private investors in the field, such as waiver of import duties, grant of tax holidays on dividends and interest incomes could also be considered. These measures will contribute to make tariff affordable to consumers as a result of lowering the electric production cost. Complementary policy measures which have already been started to encourage households and enterprises to engage in energy efficiency and conservation activities have to be strengthened and continued.

References

1. Rumman H, Guangul FM, Abdu A, Usman M, Alkharusi A (2019) Harvesting electricity using piezoelectric material in malls. In: 2019 4th MEC international conference on big data and smart city (ICBDSC), 2019, pp 1–5
2. Guangul FM, Chala GT (2020) A comparative study between the seven types of fuel cells. Appl Sci Eng Progress 3
3. Guangul FM, Chala GT (2019) SWOT analysis of wind energy as a promising conventional fuels substitute. In: 2019 4th MEC international conference on big data and smart city (ICBDSC), 2019, pp 1–6
4. Guangul FM, Chala GT (2019) Solar energy as renewable energy source: SWOT analysis. In: 2019 4th MEC international conference on big data and smart city (ICBDSC), 2019, pp 1–5
5. Guangul F, Sulaiman S, Ramli A (2013) Temperature profile and producer gas composition of high temperature air gasification of oil palm fronds. In: IOP conference series: earth and environmental science, 2013, p 012067
6. Chala GT, Guangul FM, Sharma R (2019) Biomass energy in malaysia-A SWOT analysis. presented at the 2019 IEEE Jordan international joint conference on electrical engineering and information technology (JEEIT), Jordan, 2019
7. IBRD (2018, April 11) The World Bank is helping to fight poverty and improve living standards in Ethiopia. Goals include promoting rapid economic growth and improving service delivery
8. . N. P. Commission, Growth and transformation plan ii (GTP II)(2015/16-2019/20). Addis Ababa: Federal Democratic Republic of Ethiopia, 2016
9. Adamiyatt C, Kang D, Montoya-Olsson M, de Planta R, Song A, Shin J (2018) Diversifying Ethiopia's energy portfolio with geothermal energy: a benefit-cost analysis of the Corbetti concession's potential to offset hydroelectric overdependence. University of Chicago
10. Hutchison W, Pyle DM, Mather TA, Yirgu G, Biggs J, Cohen BE et al (2016) The eruptive history and magmatic evolution of Aluto volcano: new insights into silicic peralkaline volcanism in the Ethiopian rift. J Volcanol Geoth Res 328:9–33
11. Endeshaw A (1988) Current status (1987) of geothermal exploration in Ethiopia. Geothermics 17:477–488
12. Kebede S (2016) Country update on geothermal exploration and development in Ethiopia. In: Proceedings of the 6th African Rift geothermal conference Addis Ababa, Ethiopia, 2016, pp 1–8
13. Teklemariam M, Beyene K, AmdeBerhan Y, Gebregziabher Z (2000) Geothermal development in Ethiopia. In: Proceedings, world geothermal Congress 2000, 2000, pp 475–480
14. Councel WE (2013) World energy resources 2013 survey. World Energy Council
15. Manzella A (2017) Geothermal energy. presented at the In EPJ Web of Conferences 2017
16. Lund JW (2016) Geothermal energy utilization. Renewable Energy Syst
17. Maziarz AM, Mercer BC, Brown JL (2010) Geothermal energy: a feasibility study on the application of ground source heat pumps
18. Daware K (2016, April 16) Geothermal energy and geothermal power plants. Available: https://www.electricaleasy.com/2015/12/geothermal-energy-and-geothermal-power-plant.html
19. Al-Douri Y, Waheeb S, Johan MR (2019) Exploiting of geothermal energy reserve and potential in Saudi Arabia: A case study at Ain Al Harrah. Energy Reports 5:632–638
20. Zhu J, Hu K, Lu X, Huang X, Liu K, Wu X (2015) A review of geothermal energy resources, development, and applications in China: Current status and prospects. Energy 93:466–483
21. Councel WE (2016) World energy resources 2016. World Energy Council
22. Li K, Liu C, Jiang S, Chen Y (2020) Review on hybrid geothermal and solar power systems. J Clean Prod 250:119481
23. Shiferaw A (2017) Productive capacity and economic growth in Ethiopia. United Nations, Department of Economics and Social Affairs
24. ITA (2018, April 14) Ethiopia - Energy. Available: https://www.export.gov/article?id=Ethiopia-Energy

25. Meyer S (2017) The government of Ethiopia and Ethiopian electric power sign key agreements with Corbetti Geothermal PLC. Corbetti Geothermal
26. U. F. C. Service (2016) ETHIOPIA: Power Sector Market. US Foreign Commercial Service
27. Kebede S (2013) Geothermal exploration and development in Ethiopia: Status and future plan
28. Corti G (April 19) The Ethiopian rift valley: geography and morphology. Available: https://ethiopianrift.igg.cnr.it/rift%20valley%20geography.htm
29. Bertani R, Cavadini M, Everett M, Geothermal power opportunities in the Rift Valley. RES4AFRICA
30. Kombe EY, Muguthu J (2019) Geothermal energy development in East Africa: barriers and strategies. J Energy Res Rev, pp 1–6
31. HafnerEmail M, Tagliapietra S, d Strasser L (2018) Prospects for renewable energy in Africa. In: Energy in Africa challenges and opportunities. Springer, Berlin
32. Mondal MAH, Bryan E, Ringler C, Mekonnen D, Rosegrant M (2018) Ethiopian energy status and demand scenarios: Prospects to improve energy efficiency and mitigate GHG emissions. Energy 149:161–172
33. Dalla Longa F, Strikkers T, Kober T, van der Zwaan B (2018) Advancing energy access modelling with geographic information system data. Environ Model Assess 23:627–637
34. Guta DD, Börner J (2015) Energy security, uncertainty, and energy resource use option in Ethiopia: A sector modelling approach. Uncertainty, and Energy Resource Use Option in Ethiopia: A Sector Modelling Approach (July 2015)
35. Mwagomba T (2015) Opportunities and challenges of developing geothermal in developing countries: a case of Malawi. In: Proceedings world geothermal Congress 2015, Melbourne, Australia
36. Kebede S (2014) Opportunities and challenges in geothermal exploration and development in Ethiopia. Geothermal Resources Council, vol 38
37. Barnes DF, Golumbeanu R, Diaw I (2016) Beyond electricity access: output-based aid and rural electrification in Ethiopia. World Bank
38. Richter A (2015, October 19) Ethiopia considering higher electricity tariffs to attract investments. Available: https://www.thinkgeoenergy.com/ethiopia-considering-higher-electricity-tariffs-to-attract-investments/
39. Cost To Travel (2015) Electric price in Ethiopia. Available: https://www.costtotravel.com/cost/electricity-in-ethiopia
40. Daware K (2016) Electricity rates or Tariff. Available: https://www.electricaleasy.com/2016/01/electricity-rates-or-tariff.html

Tidal and Wave Energy Potential Assessment

Girma T. Chala, M. I. N. Ma'arof, and Fiseha M. Guangul

Abstract The utilization of renewable energy sources has significantly increased in the last decade to combat the dire impact of fossil fuel emissions and sustain green energy sources in the future. Consequently, various renewable energy sources have been considered for energy production to fulfill the growing energy demands. In this modern era, technologies with the capacity of generating electricity using tidal and wave energy have been realized and now existed in many countries. This chapter presents the potential of producing offshore renewable energy from waves and tides at the ocean. Tidal and wave energies could be assets in the overall global energy production as they have high energy densities and bring many benefits, particularly in alleviating issues related to the extended use of fossil fuels. The enhanced utilization of wave and tidal energy in the future would also bring opportunity in sustaining green energy supplies.

1 Introduction

The world energy demand has increased swiftly in the last decade due to a surge in world population [1–3]. An increase in the utilization of renewable energy to contribute to the world energy production has also been observed in the same years.

G. T. Chala (✉)
Department of Mechanical (Well) Engineering, International College of Engineering and Management, P.O. Box 2511, C.P.O Seeb, P.C. 111, Muscat, Oman
e-mail: girma_tade@yahoo.com

M. I. N. Ma'arof
Department of Mechanical Engineering, INTI International University, Negeri Sembilan, Persiaran Perdana BBN, Putra Nilai, 71800 Nilai, Malaysia
e-mail: muhammad.izzat87@gmail.com

F. M. Guangul
Department of Mechanical Engineering, Middle East College, PB 79, PC 124, Al Rusayl, Muscat, Oman
e-mail: fiseham2002@yahoo.com

Hydropower, biomass, solar, wind energies have received much attention in renewable energy developments [4–6]. Though tidal and wave energies have higher energy densities as opposed to most of the renewable energy resource, it was reported that power generated from these sources is found insignificant. Nowadays, technologies with the capacity of generating electricity through the use of tidal and wave energy have now existed in many countries, which could significantly contribute to the future global energy production since around 70% of the earth is covered with the ocean. Marine energy resources include tidal current, tidal range, wave energy, ocean current, ocean thermal energy, and salinity gradient [7].

The oceans can mainly produce two types of energy: thermal energy due to a temperature gradient, and potential and kinetic energies from tides and waves. The gravitational force is the dynamism that allows for a large volume of water in the ocean to actuate, which in turn creates tides. The kinetic energy of the tides has the potential to drive a turbine to produce power [8]. As a result, tidal energy could be generated either from tidal range or tidal current. Waves, on the other hand, are created using energy from the wind hitting on the ocean's surfaces. It was estimated that the global wave energy resource could be more than 2 TW, with around 4.6% extractable through the wave energy converters [9]. From this standpoint, wave energy relates in response to the wind and highly fluctuates with seasons [10, 11]. There are different equipment made available to capture powers that could be generated from tides and waves. These include buoys, turbines, and several other types of related technologies [12].

In technologies related to wave energy, there are four common types. One of the technologies requires the use of aqua buoy and power buoy. These buoys are known as point absorber devices [13]. The buoys are used to generate power by driving the hydraulic pumps through the rise and fall of the seawater level. The other technology is the Oscillating Water Column (OWC) device [14]. This device uses the waves force to push air back and forth similar to a piston through an opening integrated with a turbine that in turn moves to generate electricity. The third technology is the overtopping device that can be located around the shores. The device focuses on accumulating the waves, leading it into the reservoirs. The accumulated waves would then generate power by driving the turbines, in almost similar manner with that of the second type. The fourth technology is the attenuator. The attenuator has a number of long multi-segmented structures that float and are placed parallel to the direction of the waves [15]. It causes flexing to be found at the connected segment and the flexing is all connected to pumps that capture the power created by waves.

There are also four common technologies available for tidal energy [16]. These include single basin single effect, single basin double effect, single basin with linked basin operation, and double basin with paired basin. All the four technologies of tidal energy require the use of barrages, dams, and a narrow bay/estuary with a sluice across it. The four types of tidal energy power plants can be generally viewed as a step-by-step progression made by engineers to further improve the technology over time [17].

The single basin single effect plant is considered the oldest type for tidal power generation [18]. The technology works by filling the basin through a sluice whenever the tide rises. When the basin has reached its highest water level, it provides a water head towards a turbine and the sluice gate closes. The turbine begins rotating only when the tide falls. Hydraulic head is formed around the barrage following an appropriate time of water release in the basin, which then passes through a generating unit found in the powerhouse. Electric power generation through this technology continues until the water head reaches the lowest possible operating level.

The single basin double effect plant is an improved version of the first type due to the fact that it has a reversible turbine to allow for a flow in both directions [19]. Although it is to a lesser extent than in the cases of power plants with unidirectional flow for the third technology available, it is a linked basin operation between a high and low basin. The upper basin opens during high tide when the level reaches above it. This applies to the lower basin only when the low tide reaches a level lower than the basin.

The double basin paired basin combines two basins which are located away from each other [20]. Both basins follow the concept of the single basin single effect tidal plant. Each basin generates electricity through filling and emptying. Although the operation is continuous, power generated may not be regular, requiring attention to alleviate this setback. This chapter presents the potential utilization of tidal and wave energy as an offshore renewable energy source. This would highlight the strength and setbacks encountered while generating power so that efficient exploitation could be possible for future enhanced green energy productions.

2 Tidal and Wave EnergyPotential

Tidal and wave energies have huge potentials of generating power globally [21–23]. Tidal energy could provide a predictable power source, with global capacity of around 1 terawatts (1 TW) [24]. Figure 1 depicts the formation of high and low tides

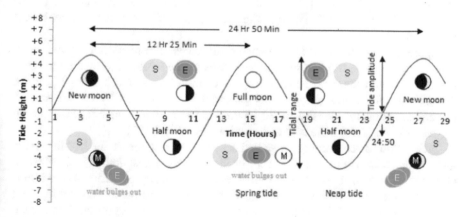

Fig. 1 Formation of high and low tides at oceans and sea [17]

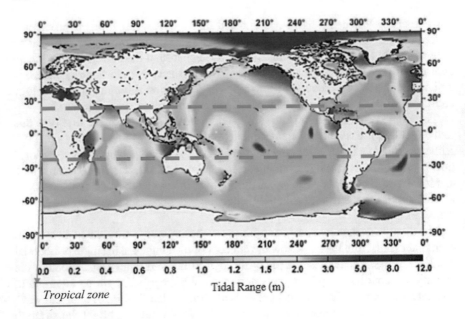

Fig. 2 Global tidal ranges at oceans [26]

at the ocean and sea surfaces with a consistent of two high and low tides formation in a day, making it predictable as opposed to that of wave energy. As the solar month is longer than the lunar month by 50 min, the two tides complete in 24-h and 50 min [17]. Figure 2 depicts tidal ranges at different locations worldwide. Tidal range could reach a maximum of 12 m or a bit higher. This could also be evidenced for tropical countries with the tidal range reaching around 12 m. The higher tidal ranges correspond to higher energy production. It can be seen that the tides vary with geographical location. Tides change water height and current at the ocean with an intense effect on the coast [25]. A 12 m range would have approximately a power production potential of 10 MW/km. The power produced relates to the water height of the places.

Figure 3 shows the world wave energy resources. The maximum wave energy density could reach 125 kW/m in some places of the world. The maximum wave energy density in some parts of the tropical countries could reach up to 70 kW/m. Accordingly, the world wave energy potential was estimated by considering an average wave energy density of 30 kW/m, and this gives a global wave energy capacity of 500 GW for the conversion efficiency of 40% [27].

Table 1 shows ocean energy installed capacity by continent in the last five years and its comparison with that of 2010. The growth was almost double in 2019 when compared with that in 2010. The global installed capacity was 250 MW, while it reached 531 MW in 2019, for which tidal barrage has contributed significantly. The advancement made in the technology of trapping ocean energy in the last decade made higher production of offshore energy on many continents. The energy density

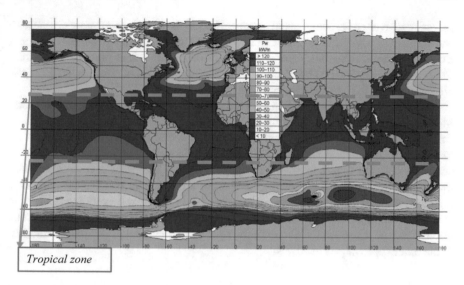

Fig. 3 World wave energy resources, in kW/m [28]

Table 1 Marine energy installed capacity (MW) by region [29]

Region	Installed capacity in MW					
	2010	2015	2016	2017	2018	2019
World	250	513	524	528	529	531
Africa		0	0	0	0	0
Asia	5	259	259	259	259	260
Europe	222	231	241	245	246	248
North America	20	20	20	20	20	20
Oceania	1	1	1	1	1	1
South America		0	0	0	0	0

of tidal and wave energy is higher than other renewable energy sources, which allows for a higher capacity of energy to be stored. Moreover, the technologies require to be installed stationary at the desired location with a proper connection to the power plant to operate without any additional work for a very long extended duration.

2.1 Economic Impact

Tidal and wave energies have many advantages [30]. Power production cost of tidal and wave energy is lower than that of fossil fuels in regard to labor cost, transportation of raw materials, and processing of the fuels. Furthermore, tidal and wave energies are competitively priced sources of energy after paying off the investment cost. Tidal and wave energies could be considered as cost-effective due to the fact that they are free of pollution, renewable, and available for usage in various parts of the world. The cost of wave power was found relatively higher than that of tidal power: 175,000 £/MW and 165, 000 £/MW, respectively [31].

The primary barrier towards an organization is the required financial capital. Most investors would not find this investment appealing due to the long payback periods. Although it has a long payback period, it is still beneficial for long-term as maintenance is minimal and is performed approximately after 30 years [32]. With adequate maintenance, tentatively, the power plant's life cycle is in fact infinite, making the economics for building the barrage a good investment in the end. The maintenance cost for tidal and wave energy could also be relatively low, due to a new discovery of a metal coating preventing metal from getting damaged by the impurities from the ocean. Consequently, the energy plant could last longer relative to the other types of power plants.

2.2   Flood Protection

Tidal power plant has the potential of protecting certain areas against flood while operating normally. The world's largest commercial tidal plant in Easter Scheldt storm surge barrier in Netherland has played a big role in protecting flood as almost 50% of plants in the country are found below sea level [33]. The dimensions of the barrier ranged from 50 m long with 20 m wide. The capacity of power generated by this tidal power plant is able to house for 1,000 households. The wave power plant found in Scotland is considered the biggest. The power plant could be able to generate around 40 MW of power for almost 30,000 households. It also provided 70 job opportunities for the citizens residing nearby, giving substantial gains to the citizens [34].

2.3   Social Impact

The development of tidal and wave energy would bring benefits to the social welfare. Social implications could be occasioned during the construction of tidal barrage. It is expected that there would be an increase in the population around the tidal and wave energy plants as the construction of the barrage progresses. For instance, La Rance tidal barrage was built in the duration of 5 years, which has then quickly become a tourist attraction place and enhanced the country's economy [35]. The barrage building also served as a road, saving time for those crossing over the bay, and this simultaneously generated income. Tidal and wave energy plants are beneficial towards the people living near the harbors as their location is closer to the power plant, giving them easy access to the energy produced.

2.4 The Straits of Malacca and Its Potential: Example from Malaysia

The coast to area ratio in Malaysia was found to be 14–1. The coast covers 4675 km long with the country's land area of about 328,550 km^2. The strait of Malacca is one of the longest straits. Figure 4 shows the location of the straits of Malacca, which has been one of the famous transportation waterways associated with the Indian and Pacific oceans. The ocean flow rates were approximated between 0.5 and 4 m/s. The ocean depth along the straits was found to be 40 m, having a 2 m/s current speed [36]. It was observed that the turbine could potentially function under these conditions.

Sakmawi et al. [36] proposed the Straits of Malacca as a potential site during a group assembly with The Marine Renewable Energy at the University of Malaya. Following the Acoustic Doppler Current Profiler (ADCP)'s information, an extreme speed of 0.48 m/s was observed at coordinates of 4°13′, 37.13″N, 100°32′, 09.02″E. A gadget was then placed in the seabed under 8 m depth, where the stream speed was found higher as compared to other areas. It was reached into decision that the Straits of Malacca have the capability of supplying energies to the capital of the country, Kuala Lumpur with the existing distance coverage [37]. Although the implementation is still in consideration, the progress depends on Malaysia's ecological contemplations. Furthermore, motivations and surveys should be conducted more frequently to spread the idea so there would be more commitment and efforts in utilizing ocean renewable energies.

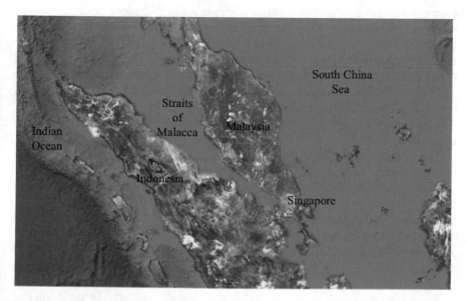

Fig. 4 Location of Straits of Malacca [37]

3 Opportunity of Tidal and Wave Energy

There are numerous opportunities that could be enlisted in the global utilization of tidal and wave energy [38]. As an alternative means of energy source, tidal and wave energies are capable of providing an efficient and everlasting supply of clean energy [39]. This phenomenon takes place twice a day without the need for human/manual intervention. Therefore, the electricity provided is naturally occurring and self-regulating.

Tidal and wave energies could also be utilized to power large remote areas i.e. far away from major cities. For instance, the Baltic Sea was used to supply energy to population within the vicinity [40]. Since the tidal and wave power plant would have various types of turbines, it provides power to these places where supplying directly from the capital possesses a significant challenge from both financial and engineering standpoints.

Tidal and wave energies could spark potential interest of investors into the country for financial support, if the threats towards the environment could be adequately measured, managed, and controlled. Environmental-safe-keeping policies such as the act on renewable energy source enhance the utilization of renewable power plants [41]. This could also initiate the installation of power plants in various places, which are away from the city that eventually gives a substantial return of investment to the investors [40].

An enhanced tidal and wave energy plants could attract more investments in the future, creating opportunities for continuous research to make tidal and wave energy more efficient and less threatening to the environment [42]. As tidal and wave energies produce sensible power, many countries are using this opportunity to support their power demands. The cost incurred in tidal plants mainly includes building a dam with turbine to generate power. Subsequently, it was highlighted that countries could afford building tidal and wave power plants. Cost of construction could also be significantly higher than other low carbon-producing sources. For instance, the tidal station for Severn Estuary in Britain costs 23 billion pounds for construction [43]. The area was selected to generate additional income besides generating power for the grid.

The development of tidal plants usually depends on the estimate of the control plant, tidal run, and geographical condition for the establishment. The Sihwa Lake tidal plant is the world's biggest tidal control plant within the city of Siheung in Gyeonggi province of South Korea. It was built for $560 million with the 12.5 km long seawall to generate a yearly estimate capacity of 552.7 GWh with tidal influx into every 30 km^2 bowl with the help of ten 25.4 MW submerged bulb turbines [43]. As wave energy uses wind energy to create sea waves, the energy produced is less than that of tidal energy. However, the wave can also be created when the ship is sailing on the sea, bringing more waves to drive the turbine for enhanced power.

Recent technologies have greatly facilitated in acquiring fossil fuel, yet, fossil fuel is significantly depleting. The idea of producing power from tides and waves was put in place to take a share in the substitution of fossil fuels. This brings opportunities for many countries and companies to invest as it typically alleviates the issues of total dependence on conventional fuels in the coming years.

Developments of wave and tidal energy in many developed countries have taken a further step to provide more efficient and consistent energy to their commonwealth. Indeed, in the future, tidal and wave would become significant and major power source fulfilling the needs of society. The various designs of the tidal and wave applications could be considered as an opportunity to provide higher efficient power. Currently, there are many different designs built to collectively utilize ocean energies.

The progress made in the developments of ocean energy technologies in the last decade has increased power productions from these sources. Table 2 shows the existing technologies for utilizing ocean renewable energy. Wave and tidal current share the biggest percentage of the existing technologies with 41.45% and 47.07%, respectively [44]. Horizontal axis turbine is a commonly used device to trap energy from tidal current followed by a vertical axis turbine. A horizontal axis tidal current turbines with a modified design could potentially enhance power production from

Table 2 Existing technology for utilizing ocean energies [44]

Ocean energy technologies	Usage (%)
Wave energy converters	
Point absorber	19.20
Oscillating water column	5.85
Oscillating wave surge converter	5.62
Attenuator	5.39
Rotation mass	1.64
Submerged pressure differential	0.23
Others	2.58
Tidal current energy converters	
Horizontal axis turbine	32.55
Vertical axis turbine	11.48
Tidal kite	1.87
Oscillatory hydrofoils	0.94
Venturi effect turbine	0.23
Tidal range converters	
Tidal barrage	4.92
Tidal lagoon	2.34
Ocean thermal energy conversion	
Closed cycle	2.11
Open cycle	0.94
Hybrid	0.47
Salinity gradient	
Membrane	0.70

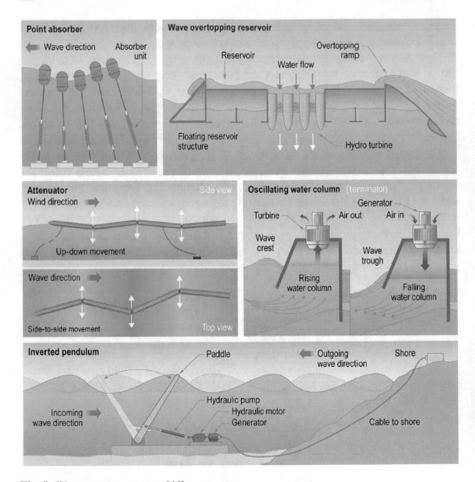

Fig. 5 Wave energy converters [46]

tidal current [45]. Point absorber is common in utilizing energy from the ocean waves. Tidal barrage and tidal lagoon are the two existing technologies available for utilizing energy from tidal range. The commonly used wave energy converters such as point absorber, attenuator, oscillating water column, and overtopping are depicted in Fig. 5.

3.1 Green Energy Source

Tidal and wave energies are considered clean and green energy sources. They can generate high amount of energy and help reduce air pollution generated from using conventional fuels in thermal power plants. On the other hand, tidal and wave energies

Fig. 6 World largest tidal plant, Sihwa Lake, South Korea [47]

can replace nuclear power by cutting off the risk of radiation leakage as emission of the radiation from nuclear power plants is colorless and nuclear pollution is invisible with high risk. Figure 6 shows the global biggest tidal power station found in South Korea.

3.2 Job Opportunities

Tidal and wave energies can offer significant economic opportunities to society. There can be many employment and entrepreneurship prospects for the society. When tidal and wave energy workforce is launched, it requires a continuous follow-up on data collection and maintenance of the devices. Thus, high labor force would make it a sustainable and efficient source of renewable energy in the future. In the years ahead, the replacement of conventional energies by renewable energies offers higher utilization of tidal and wave energies, and these later enhance job opportunities.

4   Challenges of Tidal and Wave Energy

Tidal and wave energies do not affect the ozone by draining substances such as carbon dioxide and nitrogen oxides into the atmosphere [30]. There are a couple of vital reasons for using these power resources to the extent of their impact on energy capability and natural change, and furthermore the security of having a long-term energy supply [48]. Regardless, the drawbacks to the tidal and wave energies are no exclusion. The Achilles heel of the tidal and wave energies is that the demand appears to be limited as the power production is not consistent and continuous, mainly due to the high and low tides happening twice a day, and the dependence of sea wave on the wind speed [16]. Besides, it may require a huge cost in advancing with the project. On the other hand, turbines can in fact cause a high impact on the marine ecosystem. In order to improve the design of the tidal and wave power plants, various prototypes were engineered and being continuously evaluated until today [49].

4.1 Locations

The principal drawback of tidal energy is the location of the power plant. As tidal power plants need to be built near the land, it was highlighted that finding the right location is a challenge as some waterfronts are too confined. High forthright expenses would be hampered towards tidal energy advancements. Furthermore, it appears as not a convenient source of energy for urban areas, as most urban areas are located far off from the ocean causing the metropolis to lean onto other sources of energy, and this requires practical innovations to make this approach feasible.

4.2 Influences on Marine Life

Although tidal and wave energies may sound like one of the best sources of renewable energy, it may inflict hazards towards marine life. A mega mechanical construction could induce an impact towards the fishes' movement and also affects the environment chases for land crustaceans such as crabs. In addition, there might be certain chemicals that could spill during the wave vitality stages causing the nearby water to be tainted and contaminated. Moreover, tidal and wave energy devices emit a certain electromagnetic field which is similar to the earth's electromagnetic field [50]. Hence, the marine life that uses the earth's electromagnetic field to migrate or search for food would be affected as the devices could cause a disruption in the fishes' natural instincts.

4.3 Power Production Performance in Different Climate Conditions

It was reported that the occurrence of waves is greatly influenced by the weather and general climate. When the weather is gloomy, the execution of wave energy decreases. Therefore, for long-term dependency on this energy source, the technologies utilized need to be able to endure climate variations. Moreover, wave energy is unpredictable in contrast with tidal energy [51]. This in return could eventually cause technical irregularities in the power plant. The period of electricity or potential energy generation could sustain a serious impact when affected. As the tides occur twice a day, the power generation is expected during that period, resulting in intermittent power production.

4.4 Noise and Visual Pollution

Tidal and wave energies have numerous pros and cons with an immense assumption utilized in building a control plant. Though tidal and wave power plants are regarded as air or water pollution-free, the generator along with the wave sound could initiate noise pollution [52]. Moreover, the devices located near the seashore could embrace visual pollution, particularly for the places of high tourist areas and this, in some instances, causes less people to be attracted to that area, resulting in some economic impacts on certain countries.

4.5 Common Environmental Impacts of Wave Energy

An alteration of currents and waves may influence coastal erosion. Wave amplitude, tidal velocities and water flow might bring an alter in proportion to the scale of the array. Many wave energy devices are secured to the ocean floor using concrete blocks, anchors, pilings, and chains. There may also involve dredging and scouring of the seabed when installing electrical cables, and this could cause ocean bottom disturbance [53]. Navigation hazard could be possible if wave energy devices are not illuminated at night or if their moorings break away during storms. Moreover, the quality of water might also be affected due to a potential oil spill from increased boat traffic in the area during maintenance and repair. The moving turbine can cause some reaction in the seawater creating sediment. The sediment eventually flows to the ocean and could initiate pollution endangering marine life.

4.6 Environmental Effects of Tidal Energy

Tidal range takes the concepts of conventional hydropower [54]. Though tidal energy gives a lot of positive impacts on the nation, society, and the environment, its negative impacts should not be neglected [55]. It was highlighted that the construction of tidal barrage would have ecological impact on bird feeding areas at the coastal estuaries [56]. In order to determine the impacts in detail, researchers investigated closely on the situation through pilot-scale deployment [14, 57]. Tidal project developers and researchers should work together on collecting and monitoring data to study the impacts on the environments so that the magnitude of the impacts could be minimized [58–60]. Hence, any negative impacts associated with tidal waves could be managed and at best reduced.

4.6.1 Foreseeable Effects

Tidal energy requires a huge barrage to be installed together with other equipment. These installations put the habitats in the marine under danger. Although it is only a small percentage compared to the entire marine creatures around the world, the significance of interruption of biodiversity could not be neglected. When the barrage is about to be filled with ocean water, some of the marine life might be trapped inside and eventually killed when passed through the turbine. Some of the marine creatures could be very small and pass through any barrier that is supposed to trap any marine life from entering the barrage [61]. The energy generation will also create electromagnetic interference affecting some of the sensitive marine creatures [62]. There are also potential strikes of any moving parts of the tidal energy devices which might hit and endanger marine creatures.

4.6.2 Emission of Greenhouse Gases

The horizontal axis turbine has a rotor fixed on a horizontal axis to rotate [63]. The material input for mooring and foundations causes an environmental impact. Mooring and foundations contribute more than 40% of the greenhouse gas emission. The amount of greenhouse gas emission is from 15 to 105 g of carbon dioxide [64]. Therefore, this form of threat has to be taken into consideration in the case of mega implementation of tidal energy.

4.7 Cost Impact Based on BNEF

According to the Bloomberg New Energy Finance (BNEF) analysis, it was found that there would be a significant difference in the cost of wave or tidal energy and

Table 3 Cost comparison of different power plants

Power plant	Cost k $/kW-hr
Coal	0.120–0.130
Natural gas	0.043
Nuclear	0.093
Wind offshore	0.106
Wind onshore	0.037
Solar PV	0.038
Solar thermal	0.165
Geothermal	0.037
Biomass	0.092
Hydro	0.039

other energy sources. It was estimated that the levelized cost of electricity (LCOE) of wave energy was $500/MWh while for tidal was $440/MWh. There was a slight difference between the costs of wave energy and tidal energy. Tidal and wave energy cost showed the highest when compared with other ocean energy sources such as offshore wind ($174/MWh), crystalline silicon solar PV ($122/MWh), onshore wind ($83/MWh), and large hydro ($70/MWh).

4.8 Cost Comparison in Reference to Annual Energy Outlook 2019

Most of the renewable energy sources cost significantly higher compared to non-renewable energy sources. This is because the cost of installation is higher compared to energy extraction due to the lower energy density of renewable energy. Table 3 compares the normal levelized electricity cost of different power plants in the US, which was referred from US EIA insights and investigation from Annual Energy Outlook 2019 [65]. There are alternate limit factors for each energy source. It can be observed that the cost for coal energy is higher compared to other sources, and this might be double ahead as the natural resources are depleting and many countries are projected to increase the cost of CO_s emanations in the future.

It was also reported that the levelized cost of energy for tidal and wave energy technologies is higher than that of the conventional energy sources. Figure 7 shows the levelized cost of energy for marine energy converters in the US for the year 2016. The levelized costs for river current turbine, wave point absorber, tidal current turbine, ocean current turbine, oscillating surge WEC, and Oscillating water column WEC at the capacity of 10 MW were approximated to be $0.31, $0.48, $0.42, $0.98, $0.98, and $1.47 per kW-h, respectively, which were higher than that of convention energy sources. In this regard, efforts have been put in place to make these energy sources

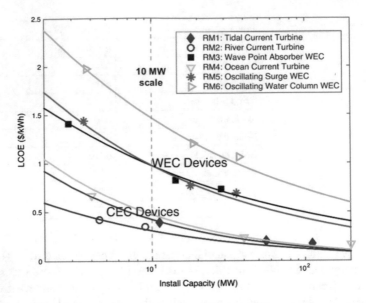

Fig. 7 Levelized cost of energy for marine energy converters [67]

competitive ahead by reducing CAPEX cost and improving device performances [66].

5 Summary

The rapid stride towards the utilization of renewable energy sources has begun since a few decades ago. This was fuelled following the depletion of fossil-based energy sources and the dire impacts from its usage. In this modern era, technologies with the capacity of generating electricity through the use of tidal and wave energy have existed, yet holistic implementation of these technologies remains to be extensive. Tidal and wave energies are viable alternative sources of energy possessing substantial amount of benefits as opposed to the conventional fuels. The implementation of tidal and wave energy enhances the economic growth of the country via investment and job opportunity to the society. The marine ecosystem can be protected as the structure of the tidal and waves can be made environmentally safe and pollution-free. Subsequently, it can prevent the huge wave from destroying the bay areas, protecting the residential areas within the vicinity of the plant. Moreover, the development of tidal and wave energies may attract tourists and improve the economy and social implications of the country.

Nevertheless, its setbacks such as the impact on the environment and that this form of energy is influenced by climate and weather should be taken into serious consideration. High attention should be given when developing both energy sources as the construction of the systems may affect animal habitats and the ecosystem, including the chemical effects of machinery defects on the ecosystem during the operation of tidal and wave plants. Besides, both energy sources might turn weak due to climate conditions. Thepower output by the plant is strongly influenced by the change of tides, hence if the climate within the area of the plant is irregular or inconsistent, the power generated by the plant may not be substantial. Though any setbacks in regard to the implementation of tidal wave energy could be resolved via continuous research and development, this is still within the confinement of its own up-hill challenges. Unceasing drive towards unremitting improvement and innovations would greatly facilitate the goal of total renewable energy. In this regard, more researches in this area are expected in an effort to utilize these energy sources efficiently to meet future green energy demands.

References

1. Barreto RA (2018) Fossil fuels, alternative energy and economic growth. Econ Model 75:196–220
2. Chala G, Abd Aziz A, Hagos F (2018) Natural gas engine technologies: challenges and energy sustainability issue. Energies 11
3. Ma'arof M, Chala GT, Ravichanthiran S (2018) A study on microbial fuel cell (MFC) with graphite electrode to power underwater monitoring devices. In J Mech Technol 9:98–105
4. Guangul FM, Chala GT (2019) SWOT analysis of wind energy as a promising conventional fuels substitute. In: 2019 4th MEC international conference on big data and smart city (ICBDSC), pp 1–6
5. Guangul FM, Chala GT (2019) Solar energy as renewable energy source: SWOT analysis. In: 2019 4th MEC international conference on big data and smart city (ICBDSC), pp 1–5
6. Chala GT, Guangul FM, Sharma R (2019) Biomass energy in Malaysia—a SWOT analysis. In: 2019 IEEE Jordan international joint conference on electrical engineering and information technology (JEEIT), pp 401–406
7. Uihlein A, Magagna D (2016) Wave and tidal current energy—A review of the current state of research beyond technology. Renew Sustain Energy Rev 58:1070–1081
8. Garrett C, Cummins P (2008) Limits to tidal current power. Renew Energy 33:2485–2490
9. Gunn K, Stock-Williams C (2012) Quantifying the global wave power resource. Renew Energy 44:296–304
10. Mirzaei A, Tangang F, Juneng L (2014) Wave energy potential along the east coast of Peninsular Malaysia. Energy 68:722–734
11. Cornett AM (2008) A global wave energy resource assessment. In: The Eighteenth international offshore and polar engineering conference
12. Bryden I, Grinsted T, Melville G (2004) Assessing the potential of a simple tidal channel to deliver useful energy. Appl Ocean Res 26:198–204
13. Elwood D, Yim SC, Prudell J, Stillinger C, von Jouanne A, Brekken T et al (2010) Design, construction, and ocean testing of a taut-moored dual-body wave energy converter with a linear generator power take-off. Renew Energy 35:348–354
14. Carstensen J, Henriksen O, Teilmann J (2006) Impacts of offshore wind farm construction on harbour porpoises: acoustic monitoring of echolocation activity using porpoise detectors (T-PODs). Mar Ecol Prog Ser 321:295–308
15. Soleimani K, Ketabdari MJ, Khorasani F (2015) Feasibility study on tidal and wave energy conversion in Iranian seas. Sustain Energy Technol Assess 11:77–86

16. Li JQ, Mei YY, Zhou XW, Zheng JT (2011) Review and tendency on development for Tidal Power Station. Appl Mech Mater, pp 2226–2230
17. Khan N, Kalair A, Abas N, Haider A (2017) Review of ocean tidal, wave and thermal energy technologies. Renew Sustain Energy Rev 72:590–604
18. Wilson E (1972) Tidal energy and its development. In: Ocean 72-IEEE international conference on engineering in the ocean environment, pp 48–56
19. Rourke FO, Boyle F, Reynolds A (2010) Tidal energy update 2009. Appl Energy 87:398–409
20. Etemadi A, Emami Y, AsefAfshar O, Emdadi A (2011) Electricity generation by the tidal barrages. Energy Proc 12:928–935
21. Aboobacker VM, Shanas PR, Alsaafani MA, Albarakati AMA (2017) Wave energy resource assessment for Red Sea. Renew Energy 114:46–58
22. Sleiti AK (2017) Tidal power technology review with potential applications in Gulf Stream. Renew Sustain Energy Rev 69:435–441
23. Jahanshahi A, Kamali M, Khalaj M, Khodaparast Z (2019) Delphi-based prioritization of economic criteria for development of wave and tidal energy technologies. Energy 167:819–827
24. Kempener R, Neumann F (2014) IRENA ocean energy technology brief-tidal energy. International Renewable Energy Agency
25. SEOS. Ocean current. Available: https://www.seos-project.eu/oceancurrents/oceancurrents-c05-p01.nl.html. Accessed on 10 Dec 2019
26. Esteban MD, Espada JM, Ortega JM, López-Gutiérrez J-S, Negro V (2019) What about marine renewable energies in Spain? J Marine Sci Eng 7:249
27. Kempner R (2014) Ocean energy technology brief 4. International Renewable Energy Agency–IRENA
28. Alcorn R (2014) Wave energy (Chap. 17). In: Letcher TM (ed) Future energy, 2nd ed. Elsevier, Boston, pp 357–382
29. I. R. E. A. (IRENA) (2020) Renewable capacity statistics 2020. Available: https://www.irena.org/publications/2020/Mar/Renewable-Capacity-Statistics-2020. Accessed on 10 March 2020
30. Iyer A, Couch S, Harrison G, Wallace A (2013) Variability and phasing of tidal current energy around the United Kingdom. Renew Energy 51:343–357
31. Fanning T, Jones C, Munday M (2014) The regional employment returns from wave and tidal energy: A Welsh analysis. Energy 76:958–966
32. Ecofriend (2017) Eco friendly tidal energy: The good, the bad and the ugly. Available: https://ecofriend.com/tidal-energy-the-good-the-bad-and-the-ugly.html. Accessed on 10 Dec 2019
33. (2015) Netherlands tidal power array installed. Available: https://www.powerengineeringint.com/renewables/netherlands-tidal-power-array-installed/. Accessed on 10 Sept 2019
34. Energy.novascotia.ca. (2017) Top 10 things you need to know about tidal energy in Nova Scotia
35. E. s. a. u. (2017) Tidal power. Available: https://www.esru.strath.ac.uk/EandE/Web_sites/01-02/RE_info/Tidal%20Power.htm. Accessed on 19 Nov 2019
36. Sakmani AS, Lam W-H, Hashim R, Chong H-Y (2013) Site selection for tidal turbine installation in the Strait of Malacca. Renew Sustain Energy Rev 21:590–602
37. Chong H-Y, Lam W-H (2013) Ocean renewable energy in Malaysia: the potential of the Straits of Malacca. Renew Sustain Energy Rev 23:169–178
38. Drew B, Plummer AR, Sahinkaya MN (2009) A review of wave energy converter technology. Sage, London
39. Alonso R, Jackson M, Santoro P, Fossati M, Solari S, Teixeira L (2017) Wave and tidal energy resource assessment in Uruguayan shelf seas. Renewable Energy 114:18–31
40. L. KS. (2015) Outlook for ocean energy development in Korea. In: the East Asian Seas Congress. Available: https://eascongress.pemsea.org/sites/default/files/file_attach/PPT-S3W3-10-Lee.pdf. Accessed on 17 Jan 2019
41. Edenhofer O, Pichs-Madruga R, Sokona Y, Seyboth K, Kadner S, Zwickel T et al (2011) Renewable energy sources and climate change mitigation: special report of the intergovernmental panel on climate change. Cambridge University Press, Cambridge
42. Neill SP, Vögler A, Goward-Brown AJ, Baston S, Lewis MJ, Gillibrand PA et al (2017) The wave and tidal resource of Scotland. Renew Energy 114:3–17

43. P. Technology (2017) Tidal giants - the world's five biggest tidal power plants. Available: Available at: https://www.power-technology.com/features/featuretidal-giants-the-worlds-five-biggest-tidal-power-plants-4211218/. Accessed on 15 May 2019
44. Shadman M, Silva C, Faller D, Wu Z, de Freitas Assad LP, Landau L et al (2019) Ocean renewable energy potential, technology, and deployments: a case study of Brazil. Energies 12:3658
45. Bai G, Li J, Fan P, Li G (2013) Numerical investigations of the effects of different arrays on power extractions of horizontal axis tidal current turbines. Renew Energy 53:180–186
46. Neill SP, Hashemi MR (2018) Wave energy (Chap. 5). In: Neill SP, Hashemi MR (eds) Fundamentals of ocean renewable energy. Academic Press, pp 107–140
47. Bae YH, Kim KO, Choi BH (2010) Lake Sihwa tidal power plant project. Ocean Eng 37:454–463
48. Magagna D, MacGillivray A, Jeffrey H, Hanmer C, Raventos A, Badcock-Broe A et al (2014) Wave and tidal energy strategic technology agenda. SI Ocean 44:1–44
49. Magagna D, Uihlein A (2015) 2014 JRC ocean energy status report. European Commission Joint Research Centre
50. Slater M, Schultz A, Jones R, Fischer C (2010) Electromagnetic field study
51. de O Falcão AF (2010) Wave energy utilization: a review of the technologies. Renew Sustain Energy Rev 14:899–918
52. Ghosh S, Chakraborty T, Saha S, Majumder M, Pal M (2016) Development of the location suitability index for wave energy production by ANN and MCDM techniques. Renew Sustain Energy Rev 59:1017–1028
53. A. E. Tutorials. Environmental impact of wave energy devices. Available: https://www.alt crnative-energy-tutorials.com/energy-articles/environmental-impact-of-wave-energy.html. Accessed on 10 March 2020
54. Chala GT, Ma'Arof M, Sharma R (2019) Trends in an increased dependence towards hydropower energy utilization-A short review. Cogent Eng, p 1631541
55. Shields M, Ford A, Woolf D (2008) Ecological considerations for tidal energy development in Scotland. In: 10th world renewable energy conference
56. Frid C, Andonegi E, Depestele J, Judd A, Rihan D, Rogers SI et al (2012) The environmental interactions of tidal and wave energy generation devices. Environ Impact Assess Rev 32:133–139
57. Nunneri C, Lenhart HJ, Burkhard B, Windhorst W (2008) Ecological risk as a tool for evaluating the effects of offshore wind farm construction in the North Sea. Reg Environ Change 8:31–43
58. Michel J, Dunagan H, Boring C, Healy E, Evans W, Dean J et al (2007) Worldwide synthesis and analysis of existing information regarding environmental effects of alternative energy uses on the outer continental shelf. US Department of the Interior, Minerals Management Service, Herndon, VA, MMS OCS Report, vol 38, p 254
59. S. E. ASSESSMENT and O. M. R. ENGERY (2007) Collision risks between marine renewable energy devices and mammals, fish and diving birds
60. Polagye B, Copping A, Kirkendall K, Boehlert G, Walker S, Wainstein M et al (2010) Environmental effects of tidal energy development: a scientific workshop. University of Washington, Seattle, Seattle, WA, USA, NMFS F/SPO-116, NOAA
61. Gill AB (2005) Offshore renewable energy: ecological implications of generating electricity in the coastal zone. J Appl Ecol 42:605–615
62. Gill A, Bartlett M, Thomsen F (2012) Potential interactions between diadromous fishes of UK conservation importance and the electromagnetic fields and subsea noise from marine renewable energy developments. J Fish Biol 81:664–695
63. Block E (2008) Tidal power: an update. Renew Energy Focus 9:58–61
64. Uihlein A (2016) Life cycle assessment of ocean energy technologies. Int J Life Cycle Assess 21:1425–1437
65. Rozenblat L (2017) Renewable energy sources: cost comparison. Available: https://www.ren ewable-energysources.com/. Accessed on 11 Apr 2020

66. Magagna D, Monfardini R, Uihlein A (2016) JRC ocean energy status report 2016 edition. Publications Office of the European Union
67. Jenne DS, Yu Y-H, Neary V (2015) Levelized cost of energy analysis of marine and hydrokinetic reference models. National Renewable Energy Lab. (NREL), Golden, CO (United States)

Potential of Offshore Renewable Energy Applications in the United Arab Emirates

Sharul Sham Dol, Abid Abdul Azeez, Mohammad Sultan Khan, Abdulqader Abdullah Hasan, and Mohammed Alavi

Abstract The United Arab Emirates (UAE) is pushing forward with investment in alternative energy sources. The UAE is well-positioned to meet its energy needs and energy projects and the fuel sources in the future. The UAE should go ahead and invest in renewable energy, which is seeking to develop alternative energy sources. Whilst solar energy availability remains essential in the region, the other renewable technologies such as wind and ocean energies are showing promising potential solutions despite many challenges. The addition of sinusoidal leading-edge propeller blades and vortex generators to wind turbines can enhance its aerodynamics performance and produce less wasted energy in the form of noise. The generation of vortex-induced vibration from finite cylinders produces energy at low ocean current speed. This chapter provides discussion and review on the potential of offshore renewable energy in the country including the conceptual novel designs to integrate such applications to the existing offshore facilities to meet the continuous energy demand.

1 Introduction

Due to the rapid growth of development in the infrastructure and investments in the Abu Dhabi Economic Vision 2030, the electrical power consumption in Abu Dhabi especially has increased dramatically. The United Arab Emirates (UAE) in general

S. S. Dol (✉) · A. A. Azeez · M. S. Khan · A. A. Hasan · M. Alavi
Department of Mechanical Engineering, College of Engineering, Abu Dhabi University, PO Box 59911, Abu Dhabi, United Arab Emirates
e-mail: sharulshambin.dol@adu.ac.ae

A. A. Azeez
e-mail: abid.azeez@adu.ac.ae

M. S. Khan
e-mail: mohammad.khan@adu.ac.ae

M. Alavi
e-mail: mohammed.alavi@adu.ac.ae

© The Author(s), under exclusive license to Springer Nature Singapore Pte Ltd. 2021
S. A. Sulaiman (ed.), *Clean Energy Opportunities in Tropical Countries*,
Green Energy and Technology, https://doi.org/10.1007/978-981-15-9140-2_12

is ranked as one of the largest per capita energy users in the world. Over the past six years, energy consumption in the UAE has been on the upsurge at an annual average of 5%. Generally, energy usage in the UAE has more than doubled over the last 10 years. This provides a test if appropriate actions are not engaged to offer alternative energy in the future. The cost of natural gas power in 2017 ranges between 0.07 and 0.10 USD/kWh depending on on-peak and off-peak hours [1]. The expected demand for electricity by the market will expand even more over in the coming few years, in its quest to achieve economic diversification in the consumption of energy-intensive areas such as petrochemicals and heavy industry and prepares to build new residential towns. It is expected that hotels number will be increased in the UAE and it may be doubled by the end of 2025. Abu Dhabi for example is also planned to have big residential and commercial developments and a hyperloop network.

At present, solar power is the main available renewable energy sector in the UAE [2]. Noor Abu Dhabi solar power plant at Sweihan has a capacity of 2 GW and it is the world's biggest single-site solar project. The Mohammed bin Rashid Solar Park in Dubai is anticipated to produce 5 GW of electricity by 2030 and is expected to generate up to AED 50 billion in investments. However, in the desert environment, dust buildup is one of the main issues that may cause a substantial decline of solar photovoltaic (PV) efficiency for onshore solar farm set-up.

Wind energy is also one of the main potential resources of renewable energy [3]. The potential of associating wind turbines into offshore structures is extensive, where it is windier. In the UAE, and specifically considering the area facing the Arabian Gulf, the onshore wave height does not look promising to harness for wave energy due to the insufficient wave height and water depth, but the further investigation can be carried out especially at the Strait of Hormuz connecting the Gulf to the Sea of Oman and wider Indian Ocean, where the depth can exceed 100 m [4]. The tidal circulation in the Strait of Hormuz is substantial, which can significantly influence coastal currents [5]. Therefore, there is a potential for vortex-induced vibration clean energy system application to convert ocean kinetic energy over the area.

2 Review of Solar Energy Basic Concept

The sun is undoubtedly a key source of renewable energy for the earth. This source of energy is responsible for wind energy, fossil fuels, energy in organic matter, tidal waves, etc. It is the source of heat that is very much essential for life on earth. Solar energy reaches the earth in the form of electromagnetic radiation. These electromagnetic waves have a range of wavelengths, each related to a certain energy level. Sunlight, or the solar radiation spectrum, includes bands between 100 nm and 1 mm, which encompasses ultraviolet, visible, and infrared radiation. The range of solar radiation wavelength is known as the electromagnetic spectrum. The 380 nm to about 700 nm range is visible to the human eye and is called the visible spectrum. Thermal radiation is electromagnetic radiation in the wavelength range from 0.1 to 100 μm. It includes a portion of the ultraviolet and all of the visible and infrared range of spectrum. The waves in the higher wavelength spectrum, carry less energy as compared to the shorter wavelengths.

The source of the sun's energy is fusion reactions. The energy in the gamma rays is then converted into heat and is viewed as light by a human eye on earth. On a specific area at a given time the amount of available daylight is an important parameter and is measured as solar radiation with units of kW/m^2 using pyranometers and pyrheliometers. The solar radiation varies during the time of day and is minimum during the nighttime and maximum during the day. The solar energy received from sun in a specific area in a certain time duration is known as solar insolation with typical units of kWh/m^2 per day. The two terms solar radiation, also known as irradiation, and insolation are used interchangeably. The amount of these depends on the location on earth, time of the day, latitude, season, weather, and the translucence level of the atmosphere [6].

The atmosphere creates a hindrance to the solar energy reaching earth. Above the earth's atmosphere, the solar energy that could fall on earth is known as extraterrestrial solar radiation. The total radiation reaching earth is either in the form of a direct beam, in a direct line from the sun to a certain location and the rest of it is diffused. Diffused radiation the solar radiation that gets scattered due to the atmosphere. The atmosphere scatter is typically due to the clouds, any suspended substances, water vapor, etc. The distance traveled by the direct beam to reach a location at sea level on earth depends on the zenith angle. The beam has to pass through a certain mass of atmosphere. Depending on the zenith angle, θ_z, the distance and hence the mass may vary. The increase in path is given by a factor known as the air mass ratio (equal to $\sec\theta_z$) [7]. The atmosphere results in absorption and scattering of the solar radiation.

On clear sunny days, most of the solar radiation is a direct beam and less than 20% is due to diffuse. However, on an overcast day, almost all the solar radiation reaching the earth's surface is due to diffuse radiation. Typically, at mid-day the solar radiation can be above $1 \ kW/m^2$ and on an overcast day less than 10% of that [8]. The solar flux, which depends on the location, time, and weather conditions may vary between 3 and 30 MJ/m^2 per day. This can be used as a source of thermal energy to run a heat engine or to produce photovoltaic power.

The extraterrestrial solar radiation also changes due to the earth's elliptical orbit around the sun. Extraterrestrial radiation increases by about 7% from July 4 to January 3, at which time the earth reaches the point in its orbit closest to the sun. The earth rotates around its axis which is makes an angle of 23.5° to its orbit while circling around the sun. The northern hemisphere experiences longer days from March 23 to September 22 due to this tilt of 23.5°. During the other half of the year, the southern hemisphere experiences longer days. In order to estimate the extraterrestrial radiation, it is required to know the location of sun in the sky at a specific location and at a given time. This requires knowledge of solar geometry. The parameters required are latitude, hour angle, and sun's declination. The hour angle at a certain location is the time of the day with respect to the solar noon. The sun's declination is the angle between a projection of the line connecting the center of the earth with the center of the sun and the earth's equatorial plane. It varies from −23.5° on the winter solstice (December 21) to +23.5° on the summer solstice (June 22) [8].

The solar energy received in different parts of the world and at a different time is not the same. It depends on the geographic location, the day of the year, time of the day, and the angle of the collecting surface with the horizontal. It also depends if it clear sky or there are clouds in the sky. Instead of taking into account all these factors, solar radiation is typically averaged over the whole year. These yearly averaged values of solar energy (W/m^2) provides important information on the available solar energy at a particular location. This helps in estimating the potential of solar energy which can be either used as either solar thermal or for power generation purposes.

The map shown in Fig. 1 displays the average solar irradiation available in the UAE. Such locations, including offshore, of the UAE, are considered as the best candidates to utilize solar energy availability. This is due to the fact that the locations have more clear sky days and less rainfall. This is the reason that solar energy availability in such a location is higher than in many other parts of the world. Since the potential is there, it is therefore considered to be one of the best locations for solar energy utilization.

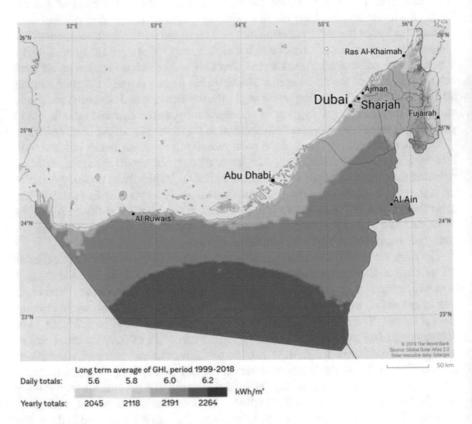

Fig. 1 Global horizontal irradiation for the United Arab Emirates. Figure courtesy of Solargis. GHI Solar Map © 2017 Solargis

3 Solar Energy Applications

The incident solar energy can be converted to beneficial forms of power or utilized as it is. The use of solar energy for non-power generation applications could be for heating/cooling, desalination, direct solar drying, furnace application, etc. On a smaller scale, it can also be used for providing energy for cooking purposes. Solar space heating systems are not very expensive but in this part of the world, the major use of energy is for space cooling and refrigeration purposes. The use of solar energy to achieve this purpose can either be in a vapor absorption system or vapor compression system. Source of heat is required to run the generator of a vapor absorption system as shown in Fig. 2. Typically, this source could be waste heat from an industrial application. However, solar energy can also be used as a free source of energy especially when waste heat is not available. Mechanical power is required to run a compressor driven vapor compression cycle and electricity is typically used to run the motor of the compressor. Solar energy must first be converted to mechanical energy in order to run the compressor.

Due to the scarcity of drinking water desalination of seawater is carried out. This requires distilling of impure water and is an energy-intensive process. Solar energy is directly used to evaporate water, as shown in Fig. 3. The evaporated water is condensed and collected for use as potable water.

The solar power generation systems, which convert solar energy to mechanical and electrical power can be of two types. The thermal energy from the sun can be converted to electric power in solar thermal system or the sunlight is directly converted into electricity energy in a photovoltaic system. In a solar thermal power system, a Rankine cycle or Organic Rankine cycle is used, as shown in Fig. 4. In a Rankine cycle the working fluid, water is sent through the steam generator to run a steam turbine as shown in Fig. 4. The boiler in a solar thermal power system gets energy from the sun to generate steam. In a typical power plant, fossil fuels are used to provide heat to the steam generators. Heat from combustion of fuel is used to change the phase of water from liquid to steam. In solar power systems this heat

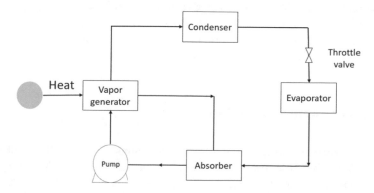

Fig. 2 Solar vapor absorption system

Fig. 3 Schematic of
desalination system
requiring solar energy

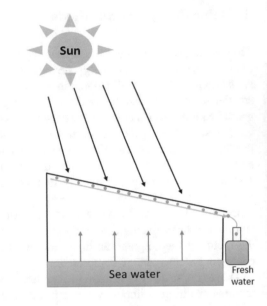

Fig. 4 Schematic of rankine
power cycle

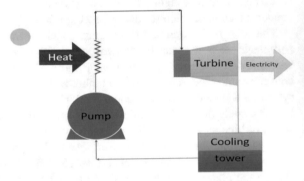

source is replaced by solar energy. The solar energy can be stored during the day so
that it can be used during the night time. However, no such large commercial solar
powerplant with storage is in operation yet. In the Organic Rankine cycle, instead of
water, a working fluid that has a boiling temperature lower than water is used.

4 Types of Solar Energy Generators

Solar power plants typically require solar concentrating collectors. The typical efficiencies of fossil fuel Rankine cycle power plants are less than 50%. When solar heat
is used in the steam generator, additional thermal losses occur and not all solar energy
is harnessed. The overall thermal efficiency can thus be less than 30%. In a typical

power plant instead of thermal efficiency, the term used to quantify the performance is heat rate. Heat rate is the cost of producing a unit of electrical energy. The units of heat rate could thus be $/kWh. It is significant to realize that solar energy is free and does not cost like the cost of fossil fuels to run a power plant. However, the initial cost of the solar power plant system is to be considered. For average solar radiation of less than 300 W/m² a very large solar receiver area would be required. Therefore, different methods of concentrating solar radiation are utilized. Figure 5 shows some typical solar collectors.

Furthermore, not all energy from the sun is in the form of beam and some are diffuse. A large number of solar reflectors are sometimes used to reflect a central receiver. The number of reflectors depends on the capacity of power generation and maybe in thousands. These reflectors are not necessary stationary and may be required to follow the motion of the sun, through a tracking system. The receiver then receives energy and transfer it to the working fluid. The receiver is thus a heat exchanger as shown in the schematic below. The receiver may transfer heat to run a gas turbine. The exhaust of the gas turbine can thus be used to transfer heat to the steam generator and hence a combined cycle power plant, as shown in Fig. 6.

In an offshore application, solar energy has several benefits. The solar power plant at site does not require supply of fuel and hence the associated pollution too is not produced. This is ideal for off-grid locations. However, their use is current limited to the times during the day when sunlight is available. The solar cells or PV cells directly convert sunlight into electricity, as shown in Fig. 7. It is a combination of both physical and chemical phenomena. The light is absorbed in the cell to generate electron–hole pairs and hence the carriers are extracted to an external circuit. N-type and P-type semiconductors form an N-P junction that is formed such that current flows as shown in Fig. 8. The flow of current requires continuous supply of energy in the form of light. Table 1 lists various types of available PV cells. The typical efficiency values of the PV cells range between 7 and 20% except for the concentrated PV cells which are much more efficient. However, the concentrated PV cells require solar tracking system to achieve high efficiency. Some low price PV cells are sensitive to high

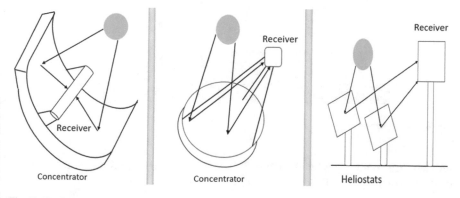

Fig. 5 Typical solar collectors

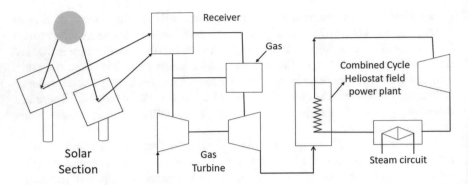

Fig. 6 Schematic of a solar combined power plant

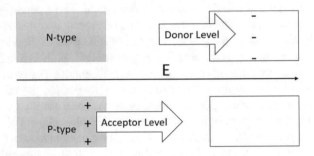

Fig. 7 PV cells

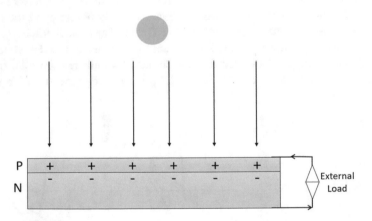

Fig. 8 P–N junction

temperature and life span of the PV cells is another factor to consider. A comparison of the various types of PV cells is provided in Table 1.

Table 1 Most common types of solar cells [10]

Solar cell type	Efficiency rate	Advantages	Disadvantages	Figure
Monocrystalline solar panels (Mono-Sl)	~20%	High-efficiency rate; optimized for commercial use; high lifetime value	Expensive	
Polycrystalline solar panels (P-Si)	~15%	Lower price	Sensitive to high temperature; lower lifespan and slightly less space efficiency	
Thin-film: amorphous silicon solar panels (A-SI)	~7–10%	Relatively low costs; easy to produce and flexible	Shorter warranties and lifespan	
Concentrated PV cell (CPV)	~41%	Very high performance and efficiency rate	Solar tracker and cooling system needed (to reach high-efficiency rate)	

5 Solar Projects in the UAE and the Potential of Offshore

Table 2 shows the status of some recent solar projects undertaken in the Gulf Cooperation Council (GCC) countries including the UAE. The capacity of the concentrated solar power and solar PV recent projects in the UAE alone exceeds 3000 MW. The price per kWh ranges from 2.4 to 7.3 US cents.

The cost US cent/kWh of such solar power projects has also been compared with fossil fuel and nuclear energy dependent power plants in Fig. 9. As can be observed the electricity generation of all fall in the same range.

Abundance of solar energy is available in the UAE as discussed above. Some of the above-mentioned projects have already been undertaken to tap the free source of energy. Similar projects can be explored for the offshore region [3].

6 Driving Forces and Challenges for Wind Energy

The demand for energy is increasing and dependents on hydrocarbon resources are putting pressure on the UAE. Over the past few decades, the UAE has been putting efforts into the expansion of renewable and alternative energy in order to meet the requirements and promote clean and sustainable development. Wind energy is considered to be one of the sources of renewable energy focus [12]. Generally, wind speed in the UAE falls below 5 m/s at most times of the year except at the mountainous and

Table 2 Results of selected auctions in GCC countries. Courtesy of IRENA [11]

Country	Project	Tech	Size (MW)	Year awarded	Year of completion	Price (US cents/kWh	status
Kuwait	KNPC, Dibdidah	Solar PV	1500	–	–	–	Bids invited
Oman	PDO Amin IPP	Solar PV	100	Nov 2018	May 2020	–	Awarded
	Ibri PV Plant	Solar PV	500	–	2021	–	Bids received
Qatar	Kahramaa	Solar PV	500	–	End 2020	–	Prequalification stage completed
Kingdom of Saudi Arabia	Sakaka	Solar PV	300	Feb 2018	End 2019	2.34	Awarded
United Arab Emirates	MBRAMSP phase 2	Solar PV	200	Jan 2015	Mar 2017	5.84	Awarded completed
	MBRAMSP phase 3	Solar PV	800	Nov 2016	End 2020	2.99	Partial completion
	MBRAMSP phase 4	CSP	700	Sep 2017	Starting in 2020	7.3	Awarded
		Solar PV	250	Nov 2018	–	2,4	–
	Noor Abu Dhabi	Solar PV	1177	Sep 2016	Apr 2019	2.94	Awarded

offshore regions as well as in the winter season that stronger winds may have speeds ranging from 5 to 10 m/s, which signifies potentiality for feasible wind power.

The UAE map shown in Fig. 10 shows that the mean wind speed available at 50-m above the sea level extracted from the National Renewable Energy Laboratory [13]. Wind speed is considered to be high in the western offshore region which can be chosen as best locations to utilize wind power.

Wind is solar-powered energy, which is clean, free, and inexhaustible. Winds arise due to uneven heating of the atmosphere by sun, irregularities of the earth's surface, and the rotation of the earth. Wind flow pattern gets changed by the buildings, land terrain, and environmental conditions. The motion of wind when harvested by the turbines helps to produce electricity. Wind turbines convert the kinetic energy inside the turbine into mechanical energy. This power can be utilized for pumping water, grinding grains, etc. Wind turbines convert the linear motion in the air into circular motion and give energy to a power generator that can supply electricity. Wind makes the blades to rotate. Low wind speed (<5 m/s) wind will result in low torque, high speed using gearbox and through a shaft which links to a generator and in turn, generates electricity. Different sizes and varying powers of wind turbines are available for a variety type of applications [14].

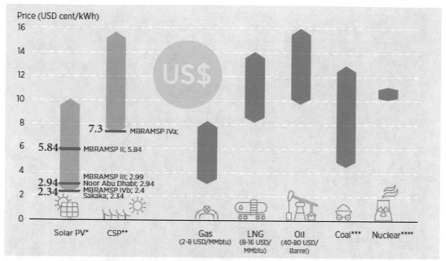

Sources: Derived from Mills, 2018; Channell et al., 2015; Manaar, 2014; Scribbler, 2015.

*   *Low = price for 300 MW Sakaka solar PV; and High = a conservative assumption based on project data and expert opinion*
**   *Low = price for 700 MW MBRAMSP IVb in Dubai; and High = price for Morocco's Noor II*
***   *Low = price for the Hassyan Clean Coal Power Plant; and High = estimate for coal with CCS*
**** *Estimated range for nuclear power based on (Mills, 2012) and (Scribbler, 2015)*

Notes: LCOE and auction/PPA prices represent one way to examine cost-competitiveness in a static analysis. These estimates are not a substitute for detailed nodal modelling, system cost tracking or analysis of factors such as backup generation requirements or demand side management. Moreover, care should be taken in comparing LCOE, auction/PPA prices and Feed in Tariff levels, as they can be very different cost metrics (see IRENA, 2018c for details). Prices for gas, LNG and oil and based on inputs from regional experts (Mills, 2018). MBRAMSP = Mohammed bin Rashid Al Maktoum Solar Park; USD cent/kWh = US Dollar cent per kilowatt-hour; PV = photovoltaic; CSP = concentrated solar power; LNG = liquefied natural gas; MMbtu = million British thermal units.

Fig. 9 Price of utility-scale electricity generation technologies in the GCC. Figure courtesy of IRENA

The aerodynamically produced noise from a wind turbine is due to the unsteady forces mainly on the surface of the wind turbine blade. Studies by Nissenbaum et al. [15] found that generated noise from wind turbines causes adverse psychological and physical health effects to the people living nearby. Therefore, it is vital to find possible ways to lower noise levels from wind turbines. There are two major sources of wind turbine noise; mechanical noise and aerodynamic noise. Mechanical noise is mostly resulted from moving parts inside the gearbox and the generator. This source of noise can be lowered through a high precision gear tooth profile designs and acoustic insulation of the nacelle. However, aerodynamic noise sources are more complex and difficult to control.

Brooks et al. [16, 17] presented two main aero-acoustic broadband noise generation mechanisms for rotating aerofoil sections. The first is the turbulent ingestion noise that is a function of inflow turbulence. The second noise mechanism is the aerofoil self-noise that is related to the blade geometry. The aerofoil self-noise production can be categorized as turbulent boundary layer trailing edge noise, laminar boundary layer vortex shedding noise, separation and stall noise, trailing edge bluntness, and vortex shedding noise and tip vortex noise. In addition to the noise formation mechanisms discussed above, wind turbines have a mechanism due to the rotation of the

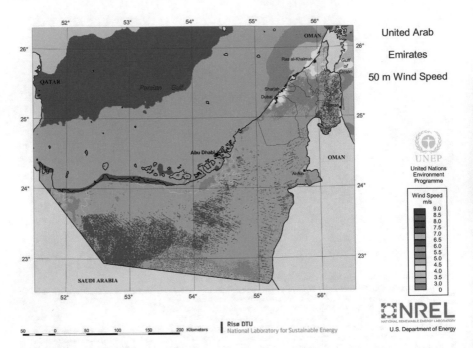

Fig. 10 Wind speed data for the United Arab Emirates. Figure courtesy of ISPRS Congress (source NREL)

blade and the interface between the wakes with the structural. This noise mechanism has a tonal character where the tones are at integer multiples of the blade passing frequency of the rotor [18]. The low frequencies in the acoustic noise spectrum are mainly due to the turbulence inflow, and are difficult to study because of the chaotic character of the turbulence flow. The offshore wind turbines installation could reduce these turbulent vortical-structural interaction effects.

7 Major Wind Energy Projects in the UAE

The Sir BaniYas Island, which is located 250 km Southwest of Abu Dhabi is the place where the region's first wind turbine was installed, as shown in Fig. 11 [19]. The wind turbine stands on a 65-m height stand and it has three rotor blades each with a 52-m wingspan, has a production capacity of 850 kilowatts per hour. This project was undertaken by Germany's Dornier Consulting. The turbine started on operation in 2005, and from the first two-year analysis which was the year 2005 and 2006, it was found that with an average wind speed of 5.1 m/s, the system produced annual energy of 1,000 MWh. The Abu Dhabi Future Energy Company- Masdar and Abu Dhabi's Tourism Development and Investment Company (TDIC) is planning to

Fig. 11 Sir BaniYas Island wind turbine. Figure courtesy of Rainharvest.co.za

develop an onshore wind farm on this same island with a capacity of up to 30 MW [12].

A company named Dusolinstalled a hybrid system in Ajman city that is a combination solar and wind energy. Wind turbine is combined with solar system best match the local conditions. The hybrid system used solar and wind system to reduce dependency on generator power. Solar and wind system charges a battery bank, which in turn supplies energy to the load at night times. If there were any changes in load capacity or insufficient solar and wind energies, the charge in the battery bank will fall below certain threshold [20]. The generator will be triggered to switch on.

8 Environmental Effects and Regulations

Wind power is the clean and environment-friendly forms of electric energy production as it does not involve any harmful forms of chemical pollutions. Moreover, the construction and operation of wind turbines do not have any environmental threat, however, there are few environmental issues, which include the following [21].

Noise Pollution

- Wind turbines produce noises especially by the parts that operate at higher rpm's and it's a limiting factor for expansion.

Bird Injuries and Mortality

- Birds that are flying in the sky may get trapped by the rotating blades of the windmill and be killed. This has got a major impact on the migrating species of birds and hence puts restrictions for the wind power development projects especially near the Gulf where the sea breeze is always strong.

Radio and TV Signal Interference

- Wind turbines operating on the top or sides of the mountains and hills interrupt the transmission of radio and television signals.

It can be ascertained that the environment-related problems that arise due to wind power are comparatively much lesser than the other renewable sources of energy and moreover there are ways to overcome these problems such as;

- Avoiding Migration Paths: The new wind development projects cooperate with the wildlife organizations to make sure the wind turbines are not affecting the migration of birds.
- Fewer and Larger Turbines: Birdlife will benefit more if the turbines sizes are increased and in-return reduce the number of wind turbines. The large size of the wind turbines has lower rpm than the small turbines making it easier for the birds to notice the blades, and dodge them.
- Site Selection: While selecting the location it should be made sure that the turbine is not situated at an area where the site is at high risk of avian population. The assessment is essential to determine if common birds of prey reside near the area where the wind turbine is to be placed.
- Radio and TV signal interference effect is solved by placing wind turbines in remote areas where less signals need to be transmitted and also by transmitting signals through fiber-optic wires in the densely populated regions.

Marine life is disturbed due to converters such as turbines, watermills, or tidal dams. The Environment Agency of Abu Dhabi has put forth two federals laws as well as two Emiri Decrees for Marine conservation [22].

9 Design Optimization for Horizontal Axis Wind Turbine

Animals provide unique encouragement in various areas including aerodynamics and hydrodynamics. Marine biologists have come across the distinct profile of the humpback whale flippers that let them manoeuvre sharp rolls and loops beneath the water surface. These flippers are characterized by the number of features including rounded tubercles on its leading-edge. All these features modify the flow pattern characteristics positively keeping the boundary layer attached to the fin maintaining lift at higher angles of attack and function as a stall control system. This aids in flattening the lift curve in the post-stall area thus helping in moving under near-stall conditions due to the unsteady flow field, which is predominantly momentous for offshore highly turbulent environment.

The performance of the modified geometry (Fig. 12) was evaluated by El Ghazali and Dol [23] by comparing the lift and drag findings to the same propeller with a straight leading-edge. Both geometries are investigated at pre-stall and post-stall conditions to determine their effect with respect to the angle of attack. An increase in the lift force and coefficient of at least 7% was linked with the addition of the sinusoidal leading-edge with improved recovery from the stall. The modified leading-edge profile affected the velocity pattern on the propeller by forming circulation and reenergising the flow to remain attached at angles of attack where separation should induce. The undetached flow is translated into reduced drag coefficient values, larger lift coefficient values, and less wasted energy in the form of noise. This novel design of propeller blades can be applied for Horizontal Axis Wind Turbine (HAWT).

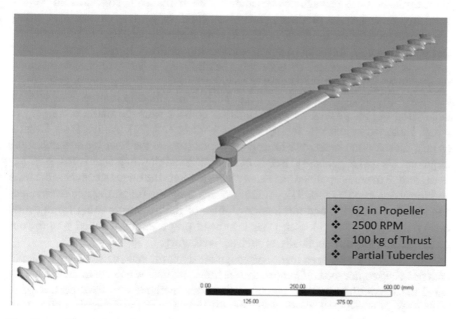

Fig. 12 Modified wind turbine propeller blades

10 Design Optimization for Vertical Axis Wind Turbine

Vertical Axis Wind Turbine (VAWT) provides a promising explanation to this specific performance issue since it is low cost and simple construction rather than HAWT, which requires high tech company with progressive technology to set it up. Drag base VAWT has small efficiency due to its low running speed that the value of the corresponding power coefficient C_p reaches only 50% of the one of the best fast running horizontal axis wind turbines (this is essentially due to the low aerodynamics performances of such rotors, based on the difference between the drag forces on the paddles). In the past few decades, various research in the flow field characteristics and aerodynamic performance by numerical investigation or wind tunnel experimental has been conducted and great milestones have been acquired. VAWTs, both drag and lift-driven [24, 25], is gaining acceptance in view of similar applications, since they can work efficiently even in presence of low-speed and unstructured flows with low noise emissions and high consistency [26].

The pressure difference between the upper and lower surfaces of a wind turbine rotor blade results in a leakage of the flow from the lower to upper surface at the blade tip. When this leakage flow encounters the mainstream, concentrated vortical structures or wakes get produced. These so-called tip vortices can cause a variety of performance losses for VAWTs. In addition, these vortices can induce structural and performance issues due to vortex-turbine interactions. In order to reduce such problems, controlling these vortices could be obtained by active or passive methods. Passive Flow Control (PFC) methods do not adjust to modifications in flow conditions. Nevertheless, PFC methods have found application in real systems such as delta-wing vortex generators for separation control near the blade roots [27] or winglet like tip extensions for tip leakage control [28, 29]. Active Flow Control (AFC), on the other hand, can be applied depending on the flow settings and force requirements and is generally used to control the local flow characteristics. AFC has been studied in many applications mostly to manipulate the boundary layer separation and transition profiles [30, 31]. AFC for controlling the tip leakage characteristics and the size, vorticity, and turbulence profiles of the tip vortex using tip injection (also referred to as tip blowing) has also been proposed and studied by previous researchers in various applications such as fixed wings [32].

Several efforts have been made to manipulate dynamic stall, particularly for applications in helicopter rotors. Comprehensive literature review was conducted to understand the feasibility and effectiveness of various methods that have been applied to control dynamic stall, particularly the lift linked with the dynamic stall vortex. Gardner et al. [33] investigated the effects of fluid injection into the flow-through different jet configurations and noted considerable. On the other hand, passive flow-control means, such as vortex generator (VG), have recently been studied as a means to manipulate dynamic stall. These means have a zero net-mass flux and, therefore, are easier to apply. It has generally been observed that an increase in the VG height can result in a decrease of the lift hysteresis [34]. Furthermore, VG configurations

forming counter-rotating streamwise vortices have been found to be more beneficial for reducing the lift hysteresis compared to the configurations that generate
co-rotating streamwise vortices [35, 36].

The findings on the recent research efforts on the aerodynamically produced noise
from VAWT implied that the higher solidity and higher tip speed ratio rotors are noisier
than the normal turbines [37]. The outcome of spacing between the airfoils in every
blade at different tip speed ratios has been investigated. The results found that the
60% spacing is the best configuration of the double-airfoil from the noise reduction
point of view.

The modification of H-rotor type VAWT (Fig. 13) to have stable and efficient
operation for offshore operation is an important target for the project. Several passive
flow control techniques are considered (Fig. 14). A new active flow control at the
turbine rotor tip is also taken into consideration. The flow at the blade leading edge
is directed through a small channel to the blade tip as shown in Fig. 15. The flow
jet at the tip interacts with the tip vortex and reduces the tip vortex strength. The
triangular vane vortex generators can be used for enhancing the turbine performance

Fig. 13 VAWT proposed
model

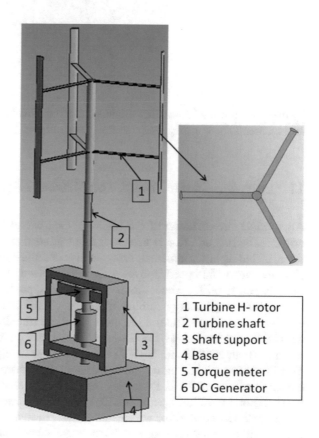

1 Turbine H- rotor
2 Turbine shaft
3 Shaft support
4 Base
5 Torque meter
6 DC Generator

Fig. 14 Tip vortex passive
model

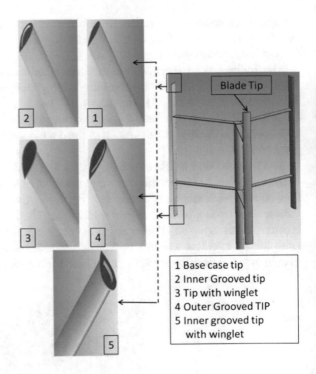

by making control of the vortex shedding that happened after the turbine blades. The
arrangements for vortex generators are shown in Fig. 16.

11   Potential for Buildings and Offshore Applications

Appropriate site selection of our offshore wind energy turbine is assessed on two
criteria. The first criterion is used to create excluded areas that have a set of conditions, including environmental, safety, and economic parameters. The second criterion involves wind speed that its distribution in the UAE has been mentioned in
Sect. 6. Based on the above criteria, the most appropriate potential locations for
the offshore wind energy turbine would be off the coastal shores of the Emirate
of Abu Dhabi. Upper Zakum Offshore Oil Field (the description on the oilfield is
described in Sect. 12) would be the potential site for this technology. The wind
turbines (both HAWT and VAWT) can be erected using the same facilities as the
oilfield platform. The maintenance and transport concerns can be minimized if the
existing facilities are used. Producing the electricity on-site also reduces the need for
transmission for oilfield operations. This in turn reduces transmission losses as well
as the materials needed for wiring and poles. However, it becomes massive test to
design suitable wind turbine, which can harness wind energy and be used on small
scale especially over buildings and platforms. The HAWT design in the previous

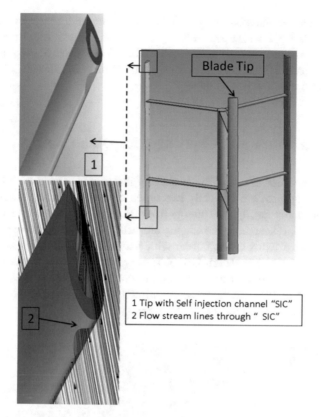

Fig. 15 Tip vortex active model

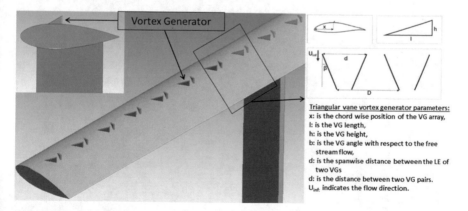

Fig. 16 Triangular vane vortex generator parameters

section can help on manufacturing a simple wind turbine that works efficiently on building and platform, with no need for special or big companies for the installation process. The designed product will help to reduce the consumed electricity and the loads on the electrical national grid. The layout of the wind farm can be enhanced for the maximum production of energy.

12 Driving Forces and Challenges for Ocean Energy

Ocean can be regarded as one of the world's largest source of renewable energy apart from solar. It can produce basically two types of energy; thermal energy through the heat trapped and mechanical energy from tidal and waves. Apart from the large cities in the UAE, vast areas in the country are still covered by dunes (e.g. Liwa Oasis in the south that extends to the border with Saudi Arabia) and mountains (e.g. Al Hajar Mountains in the Northern UAE). There are also plenty of sandy beaches in the vicinity of Fujairah. Normally, renewable energy harnessing sites are remotely situated and all of these geographic conditions hinder the rapid development of renewable technology applications in the country.

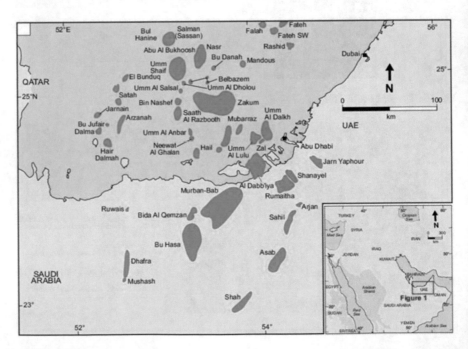

Fig. 17 Location map of the onshore and offshore fields of Abu Dhabi. Figure courtesy of GeoArabia. Copyright © 2020 Petrolink

There are numerous oilfields offshore of Abu Dhabi, as shown in Fig. 17 [38]. Upper Zakum, for example, is the second-largest offshore oilfield in the world. The field is producing 640,000 barrels of oil per day through a chain of approximately 450 oil wells connected to more than 90 different platforms [39]. So there will be a stronger appeal in the UAE to integrate the ocean energy technology to the existing offshore facility like Upper Zakum to facilitate the development and energy transport. They can share the same facility therefore the cost would be practical.

However, there are several challenges for the ocean renewable energy operators to think about before committing to the project. This includes the condition of the Arabian Gulf. Owing to its shallow depth and relative isolation [40], the extraction of tidal and kinetic energy needs further research. Feasibility study on converting ocean waves energy by Pelamis for various the UAE marine territories has been done by Houhou et al. [41] and the initial results look promising. Design of non-intrusive tidal harvester and its potential application for Arabian Gulf was proposed by Mohamad et al. [42] for tidal range as low as 1 m, which is the mean value in the Arabian Gulf in the vicinity of Dubai. If only 10% of the Arabian Gulf area was used, a power of 31.22 GW can be harnessed, which is much more than the capacity of tidal barrage at La Rance in France. However, this tidal barrage project could hugely interrupt the marine activities in the Arabian Gulf that are already crowded with trade and military vessels as well as oil and gas explorations.

13 Environmental Effects and Regulations

There are several ocean energy conversion techniques that contribute to renewable energy, however, they do have some environmental effects. The ocean energy conversion techniques utilize valuable coastal areas as they generally operate near shore or surface. Marine life is disturbed due to converters such as turbines, watermills, or tidal dams. The Environment Agency of Abu Dhabi has put forth two federals laws as well as two Emiri Decrees for Marine conservation [43]. To ensure the safe operation of any ocean energy conversion devices, certain general requirements have to be met. They are as follows: (1) The device should not obstruct navigation, (2) The availability of high energy density, (3) The device should not contribute to reducing the value of the coastal real estate, (4) Low maintenance, (5) The system should be Robust, and (6) The device should have no impact on marine life and the environment [44]. Marine fouling can also impact the efficiency of ocean energy convertors.

14 What is Vortex-Induced Vibration (VIV)?

Main renewable resources such as solar and wind energies will not be able to meet the continuous energy demand as the technologies depend much on weather conditions and the environment [45]. Likewise, hydro-energy can only be generated when sufficient water streams are available, which is scarce in the Middle East region. The construction of hydropower dams and reservoirs could have also interrupted the ecosystem. The UAE however can utilize the kinetic energy captured from the ocean current but the limitation of the Arabian Gulf lies on its low ocean current speed to generate such energy. Ocean technologies based on tidal and currents turbines can only run efficiently for current flow stronger than 2 m/s [46].

A potential solution to overcome this problem is to extract turbulent energy based on vortex-induced vibration (VIV), which can work in low-speed conditions [47]. VIV is a turbulence motion induced on bluff body that forms periodic fluctuations of lift forces and pushing it up and down normal to fluid flow. This oscillation generated by the turbulence motion can be converted to other forms of energy and can be utilized as renewable energy. Based on the previous study, VIV-powered system has the capability to generate power in the range of current speed 0.26–2.6 m/s [48]. Thus, this power generation system might be suitable in generating alternative energy as the typical ocean current (surface speed) in the Arabian Gulf is between 0.1 and 0.2 m/s but magnifies up to 0.3 m/s at the Strait of Hormuz [49].

15   Review on VIV Basic Concept

The VIV is the alternating motion of the cylinder as a result of difference in pressure distribution of the flow near the cylinder. VIV is a type of force-induced vibration (FIV). Other types of FIV are turbulence-induced vibration and vibration induced by tip vortex. Many renowned investigators in the field have published extensive reviews about the VIV mechanism (see [45, 47, 48]). The back and forth motion, which are lift and drag, is due to alternating vortex shedding, as shown in Fig. 18 at which the vortex shedding frequency, f_s was close to the cylinder natural frequency, f_n at the free stream velocity.

The St is the dimensionless frequency of the excitation force in the lift direction whereas the frequency of the excitation force in the drag direction is normally two times the lift direction. The lift force can be explained as a group of vortices shed to one side of the cylinder once per cycle and the other side. The lift force, which appears when the vortex shedding starts to occur, causes the cross-flow motion (normal to the fluid flow direction). Similarly, the drag force appears as a result of vortex shedding

Fig. 18 Vortex shedding from a finite cylinder

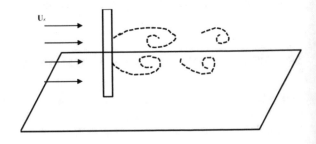

but with all vortices shed downstream. In-line motion (in the same direction as the fluid flow) of the cylinder is caused by drag force. Since all the vortices are shed downstream of the cylinder in the drag direction, the drag force associated with vortex shedding occurs at twice the frequency of the lift force [47].

Lock-in happens when the natural frequency of the structure is in proximity to the vortex shedding. Both frequencies synchronize and amplified amplitude vortex-induced structural vibration can occur. The vortices in the lock-in condition can produce a colossal amount of energy. Besides the increase in circulation, consequences of lock-in also increase correlation length, in-line drag force, and lock-in bandwidth; all of which result in the increase of maximum amplitude.

When the body oscillating frequency is synchronized with the periodic vortex mode, the response displacement, $y(t)$ can be expressed by the harmonic response of a linear oscillator;

$$y(t) = A\sin(2\pi f t) \tag{1}$$

where A represents the harmonic amplitude and f represents the oscillating frequency due to the shedding of vortices. When the fluid force and the body response oscillate at the same frequency, which is roughly close to the natural frequency of the system, the lift force, $F_L(t)$ can be expressed by;

$$F_L(t) = F_L\sin(2\pi f t + \phi) \tag{2}$$

where ϕ is the phase angle between the displacement and the fluid force. The phase angle is very important for body excitation to occur.

When lock-in in the in-line direction happens, the in-line vibration is dominated by vibration at twice the cross-flow frequency due to the drag force which occurs at twice the vortex shedding frequency. Similarly, when the vortex shedding frequency is close to the cylinder natural frequency, lock-in in the cross-flow happens. The drag coefficient for vibrating cylinder can reach few times greater than the stationary cylinder.

16 Current Latest Work on VIV

A dual serial vortex-induced energy harvesting system consisting of two identical cantilevers based piezoelectric harvester with a specific spacing is tested. Using dual harvesters proved to be better than the traditional single harvester and is capable of generating between 2.67 and 6.79 times the output power [50]. The working range of VIV generators can be enhanced using a bi-stable spring. An OGY (Ott, Grebogi and Yorke) controller has been designed and developed to stabilize the chaotic responses that may be generated by the bi-stable spring [51]. The shape of the cylinder used for VIV application has an effect on the energy harvested. The output power increases with the use of an elliptical cylinder with greater major axis than the minor axis.[52]. The modification of a circular cylinder by adding a slit and also a concave rear surface enhanced the performance of piezoelectric energy harvester using vortex-induced vibration [53]. The effect of cylinder diameter on the performance of VIV was tested. It was found that a cylinder diameter of 2 in. gave the highest amplitude of oscillation motion of 0.0065 m within the Reynolds number of between 300 and 300,000 [54]. The shape, size, and orientation or arrangement of the cylinders play a significant role in designing an efficient vortex induced vibration energy generator.

17   Potential for the Offshore Applications and Its Challenges

There are two principal jurisdictions that consist of Offshore companies in the UAE. Ras Al Khaimah (RAK) and the Jebel Ali Free Zone (JAFZA) are the two Offshore jurisdictions. Also, Abu Dhabi National Oil Company (ADNOC) has a dedicated offshore arm for the development of oil and gas resources in Abu Dhabi Waters. They have the capacity to produce about 2 billion cubic feet of gas per day and about 2 million barrels of oil per day [55]. There is a strong potential to integrate renewable energy technologies into these offshore facilities but to do that without disrupting the oil and gas operations, remains a huge challenge.

The sea current data for Jebel Ali Offshore Station is available on the Dubai coast website [56]. Figure 19 shows the ocean current data for the month of June 2020. It can be noticed that the average current falls in the range of 0.22–0.24 m/s. This is the typical current speed required to power the Vortex-Induced Vibration Device.

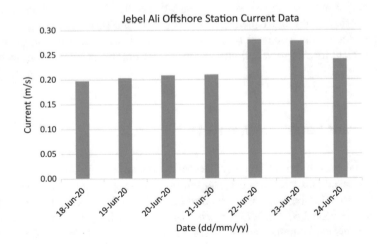

Fig. 19 Seacurrent data (m/s) at Jebel Ali Offshore Station

18 Offshore-Integrated Renewable Energy Applications

The integration of VIV into offshore platform design can be developed by taking into account the feasibility of offshore installation, multi-cylinder arrangements, and arrays, and by estimating the output power that can be generated. The idea is to combine all of these renewable technologies as an integrated system to be installed at the platform to compensate for the continuous energy demand. The type of wind turbine chosen to be built on the offshore platform was the horizontal wind turbine with variable speed and three blades which can work in the low wind speed which is typically around 4–6 m/s. One such turbine is the turbine produced by the manufacturing Vestas. The V90 model of Vestas gives the rated power of 3 MW with a cut in speed of 4 m/s. The rotor diameter is 90 m and the swept area is 6362 m^2 [57].

The integrated platform conceptual design is shown in Fig. 20. In the conceptual design, there are three platforms, which are connected together and each platform has four HAWT Vestas V90 and set of solar panels. VIV generators are integrated below the platform. In the design the solar panels could also be placed even on the

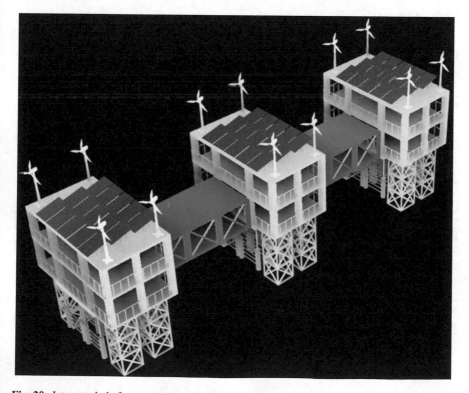

Fig. 20 Integrated platform conceptual design

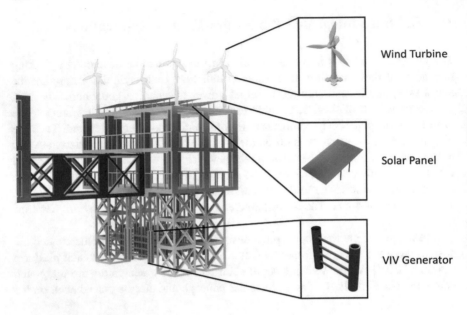

Fig. 21 Labelled platform

bridges, however, it will be less effective as it will not receive sufficient solar power as the bridge is placed lower than the platform.

The labeled single platform detailed view is depicted Fig. 21. The detailed view shows the location of wind, solar cells, and VIV systems. The battery can be installed on the top and bottom surface of the bridge that connects the platforms.

19   Summary

Plans for having renewable energy sources in the United Arab Emirates have been pointed out as an important objective for residential and industrial growths. The potential of offshore renewable energy in the country including the conceptual designs are discussed. There is a strong potential for future investment to integrate renewable technologies into the existing oilfield offshore facilities of Abu Dhabi. The technology advancement in solar cells, wind turbines, and VIV are proposed to incorporate local environment and conditions. This chapter can set as a platform for future applied research collaboration as well as project consultancies with industry that has significant economic, environmental, and industrial aspects for the UAE.

References

1. Abu Dhabi Distribution Co. (2017) Water and Electricity Tariffs 2017, Abu Dhabi
2. Mezher T, Dawelbait G, Abbas Z (2012) Renewable energy policy options for Abu Dhabi: drivers and barriers. Energy Pol 315–328
3. Al Qubaisi AA, Dol SS, Khan MS, Azeez AA (2018) Feasibility study and design of floating offshore wind turbine farm in United Arab Emirates. In: 2018 advances in science and engineering technology international conferences (ASET). IEEE, pp 1–5
4. Dubai Municipality (2020) Dubai coastal zone monitoring and forecasting programme. Available https://www.dubaicoast.ae/. Accessed on 17 May 2020
5. Vaughan GO, Al-Mansoori N, Burt JA (2019) The Arabian Gulf. In: Sheppard C (ed) World seas: an environmental evaluation, 2 ed, vol 2. Academic Press, pp 1–23.https://doi.org/10.1016/B978-0-08-100853-9.00001-4
6. Bhatia SC (2014) Advanced renewable energy systems, New Delhi Woodhead Publishing India PVT LTD
7. Twidell J, Weir T (2015) Renewable energy resources, 3rd edn. Routledge Taylor & Francis Group
8. Kreith F, Goswami DY (2007) Handbook of energy efficiency and renewable energy, Taylor & Francis Group
9. Solargis Admin (2019) Solar resource maps of United Arab Emirates, Solargis. https://solargis.com/maps-and-gis-data/download/united-arab-emirates. Accessed on 20 May 2020
10. Greenmatch Admin (2020) 7 different types of solar panels explained, Greenmatch. https://www.greenmatch.co.uk/blog/2015/09/types-of-solar-panels. Accessed on 20 May 2020
11. IRENA (2019) Renewable energy market analysis: GCC 2019. IRENA, Abu Dhabi
12. Bachellerie IJ (2012) Renewable energy in the GCC countries resources, potential, and prospects. Gulf Research Center, Jeddah
13. United Arab Emirates 50 WindSpeed, NREL. https://www.nrel.gov/wind/images/map_united_arab_emirates_speed.jpg. Accessed on 15 May 2020
14. Sumathi S, Kumar L, Surekha P (2015) Solar PV and wind energy conversion systems. Springer, Cham, Switzerland
15. Nissenbaum MA, Aramini JJ, Hanning CD (2012) Effects of industrial wind turbine noise on sleep and health. Noise Health 14(60):237
16. Brooks TF, Schlinker RH (1983) Progress in rotor broadband noise research. Vertica 7:287–307
17. Brooks TF, Pope DS, Marcolini MA (1989) Airfoil self-noise and prediction, vol 1218. National aeronautics and space administration, Office of Management, Scientific and Technical Information Division
18. McAlpine A, Kingan MJ (2012) Far-field sound radiation due to an installed open rotor. Int J Aero Acoust 11(2):213–246
19. Admin GreenEnergy (2010) The use of wind turbines vs solar power in South Africa, RainHarvest.co.za. https://www.rainharvest.co.za/2010/05/the-use-of-wind-turbines-vs-solar-power-in-south-africa/. Accessed on 25 May 2020
20. "Solar+Wind+GenSet Hybrid," DuSOL. https://www.dusol.ae/projects.php?pid=0090. Accessed on 30 May 2020
21. Michaelides EES (2012) Alternative energy sources. Springer, Berlin, Heidelberg, Fort Worth, TX
22. Environmental Agency (2019) Laws and policies, Abu Dhabi
23. El Ghazali AF, Dol SS (2020) Aerodynamic optimization of unmanned aerial vehicle through propeller improvements. J Appl Fluid Mech 13(3):793–803
24. Tjiu W, Marnoto T, Mat S, Ruslan MH, Sopian K (2015) Darrieus vertical axis wind turbine for power generation I: assessment of Darrieus VAWT configurations. Renew Energy 75:50–67
25. Bianchini A, Ferrara G, Ferrari L (2015) Design guidelines for H-Darrieus wind turbines: optimization of the annual energy yield. Energy Convers Manage 89:690–707

26. Bhutta MMA, Hayat N, Farooq AU, Ali Z, Jamil SR, Hussain Z (2012) Vertical axis wind turbine—a review of various configurations and design techniques. Renew Sustain Energy Rev 6(4):1926–1939
27. Barrett R, Farokhi S (1996) Subsonic aerodynamics and performance of a smart vortex generator system. J Aircr 33(2)
28. Gaunaa M, Johansen J (2007) Determination of the maximum aerodynamic efficiency of wind turbine rotors with winglets. The science of making torque from wind (TORQUE2007). J Phys Conf Ser 75:012006
29. Shimizu Y, Ismaili E, Kamada Y, Maeda T (2003) Rotor configuration effect on the performance of a HAWT with tip-mounted mie-type vanes ASME J Sol Energy Eng 125:441e447
30. Smith BL, Glezer A (1998) The formation and evolution of synthetic jets. Phys Fluids 10(9), 281e2297
31. Bons J, Sondergaard R, Rivir R (2002) The fluid dynamics of LPT blade separation control using pulsed jets. ASME J Turbomach 124(1):77–85
32. Gursul I, Vardaki E, Margaris P, Wang Z (2007) Control of wing vortices active flow control. In: Notes on numerical fluid mechanics and multidisciplinary design. Springer, Berlin, pp. 137–151
33. Gardner A, Richter K, Rosemann H (2011) Numerical investigation of air jets for dynamic stall control on the OA209 airfoil. CEAS Aeronaut J 1(1–4):69–82
34. Chan HB, Yong TH, Kumar P, Wee SK, Dol SS (2016) The numerical investigation on the effects of aspect ratio and cross-sectional shape on the wake structure behind a cantilever. ARPN J Eng Appl Sci 11(16):9922–9932
35. Yong TH, Chan HB, Dol SS, Wee SK, Kumar P (2017) The flow dynamics behind a flexible finite cylinder as a flexible agitator. In: IOP conference series: materials science and engineering, vol 206, no 1. IOP Publishing, p 012033
36. Dol SS, Chan HB, Wee SK, Kumar P (2020) The effects of flexible vortex generator on the wake structures for improving turbulence. In: IOP conference series: materials science and engineering, vol 715, no 1. IOP Publishing, p 012070
37. Mohamed MH (2016) Reduction of the generated aero-acoustics noise of a vertical axis wind turbine using CFD (Computational Fluid Dynamics) techniques. Energy 96:531–544
38. Morad S, Al-Aasm IS, Nader FH, Ceriani A, Gasparrini M, Mansurbeg H (2012) Impact of diagenesis on the spatial and temporal distribution of reservoir quality in the Jurassic Arab D and C Members, offshore Abu Dhabi oilfield. United Arab Emirates, GeoArabia 17(3):17–56
39. Dol SS (2020) Aerodynamic optimization of unmanned aerial vehicle for offshore search and rescue (SAR) operation. In: IOP conference series: materials science and engineering, vol 715, no 1. IOP Publishing, p 012015
40. Kassem OM, Dol SS (2020) Feasibility study of offshore wind and solar energy technologies in United Arab Emirates. Int J Eng Res Mech Civil Eng 5(1):1–5
41. Houhou MR, Dol SS, Khan MS, Azeez AA (2018) Feasibility study on converting ocean waves energy by pelamis in United Arab Emirates. In: 2018 advances in science and engineering technology international conferences (ASET). IEEE, pp 1–5
42. Mohamad OA, Dol SS, Khan MS (2020) Design of non-intrusive tidal harvester and its potential application for Arabian Gulf. In: 2020 advances in science and engineering technology international conferences (ASET). IEEE, pp 1–5
43. Ministry of Climate Change & Environment, Environment and Green Development, Ministry of Climate Change & Environment (United Arab Emirates). https://www.moccae.gov.ae/en/knowledge-and-statistics/green-economy.aspx. Accessed on 15 May 2020
44. Zahari MA, Dol SS (2014) Application of vortex induced vibration energy generation technologies to the offshore oil and gas platform: the preliminary study. Int J Mech Aerosp Industr Mechatron Eng 8(7):1321–1324
45. Khing TY, Zahari MA, Dol SS (2015) Application of vortex induced vibration energy generation technologies to the offshore oil and gas platform: the feasibility study. Int J Aerosp Mech Eng 9(4):661–666
46. Commission of the European Communities, DGXII (1996) Wave energy project results: the exploitation of tidal marine currents, Report EUR16683EN

47. Zahari MA, Chan HB, Yong TH, Dol SS (2015) The effects of spring stiffness on vortex-induced vibration for energy generation. IOP Conf Ser Mater Sci Eng 78:012041
48. Bernitsas MM, Raghavan K, Ben-Simon Y, Garcia EM (2006) VIVACE (Vortex induced vibration aquatic clean energy): a new concept in generation of clean and renewable energy from fluid flow. J Offshore Mech Arctic Eng 1–15
49. Kämpf J, Sadrinasab M (2006) The circulation of the Persian Gulf: a numerical study. Ocean Sci 2:27–41
50. Zhou S, Wang J (2018) Dual serial vortex-induced energy harvesting system for enhanced energy harvesting. AIP Adv 10
51. Huynh BH, Tjahjowidodo T, Zhong ZW, Wang Y, Srikanth N (2018) Design and experiment of controlled bistable vortex induced vibration energy harvesting systems operating in chaotic regions. Mech Syst Signal Process 98:1097–1115
52. Azeez AA, Dol SS, Khan MS (2019) Effects of cylinder shape on the performance of vortex induced vibration for aquatic renewable energy. In: 2019 advances in science and engineering technology international conferences (ASET), pp 1–4
53. Pan F, Xu Z, Pan P, Jin L (2017) Piezoelectric energy harvesting from vortex-induced vibration using a modified circular cylinder. In: 2017 20th international conference on electrical machines and systems (ICEMS)
54. Zahari MA, Dol SS (2015) Effects of different sizes of cylinder diameter on vortex-induced vibration for energy generation. J Appl Sci 15:783–791
55. ADNOC, "ADNOC OFFSHORE," ADNOC (2019) Retrieved from https://www.adnoc.ae/en/adnoc-offshore/who-we-are/about-adnoc-offshore
56. Dubai Coast, "Oceans Map," Dubai Coast. https://www.dubaicoast.ae/OceansMap/. Accessed on 25 June 2020
57. Admin wind-turbine-models (2020) Vestas V90-3.0, wind-turbine-models.com. https://en.wind-turbine-models.com/turbines/603-vestas-v90-3.0#pictures. Accessed on 30 June 2020

Current and Future Perspectives of Integrated Energy Systems

Muhammad Yasin Naz, Shazia Shukrullah, and Abdul Ghaffar

Abstract An integrated energy system is a combination of two or more energy conversion systems. A synergistic benefit of such systems is the output that is greater than the sum of the individuals. A well-designed integrated energy system can substantially reduce the consumption of fossil fuels and boost system reliability. However, for efficient integrated energy systems, it is important to have a good understanding of the system operators responsible for energy flow and for balancing the demand and supply of energy. This chapter discusses the important aspects of the integrated energy systems running on renewable and or conventional resources. Basic knowledge and analysis of integrated energy systems are presented to decide on their merits and demerits. The focused areas include the deployment of electricity generation technologies, challenges associated with local and global optimization of integrated energy systems, and the implications of specified levels of production of renewable energy. It is concluded that a well-designed integrated energy system can substantially reduce the consumption of fossil fuels and boost system reliability. For high load applications, wind-diesel integrated systems are more attractive than wind-PV integration.

1   Integration of Energy Systems

A solution to the environmental complications that we are facing nowadays necessitates long-term action for sustainable development. Renewable resources are the most

M. Y. Naz (✉) · S. Shukrullah · A. Ghaffar
Department of Physics, University of Agriculture, Faisalabad 38040, Pakistan
e-mail: yasin603@yahoo.com

S. Shukrullah
e-mail: zshukrullah@gmail.com

A. Ghaffar
e-mail: aghaffar16@uaf.edu.pk

S. A. Sulaiman (ed.), *Clean Energy Opportunities in Tropical Countries*,
Green Energy and Technology, https://doi.org/10.1007/978-981-15-9140-2_13

well-organized and effective solution for this purpose. The definition of a 'Renewable Energy Source' is that once it has been used, it can be used again and there will always be a replacement [1]. The most important fact about a renewable resource is that it does not impact the environment excessively when compared to conventional energy sources. These energy sources provide a cleaner source of energy that helps to reverse the effects of certain forms of environmental pollution. This reflects a closer linkage between sustainable development and renewable energy. This relationship is described using practical cases, and related examples are given in this chapter. There are many renewable energy resources, including wind energy, hydropower energy, geothermal energy, biomass energy, solar energy, biogas energy, and other forms of energy [2, 3]. These renewable energy sources are better replacements for fossil fuels and they will continue to remain even after they are used constantly, and will never run out. The scientific community, besides political and environmental institutions, has been considering the role of renewable energy systems as the key to future global sustainable development. At the national level, at least 30 countries across the world generate more than 20% of their energy requirement from renewable sources. Hopefully, the conclusions and recommendations offered in the present study will be useful for energy scientists, engineers, and policymakers.

The global energy system comprises of renewable, fossil, and nuclear energy sources. It also includes electrical energy, fuel energy, and thermal energy pathways to convert and deliver energy services at diverse physical scales [1, 4]. Interaction and interdependency among the pathways and physical scales have been increasing with time. Integration of different energy systems permits real-time analysis and control of such interaction and interdependency along economic, technical, and social dimensions [1]. By integrating and optimizing energy systems across multiple pathways and scales, one can make use of the potential co-benefits, which include high reliability, better performance, low-cost, and reduced environmental impact.

As shown in Fig. 1, an integrated energy system (IES) is formed by combining two or more energy sources. The fuels from different sources for the same device are integrated to meet the shortcomings inherent in each. IESs offer the synergistic benefit of greater output than the sum of the individual components. The efficiency of IESs is typically higher than of the individual technologies. Moreover, higher reliability is possible through an IES, when dealing with redundant technologies and energy storage systems [5, 6]. A well-designed IES substantially reduces the consumption of fossil fuels and boosts system reliability [7]. Other than the diesel generator and renewable energy conversion systems, IESs also includes a control system, an energy storage system or battery bank, and system architecture for optimization of the components' performance.

IESs are a cost-effective solution to AC electricity needs in rural areas [8]. Specifically, wind-PV integrated systems are an attractive choice for low load applications (<10 kWh/day) [7]. For high load applications, wind-diesel integrated energy systems are more beneficial than a wind-PV integrated system [9]. This short review covers IESs constituted of solar energy, biomass energy, wind energy, and others [10, 11]. This review provides basic information about IESs, accompanied by an analysis

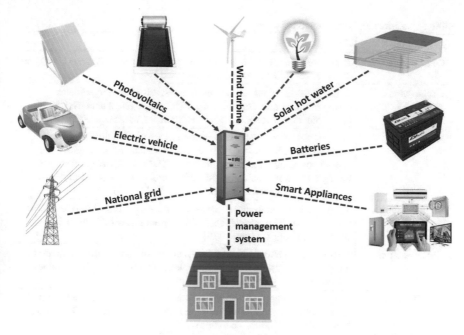

Fig. 1 A concept model of an integrated energy system

of such systems with reference to PV hybrid energy systems, wind hybrid energy systems, and many others [12].

2 Types of Integrated Energy Systems

Due to the rapidly increasing threat of global warming confronting planet Earth, renewable energy is now considered more practical and realistic. The most crucial and debated topic in recent years has been "climate change", and the most significant factor of climate change is the rising temperature of our atmosphere. In this scenario, renewable resources such as solar energy, wind energy, geothermal energy, hydro-energy, ocean, and bioenergy have found favor because of their availability and cleanliness. Renewable energies allow technological developments to meet the fast-growing energy demands without causing any harmful effects to the environment, in particular the atmosphere. This type of energy is safe for the future of our planet. IESs can combine fossil, wind, solar, battery, nuclear and hydel energy to keep the island hideaway running around the clock [13]. There are different types of IESs that are described in Table 1. The peak hours of wind and solar energy occur during different seasons or even at different times of the day in different parts of the world. An integrated system is more likely to deliver electricity as per our needs [6] and store extra energy in battery banks. A gasoline or diesel generator is generally incorporated

Table 1 Different types/combinations of IES

Types/Combinations	Description
PV-fuel cell	These are integrated designs that can be used for both systems i.e. system that is either grid-connected or standalone [14]
PV solar-wind	These PV solar-wind cells are used for HOMER (hypergeometric optimization of motif enrichment), PPA (power pinch analysis), and for the technical and economic analysis for different systems sizes [15–17]
Fuel cell-PV-wind	These types of renewable energy sources are used in different applications in Saudi Arabia as well as in Yanbu and simulations of this system are being done using software (HOMER). There are three systems through which it should pass through in order to have a complete cost analysis assessment in developing countries [18, 19]
PV solar-battery	These types of renewable energy sources are used for microgeneration by using the MIMO controller application and it basically helps in less consumption of grid energy [20]
PV solar-thermal	These types of energy sources are used for collector systems that are of air type, PVT system of liquid type along with flat plate photovoltaic thermal collector system [21, 22]
PV-wind-battery	This system only consisted of lead-acid and lithium-ion batteries. For storage purpose storage batteries are also used. It meets the demand in household applications up to 90%
Wind-diesel	These are used for different optimization techniques and their objective functions are (ISE), (ITAE) and (ITSE) are measured [23]
Fuel cell-wind	These types of hybrid energy systems can give its efficiency up to 99.28% and 99.41% by using (MPPT) [19]
Wind-thermal	Required and limited power flow by using an algorithm in MATLAB or SIMULINK by using IEEE 30 at test systems [24]
Fuel cell-battery	They are used in ESS devices which have high power and energy. ESS is basically a heterogeneous energy storage system [25]
Solar–diesel–fuel cell–hydrogen	These are the most important source of energy used for sizing of units, management of energy, cost optimization as well as modeling of renewable energy components [1, 26]
Diesel-wind-fuel cell	They can be used for every type of system connection, i.e. grid-connected, isolated or standalone system as well as in different controllers such as PID and PD controllers [27]
Wind-diesel-hydro	They do have applications in FACTS devices and STATCOM devices used for the improvement of the power factor [28]
Biomass-solar	There are two combined cycles systems are integrated. First used for thermal integration with following the second system for thermochemical hybrid routines [26]

in IESs along with renewable energy converters, to cater for days when renewable energy is not available. The key characteristics of IESs for making a selection of renewable energy technologies are summarized in Table 2 [2].

Different designs of IESs have been comprehensively investigated and several optimization methods including genetic algorithms [18] and linear programming [29]

Table 2 Summary of IES characteristics for renewable energy sources

Technology		Plant size range (MW)	Variability: characteristic time scales for power system operation (Time scale)	Dispatchibility (See legend)	Geographical diversity potential (See legend)	Predictability (See legend)	Capacity factor range (%)	Capacity credit range (%)	Active power, frequency control (See legend)	Voltage, reactive power control (See legend)
Bioenergy		0.1–100	Sesons (depending on avaliability of biomass)	***	*	**	50–90	Similar to thermal and CHP	**	**
Solar energy	PV	0.004–100	Minutes to years	*	**	*	12–27	<25–75	*	*
	CSP/thermal storage	50–250	Hours to years	**	*	**	35–42	90	**	**
Geothermal		2–100	years	***	N/A	**	60–90	Similar to thermal	**	**
Wind energy		5–300	Minutes to years	*	**	*	20–40 onshore 30–45 offshore		*	**
Hydropower	Run of river	0.1–1500	Hours to years	**	*	**	20–95	0–90	**	**
	Reservoir	1–20,000	Days to years	***	*	**	30–60	Similar to thermal	**	**
Ocean energy	Tidal range	0.1–300	Hours to days	*	*	**	22.5–28.5	<10	**	**
	Tidal current	1–200	Hours to days	*	*	**	19–60	10–20	*	**
	Wave	1–200	Minutes to years	*	**	*	33–31	16	*	*

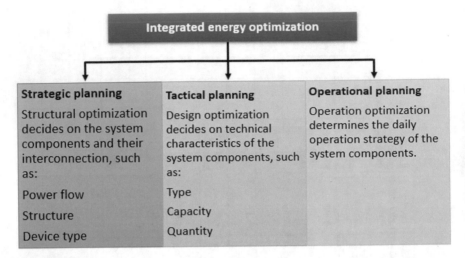

Fig. 2 Planning factors involved in structural, design, and operational optimization of IES

utilized to arrive at the most reliable and economical models of IESs. Optimization of IESs is carried out by considering the structural parameters, design parameters, and operation parameters. Figure 2 explains the factors involved in structural, design, and operational optimizations. The structural optimization is based on the system components and their interconnections, such as power flow structure and device type. The design optimization is based on technical characteristics of the system components, such as type, capacity, and quantity. On the other hand, the daily operation strategy of the system components is determined from the operation optimization.

Different software packages like HOMER [20], RETScreen [30], HOGA [31] and Hybrid [32] have also been built for the appropriate choice of power generation techniques and their scale. Such software packages make the study of IESs easier and more interesting. Some of the studies focus only on testing the reliability of IES designs [32–35], whereas others bear in mind several sizes and types of existing generation systems with a view to reduce fuel and investment costs and improve system operation [36]. An optimization-based methodology is needed, which concurrently reduces fuel and investment costs (installation and unit cost) while preserving system reliability [37].

Different energy management systems (EMSs) are used to lower energy consumption and related costs. These systems are based on systematic activities, procedures, strategy planning, implementation, organization, control, and culture, which continuously lower energy consumption. As shown in Fig. 3, an EMS uses hierarchical frameworks to promote coordination between the entities involved in the network. The use of advanced metering infrastructure (AMI) in EMSs at homes, along with data-driven decision support, many involve a large number of appliances in determining the demand response. For home EMSs an EMS could collect price signals from system operators that could be viewed as moderators between the end-users

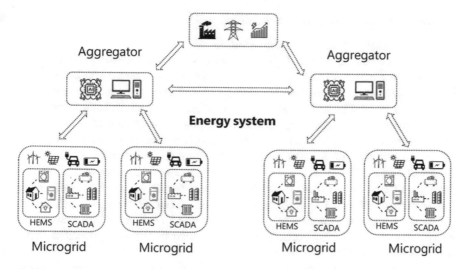

Fig. 3 Schematic of a hierarchical EMS enabled by two-way communication between SCADA, home EMS, micro-grid, and aggregator

and the grid operators. A home EMS decides on how to schedule different devices to maximize utility without compromising the normal operation.

For residential and commercial uses, IESs are formed by a combination of heat and power. These systems utilize different energy sources to produce electricity and heat simultaneously. Geothermal pumps and solar water heaters are also combined to achieve efficient power solutions for homes, multifamily buildings, and commercial areas. At this scale, IESs are typically low voltage and may range from kilowatts to megawatts. Generally, such systems are owned by individuals and are operated under local rules and regulations. These systems cannot communicate power generation and demand information to higher levels; however, mixing of the sources and high-density development permit cost-effective integration of systems. For towns, the energy management system should incorporate a variety of energy sources at an aggregated level [38].

Dispatchability:

*low partial dispatchability, **partial dispatchability, ***dispatchable.

Geographical Diversity Potential:

*moderate potential, **high diversity potential.

Predictability:

*moderate prediction accuracy (typical <10% RMS error of rated power day ahead, **high prediction accuracy.

Active Power and Frequency Control:

*good possibilities, **full control possibilities.

At the city level, the electrical systems consist of distribution and sub-transmission systems. On a national scale, IESs cover geographically isolated energy resources in utility service territories and balancing areas. These systems integrate larger-scale power generation technologies, such as large solar power plants, wind farms, pumped hydro systems, and large photovoltaic plants. Integration of power sources at the national level needs to cover multiple regulatory areas and jurisdictions, which raises adaptation issues for such systems. The designing of an IES and selection of energy sources is based on several factors and the key factors are resource accessibility and cost [14]. Figure 4 provides a cost comparison of renewable energy from different sources. The design method of IES needs the sizing and selection of the most appropriate grouping of power conditioning devices, energy sources, and energy storage systems collectively with the implementation of a proficient energy dispatch stratagem [18].

The complexity of IES also increases with an increase in size and capacity, making it more difficult to maintain stability, reliability, and economic operation. As shown in Fig. 5, several factors are involved in the design and implementation of efficient system integration. IESs are assumed to be capital-intensive and are typically designed to work for decades. However, many of the existing IESs suffer from aging infrastructure, with upgradation having been neglected. *World Energy Outlook 2011*

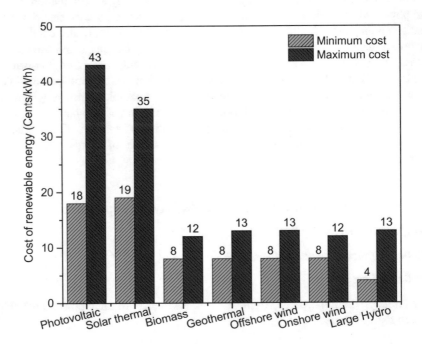

Fig. 4 A cost comparison of different renewable energy sources

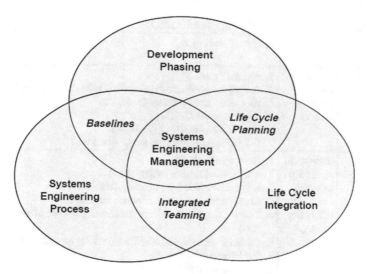

Fig. 5 Basic considerations for design, development, and implementation of an efficient IES

[39] has predicted a US$35 trillion investment in energy infrastructure over the next 25 years. This highlights the high risks associated with over-investing in the field based on weak assumptions and high uncertainty.

IESs, primarily, are complex to a degree that is difficult to appreciate. The major technical complexity is intertwined with social, economic, and political factors. Without a comprehensive approach, there is a significant risk that local optimizations may yield solutions far from a global optimum. On the other hand, global optimizations may result in a brittle solution, risking reliability and security. A group of optimal sub-systems can improve the results as well as resiliency. However, the boundaries between the sub-systems are not clearly understood and interactions between sub-systems remain undefined. Although solar and wind energies yield electricity at no fuel cost and free of emissions, they have several drawbacks that pose a serious challenge for grid operators. The energy harvesting from such systems depends on sunshine and weather conditions, which change vividly over time. Will the existing grid designs be able to manage the weather-prone renewable energy resources in a reliable and affordable manner? That remains the question.

3 Energy Sustainability Framework

Accessibility is referred to access to electricity at competitive rates in a sustainable manner. The quantitative indicators of accessibility are electricity tariff, electrification rate, average spending in the form of electricity bills, per capita electricity growth, and consumption and electricity intensity. Table 3 provides an overview of the indicators of meeting the energy sustainability objectives. The qualitative criteria are the

Table 3 Summary of indicators of energy sustainability

Energy objective	Dimension	Indicators and criteria
Accessibility	Affordable price Energy sources	1. Electricity price ($) 2. Average expenditure on electricity bills (% of income) 3. Electricity tariff subsidy (checklist) 4. Electrification rate (%) 5. Electricity intensity (kWh/GDP) 6. Electricity consumption per capita (kWh)
Availability	Short-term reliability of supply Long-term continuity of supply	1. Reserve margin (%) 2. SAIF(no. of times), SAIDI (min) 3. Reliability operating standards (checklist) 4. Cross-border supply and interconnections 5. Fuel mixing for electricity generation (% share of each type of fuel) 6. Reliance on imported fuel for electricity generation (%) 7. Fuel diversity 8. Strategy for nuclear power 9. Renewable energy polices 10. Share of renewable energy in electricity generation (%)
Acceptability	Safety greenhouse emission	1. CO_2 emission per capita (tCO_2) 2. CO_2 intensity in term of economic and electricity output

subsidies on electricity tariff. This aspect narrates the price of sustainable electricity supply to the consumers. Several countries around the world have used subsidies as one of the solutions to improve access to electricity among the lower-income classes. However, the burden of subsidy is derailing the economy and national budget of many underdeveloped and developing nations [40]. The energy security aspect is related to the availability of energy resources and their reliability. The latter is divided into long-term and short-term reliability. The short-term reliability depends on production of electricity, transmission, and distribution of electricity to the consumers. Because of the sheer complexity of reliability, many metrics are needed to assess the level of reliability of the electricity supply. The metrics include System Average Duration Index (SAIDI), System Average Interruption Frequency Index (SAIFI), and system reserve margin. SAIDI and SAIFI are widely used for the calculation of grid stability over a given period of time [43]. SAIDI calculates the total interruption time over the year, in minutes or hours per consumer. SAIDI is defined as:

$$\text{SAIDI} = \frac{\text{The sum of all customer interruption durations}}{\text{The total number of customer served}} \tag{1}$$

This index is a calculation of the time of reaction or regeneration when outages occur. SAIFI estimates the total number of occasions a consumer is affected during a year, which is calculated as:

$$\text{SAIFI} = \frac{\text{Total number of customers interrupted}}{\text{The average number of customers served}} \tag{2}$$

Together, SAIDI and SAIFI offer an accurate indicator of energy supply efficiency because they identify core aspects of the power outage spectrum faced by end-users. The reserve margin is a metric for calculating the volume of potential for the spare generation above peak demand. This test is seen as a quantitative measure of the adequacy of the method. The percentage reserve margin can be calculated using the relation:

$$\text{Reserve margin\%} = \frac{\text{Total installed capacity} - \text{Peak demand}}{\text{Peak demand}} \tag{3}$$

For long-term continuity of electricity supply, diversification of energy supplies will reduce the impact volatility in fuel price and physical supply disturbance. However, diversity is a complex subject in terms of energy security where the problems are the availability of fuel sources, fuel type, a supply of fuel from geographic regions, and technology.

4 A Review of the Reported IES Combinations

A survey of the published literature in integrated systems reveals that a number of IES combinations have been reported so far. Each combination has different efficiency and performance indicators from others. These combinations are made by considering the availability of energy sources and sustainability factors. Table 4 provides a summary of some of the IES combinations reported in the literature.

5 Energy Demand Balancing

To comprehend the efficient integration of variable energy sources into an electric system, it is vital to first know that the system operators responsible for managing the flow of electricity on a grid are exactly in balance with demand [55]. The electricity demand for cooling, heating, lighting, and electronics fluctuates substantially throughout the day and year. Therefore, the grids should be designed to handle large variations in electricity demand. System operators should balance the frequently changing demand and supply despite daily load cycles and minute to minute fluctuations. To prepare power plants for increased or reduced electricity demand, system operators forecast the demand from consumption trends, weather, and previous demand history. To regulate the variations in energy demand and supply, some of the power plants in the network are held in reserve to meet the increased energy demand at short notice. Such dispatchable sources of energy are likely to be hydropower or

Table 4 Review of different IES combinations

Ref	IES	Outcomes	Efficiency
[41]	Wind-PV-Thermal solar	• The solar PV and wind sources have slight dispatchability. The output of these sources cannot be increased but can be reduced on-demand • The challenges and cost of integration of larger amounts of solar and wind vary with a number of factors; for example, energy mixing, size of the balancing area, operational practices, and extent of geographical spread of solar and wind sources	• Thermal efficiency = 17.18% • Overall energy efficiency = 45% • Electrical efficiency = 10.01% • Exergy efficiency = 10.75%
[42]	PV-thermal solar	• The effect of the duct length on the energy efficiency of the mixed system was studied. The overall thermal and energy efficiencies decreased from 17 to 10% and 46% to 38.5%, respectively, with an increase in length of the duct from 1.2 to 6 m	• Thermal efficiency = 17.18% • Electrical efficiency = 10.01 • Overall energy efficiency = 45%
[43]	PV-thermal solar	• The impact of six selected parameters on the integrated system was evaluated. According to the first law, a glazed PV-thermal solar system was more suitable for maximization of the amount of either thermal or overall energy output • An increase in efficiency of PV cell, the ratio of water mass to the collector area, packing factor and wind velocity was supportive of the unglazed systems • An increase in ambient temperature and on-site solar radiation were favorable for glazed systems	• The conversion efficiency of PV cell was measured at about 0.13 • The glazed collector exhibited higher thermal efficiency than an unglazed collector • The thermal efficiency of glazed collector = 50.3% • The thermal efficiency of unglazed collector = 40.8% • The conversion efficiency of solar cell = 9.3%
[44]	PV-thermal solar	• An instantaneous increase in overall energy and energy efficiency of PV-thermal solar air heater was observed • The reported work revealed an increase of 2 to 3% in energy output due to thermal energy. Additionally, 12% electrical output was reported from PV-thermal integrated system. Such integration increases the overall electrical efficiency of the system by up to 15%	• Instantaneous energy = 55–65% • Exergy efficiency = 12–15% • The electrical efficiency = 14–15%

(continued)

Table 4 (continued)

Ref	IES	Outcomes	Efficiency
[45]	PV-thermal solar	• A PV-thermal solar system was tested under different environments theoretically and experimentally. The theoretical response of Tedlar back surface, solar cell, and greenhouse room was approximately the same as the experimental data • The measured and predicted responses of the Tedlar back surface, solar cell, and greenhouse room temperature was verified in terms of RMS of percent deviation (7.05–17.58%) and correlation coefficient (0.95–0.97). Both responses showed fairly good agreement • The energy analysis of PV-thermal greenhouse system predicted an energy efficiency level of 4%	• Exergy efficiency = 4%
[46]	PV-thermal solar	• The collectors were partially covered with PV modules, which combined the electricity generation and production of hot water • It was beneficial for both the consumers whose primary requirement was the production of hot water and those whose primary requirement of production of electricity • The collectors covered with PV modules were beneficial for the users of electricity	• Electrical efficiency was decreased from 0.091 to 0.08) • The instantaneous efficiency of four collectors' module was 61% • The instantaneous efficiency of 10 collector's module was 51%

(continued)

Table 4 (continued)

Ref	IES	Outcomes	Efficiency
[47]	PV-thermal solar	• A thermal model of unglazed PV-thermal heating system was studied experimentally during the summer season • The glazed hybrid PV-thermal model, in the absence of a Tedlar, provided the best performance during the summer season • No temperature difference was observed for a solar cell of unglazed PV-thermal module and without Tedlar. However, a marginal rise in outlet air temperature was observed in the absence of Tedlar • The overall efficiency of the solar cell and hybrid system was increased with an increase in airflow rate through the duct. An increase in efficiency with flowrate was attributed to a reduction in losses from the system • Conversely, the efficiency of the system was decreased with the length of the module. More losses from the system were expected with an increase in module length. However, as predicted, the efficiency of the solar cell increased with a decrease in temperature • The overall efficiency of the PV-thermal system was also increased when smaller modules were connected in series for a specified length of the system	• Overall efficiency was measured by about 55%
[48]	PV-thermal solar	• The day to day efficiency of a PV-thermal solar system was reported higher for water-based heating than the air-based heating • The thermal efficiency of the hybrid system was lower during summer when compared with system efficiency during winter • The system efficiency during summer and winter was measured by about 65% and 77%, respectively • The unglazed system in the absence of Tedlar performed well at low operating temperatures • The system glazed with Tedlar performed well at high operating temperatures • The hybrid system, utilizing water as a working fluid, resulted in a better performance with the exception of glazing without Tedlar	• The thermal efficiency of the hybrid system during the summer and winter seasons was measured by about 65% and 77%, respectively • During winter, the electrical efficiency was about 11.83% and thermal efficiency was about 29%

(continued)

Table 4 (continued)

Ref	IES	Outcomes	Efficiency
[49]	Thermal solar	• A parabolic trough collector was used in the reported work. The energy and exergy losses took place at the collector. The losses were measured about 36.2% and 70.4%, respectively • The tested thermal solar system provided higher energy conversion efficiency as compared to the conventional standalone solar-thermal systems	• The energy efficiency was increased from 10.2% to 58.0% and the exergy efficiency was increased from 12/5 to 15.2% • The overall energy efficiency of the system in winter and summer was about 63.8% and 27.3%, respectively • The overall exergy efficiency of the system during winter and summer was about 16.9% and 9.9%, respectively
[50]	Gasification-FC	• An MSW gasification unit was combined with the Stirling engine and SOFC for a decentralized CHP plant. The electric power capacity of the system was anticipated about 10 kW which was analyzed thermodynamically • Different plant efficiencies were obtained with a change in MSW composition. The overall plant efficiency ranged from 43 to 48%. About 45% of plant efficiency was possible with a mean composition • Owing to a change in mole fractions of the gaseous components in the gasifier, the plant efficiency was decreased with a rise in gasifier temperature • The maximum efficiency was possible with SOFC operating temperature of 690 °C and constant flowrate of the fuel	• The efficiency of AC/DC convertor was 0.95 • The electrical efficiency of the plant was about 48% and CPH efficiency was about 95% • The changing MSW compositions resulted in different plant efficiencies in the range of 43–48%
[51]	PV-biomass	• A grid system was integrated with PV and biomass. The findings of the study suggested that it is the most feasible combination for grid-connected systems • The cost of production of energy from this combination was about 0.143 \$/kWh. The renewable fraction was about 0.91 • The major part of the produced energy came from the PV system. The rice husk biomass contributed about 14% to energy production • The grid-connected PV-biomass system showed better results for energy cost and a renewable fraction as compared to the standalone power generation systems	• The efficiency of the biomass generator was about 59.2% • Almost 78% of the produced energy came from the PV part of the hybrid system

(continued)

Table 4 (continued)

Ref	IES	Outcomes	Efficiency
[52]	PV-wind	• The theoretical basis for comparison of the domains was developed. The efficiencies of the different energy sources were compared using different theoretical models	• The solar PV contributed 9.0% of energy • The wind turbines contributed 2.9% of energy • The batteries contributed 3.1% of energy • The capacitors contributed 21.1% of energy
[53]	PV-diesel	• The long-term solar radiation data of Dhahran city on the East Coast of Saudi Arabia was investigated for techno-economic feasibility of using PV–diesel–battery integrated buildings. A 4 kW PV facility was integrated with a 10 kW diesel power system and battery storage of 3 h, which was equivalent to 3 h of mean load. The PV penetration was about 22%	• The hybrid energy systems were developed for residential buildings with an annual electrical energy demand of 35,120 kWh • The cost of electrical energy from the integrated system was anticipated by about 0.179 $/kWh. For this typical case, the price of diesel was assumed 0.1$/L
[54]	PV-diesel-battery	• A PV–diesel–battery hybrid system was simulated. It was revealed that the integration of 80 kW PV system with 175 kW diesel power plant and battery storage of 3 h would result in a low-cost but an efficiency power production system • The emphasis of the work was mainly on unmet load, percentage fuel savings, excess electricity generation, and reduction in carbon emissions. Different scenarios such as PV–diesel with storage, PV–diesel without storage, diesel only situation and PV–diesel–battery systems were set to check the production cost and efficiency	• The PV penetration was about 26% • The cost of generating energy from the proposed hybrid system was found to be 0.149 $/kW h • For this typical case, the price of the diesel fuel was assumed 0.1 $/L)
[13]	PV-wind-diesel	• Parametric study on a PV-wind-diesel integrated system was carried out. The experimental analysis revealed that for couple of 10 kW wind turbines integrated with 3 days of battery storage and PV field of 30 m², the diesel-based back-up needs to provide 23% of the total energy demand • In the absence of a battery storage facility, the diesel-based back-up need to provide 48% of the total energy demand	• Module reference efficiency was 0.111 • Overall efficiency of wind turbine was 35% • The diesel-based back-up needs to provide 23% of the total energy demand with a battery storage facility and 48% of the total energy demand in the absence of the battery storage facility

natural gas. Nuclear and coal power plants usually work at full capacity, until they are shut down for maintenance or repair. These plants are based on more flexible energy sources to meet the changing energy demand.

Conventional plants are run with fossil fuel and are dispatchable to varying degrees. Some renewable systems, including biopower, hydropower, geothermal, and concentrated solar power plants with thermal storage are also categorized as dispatchable technologies. The output from river-based hydropower systems generally remains constant over short periods of time but varies over longer periods. Some of the ocean technologies, such as ocean current, are also capable of providing constant output. In some cases, these technologies can offer some level of dispatchability as well.

Photovoltaic and wind have little dispatchability; output from these sources can only be reduced but not enhanced on demand. The other challenges are the uncertainty and variability in the output profiles of these systems. High level of deployment of these systems can therefore be problematic for reliable grid operation. However, the requirement of balanced demand and supply should be dealt on aggregate basis. The variations and uncertainty in the output of any individual source or load do not ultimately define the integration challenges associated with renewable energy resources [55].

6 Demand Response and Flexibility

Demand response programs (DRPs) are used to meet variations in demand and balance the variations in output [39, 55]. DRPs facilitate the companies in adjusting bill payers' heating, cooling, and other energy services in exchange for monetary credits on their bill. If a utility suddenly wants extra energy due to a spike in demand or reduction in production, DRPs sends a signal to reduce the energy consumption of program participants. DRP products are tailored in a way that participants do not notice but they are still substantially beneficial in reducing the system cost, lowering emissions, and increasing the system resilience [56]. Balancing of changes in demand and energy production using DRPs is less expensive than manual adjustment of the output of dispatchable power sources. For example, demand response can be effective during infrequent events of low wind generation and it is less expensive as compared to maintaining the extra reserves year-round.

In order to increase the role of demand response in integrating power sources, the issues of lack of advanced meters and limited participation incentives for ratepayers need to be addressed. Renewable energy plants are capital-intensive but the least costly to operate. They require no fuel, which is a low-cost approach to run them, and they use dispatchable sources to adjust the variations in renewable output. Dispatchable power sources can be started and stopped to ramp the energy production up or down quickly. Natural gas based dispatchable power plants are considered more flexible than coal and nuclear power plants. However, flexibility comes at a cost; still, the overall cost of the power system can be reduced by adding more flexibility. But the

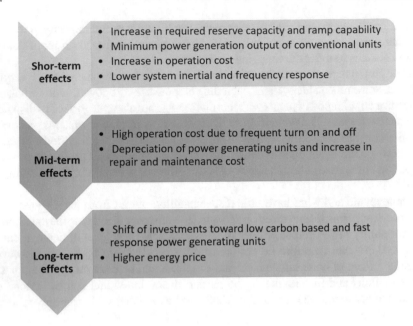

Fig. 6 Effect of energy sources on flexibility of IES

real challenge is to forecast the amount of flexibility needed and distribution of cost associated with a flexible plant. As shown in Fig. 6, the energy sources have short-term effects, mid-term effects, and long-term term effects on the flexibility of IESs. Secure and efficient operation of IESs requires a sufficient allocation of reserves to account for the fluctuations in the generation, supply, and demand. Therefore, more conventional resources are generally allocated as a buffer to provide the balance of power in the operational cycle. Since plant flexibility is beneficial for the market, policymakers should incorporate evaluation of plant flexibility into energy resource planning. The plant flexibility may also be improved by providing incentives for utilities and power plants to invest in flexibility [57, 58].

7   Problems with Hybrid Renewable Energy Systems

Although a hybrid system has a number of benefits, there are some concerns and problems associated with hybrid systems which have to be attended to:

- The stability issue: As the power production from distinct sources of a hybrid system is comparable, an abrupt change in the output power from any of the sources or a sudden alteration in the load can influence the system stability extensively.

- Due to the reliance of the renewable sources used in a hybrid system on weather outcomes, load sharing among the different sources engaged for power generation, the optimum power dispatch, and the verification of cost per unit generation are not easy.
- Distinct sources of the hybrid systems have to be functioning at a point that provides the most efficient generation. Indeed, this may not be happening due to the fact that the load sharing is often not related to the capacity of the sources. Numerous features decide load sharing, like availability of fuel, the reliability of the source, switching required between the sources, the economy of use, etc. Hence, it is preferable to assess the schemes to raise the efficiency to as high a level as possible [14].
- The reliability of power can be guaranteed by including weather unrelated sources like a fuel cell or diesel generator.
- Most hybrid systems entail storage devices, for which batteries are commonly used. These batteries need a prolonged examination, as a battery's life is limited to a few years. It is testified that the battery life should rise to around years for the cost-effective use in hybrid systems.

8 Summary

Integrated power sources can produce greater output than the sum of the individual sources. The efficiency and reliability of integrated systems are typically higher than those of individual technologies. A well-designed integrated energy system can substantially reduce the consumption of fossil fuels and boost system reliability. Specifically, wind-PV integrated systems are an attractive choice for low load applications. For high load applications, wind-diesel integrated systems are more attractive than wind-PV integration. Integrated energy systems become more complex, with an increase in size and capacity. It would be difficult to maintain stability, reliability, and economic operation with increasing system size.

Although integrated energy systems are assumed to be capital-intensive and designed to work for decades, many of the existing systems face the issues of aging infrastructure and lack of upgradation. Without a comprehensive approach, there is a significant risk that local optimization may yield solutions far from a global optimum. At the same time, global optimization may result in brittle solutions, putting at risk reliability and security. Therefore, to comprehend the efficient integration of variable energy sources into an electric system, it is vital to first know that the system operators responsible for managing the flow of electricity on a grid are exactly in balance with demand.

References

1. Bajpai P, Dash V Hybrid renewable energy systems for power generation in stand-alone applications: a review. Renew Sustain Energy Rev 16:2926–2939. 2012/06/01/2012
2. Sims R, Mercado P, Krewitt W, Bhuyan G, Flynn D, Holttinen H et al (2011) Integration of renewable energy into present and future energy systems. In IPCC special report on renewable energy sources and climate change mitigation. Cambridge University Press, Cambridge, United Kingdom and New York, NY, USA
3. Hoai Trinh T, Uemura Y A theoretical equation presenting slope in van krevelen diagram for biomass pyrolysis. Platform J Eng 3, 56–64. 2019-05-06 (2019)
4. Shivarama Krishna K, Sathish Kumar K A review on hybrid renewable energy systems. Renew Sustain Energy Rev 52:907–916. 2015/12/01/2015
5. Upadhyay S, Sharma MP A review on configurations, control and sizing methodologies of hybrid energy systems. Renew Sustain Energy Rev 38:47–63. 2014/10/01/2014
6. Wichert B PV-diesel hybrid energy systems for remote area power generation—a review of current practice and future developments. Renew Sustain Energy Rev 1:209–228. 1997/09/01/1997
7. Wandhare RG, Agarwal V (2015) Novel integration of a PV-wind energy system with enhanced efficiency. IEEE Trans Power Electron 30:3638–3649
8. Chen F, Duic N, Manuel Alves L, da Graça Carvalho M Renewislands—renewable energy solutions for islands. Renew Sustain Energy Rev 11:1888–1902. 2007/10/01/2007
9. Paiva JE, Carvalho AS Controllable hybrid power system based on renewable energy sources for modern electrical grids. Renew Energy 53:271–279. 2013/05/01/2013
10. Pradhan SR, Bhuyan PP, Sahoo SK, Prasad GRKDS (2013) Design of standalone hybrid biomass & PV system of an off-grid house in a remote area. Int J Eng Res Appl 3:433–437
11. Ma'arof MIN, Chala GT, Chaudhry MB, Premakumar BK (2020) A comparative study of electricity generation from industrial wastewater through microbial fuel cell. Platform J Eng 4: 70–75, 2020-06-30
12. Mahmoudi H, Abdul-Wahab SA, Goosen MFA, Sablani SS, Perret J, Ouagued A et al Weather data and analysis of hybrid photovoltaic–wind power generation systems adapted to a seawater greenhouse desalination unit designed for arid coastal countries. Desalination 222:119–127, 3/1/2008
13. Elhadidy MA, Shaahid SM Parametric study of hybrid (wind + solar + diesel) power generating systems. Renew Energy 21:129–139, 2000/10/01/2000
14. Bhave A (1999) Hybrid solar–wind domestic power generating system—a case study. Renew Energy 17:355–358
15. Yang H, Lu L, Burnett J (2003) Weather data and probability analysis of hybrid photovoltaic–wind power generation systems in Hong Kong. Renew Energy 28:1813–1824
16. Rozali NEM, Yahaya MSAM (2016) Study of the effects of seasonal climate variations on hybrid power systems using power pinch analysis. Proc Eng 148:1030–1033
17. Nagaraj R, Thirugnanamurthy D, Rajput MM, Panigrahi B (2016) Techno-economic analysis of hybrid power system sizing applied to small desalination plants for sustainable operation. Int J Sustain Built Environ 5:269–276
18. El Khashab H, Al Ghamedi M (2015) Comparison between hybrid renewable energy systems in Saudi Arabia. J Electric Syst Inf Technol 2:111–119
19. Fathabadi H (2016) Novel highly accurate universal maximum power point tracker for maximum power extraction from hybrid fuel cell/photovoltaic/wind power generation systems. Energy 116:402–416
20. Allison J (2017) Robust multi-objective control of hybrid renewable microgeneration systems with energy storage. Appl Therm Eng 114:1498–1506
21. Herrando M, Markides CN, Hellgardt K (2014) A UK-based assessment of hybrid PV and solar-thermal systems for domestic heating and power: system performance. Appl Energy 122:288–309

22. Chow TT (2010) A review on photovoltaic/thermal hybrid solar technology. Appl Energy 87:365–379
23. Behera S, Sahoo CP, Subudhi B, Pati BB (2016) Reactive power control of isolated wind-diesel hybrid power system using grey wolf optimization technique. Proc Comput Sci 92:345–354
24. Panda A, Tripathy M (2016) Solution of wind integrated thermal generation system for environmental optimal power flow using hybrid algorithm. J Electric Syst Inf Technol 3:151–160
25. Hemmati R, Saboori H (2016) Emergence of hybrid energy storage systems in renewable energy and transport applications—a review. Renew Sustain Energy Rev 65:11–23
26. Liu Q, Bai Z, Wang X, Lei J, Jin H (2016) Investigation of thermodynamic performances for two solar-biomass hybrid combined cycle power generation systems. Energy Convers Manage 122:252–262
27. Tah A, Das D (2016) Operation of small hybrid autonomous power generation system in isolated, interconnected and grid connected modes. Sustain Energy Technol Assess 17:11–25
28. Mohanty A, Viswavandya M, Ray PK, Mohanty S (2016) Reactive power control and optimisation of hybrid off shore tidal turbine with system uncertainties. J Ocean Eng Sci 1:256–267
29. Ekins-Daukes N (2009) Solar energy for heat and electricity: the potential for mitigating climate change. Briefing Paper no 1:1–12
30. Slotine J-JE, Li W (1991) Applied nonlinear control, vol 199. Prentice hall Englewood Cliffs, NJ
31. Cheng D, Isidori A, Respondek W, Tarn TJ (1988) Exact linearization of nonlinear systems with outputs. Theory Comput Syst 21:63–83
32. Arif J, Ray S, Chaudhuri B (2013) MIMO feedback linearization control for power systems. Int J Elcctr Power Energy Syst 45:87–97
33. Roca L, Guzman JL, Normey-Rico JE, Berenguel M, Yebra L (2011) Filtered Smith predictor with feedback linearization and constraints handling applied to a solar collector field. Sol Energy 85:1056–1067
34. Moradi H, Saffar-Avval M, Alasty A (2013) Nonlinear dynamics, bifurcation and performance analysis of an air-handling unit: disturbance rejection via feedback linearization. Energy Build 56:150–159
35. Isidori A, Ruberti A (1984) On the synthesis of linear input-output responses for nonlinear systems. Syst Control Lett 4:17–22
36. Byrnes CI, Isidori A (1991) Asymptotic stabilization of minimum phase nonlinear systems. IEEE Trans Autom Control 36:1122–1137
37. Ge J-H, Frank P, Lin C-F (1996) H∞ control via output feedback for state delayed systems. Int J Control 64:1–7
38. Inderwildi O, Zhang C, Wang X, Kraft M (2020) The impact of intelligent cyber-physical systems on the decarbonization of energy. Energy Environ Sci 13:744–771
39. World Energy Outlook (2011) https://www.iea.org/newsroom/news/2011/november/world-ene rgy-outlook-2011.html
40. Wamukonya N (2003) Power sector reform in developing countries: mismatched agendas. Energy Policy 31:1273–1289
41. Sarhaddi F, Farahat S, Ajam H, Behzadmehr A (2010) Exergetic performance assessment of a solar photovoltaic thermal (PV/T) air collector. Energy Build 42:2184–2199, 11
42. Sarhaddi F, Farahat S, Ajam H, Behzadmehr A, Mahdavi Adeli M (2010) An improved thermal and electrical model for a solar photovoltaic thermal (PV/T) air collector. Appl Energy 87:2328–2339, 7
43. Chow TT, Pei G, Fong KF, Lin Z, Chan ALS, Ji J (2009) Energy and exergy analysis of photovoltaic–thermal collector with and without glass cover. Appl Energy 86:310–316, 3
44. Joshi AS, Tiwari A (2007) Energy and exergy efficiencies of a hybrid photovoltaic–thermal (PV/T) air collector. Renew Energy 32:2223–2241, 10
45. Nayak S, Tiwari GN (2008) Energy and exergy analysis of photovoltaic/thermal integrated with a solar greenhouse. Energy Build 40:2015–2021

46. Dubey S, Tiwari GN (2009) Analysis of PV/T flat plate water collectors connected in series. Solar Energy 83:1485–1498, 9
47. Tiwari A, Sodha MS Parametric study of various configurations of hybrid PV/thermal air collector: experimental validation of theoretical model. Solar Energy Mater Solar Cells 91:17–28, 1/5/2007
48. Tiwari A, Sodha MS (2006) Performance evaluation of hybrid PV/thermal water/air heating system: a parametric study. Renew Energy 31:2460–2474, 12
49. Zhai H, Dai YJ, Wu JY, Wang RZ (2009) Energy and exergy analyses on a novel hybrid solar heating, cooling and power generation system for remote areas. Appl Energy 86:1395–1404, 9
50. Jain S, Agarwal V (2008) An integrated hybrid power supply for distributed generation applications fed by nonconventional energy sources. IEEE Trans Energy Convers 23:622–631
51. Bhattacharjee S, Dey A (2014) Techno-economic performance evaluation of grid integrated PV-biomass hybrid power generation for rice mill. Sustain Energy Technol Assess 7:6–16, 9
52. Benson CL, Magee CL (2014) On improvement rates for renewable energy technologies: solar PV, wind turbines, capacitors, and batteries. Renew Energy 68:745–751, 8
53. Shaahid SM, Elhadidy MA (2008) Economic analysis of hybrid photovoltaic–diesel–battery power systems for residential loads in hot regions—a step to clean future. Renew Sustain Energy Rev 12:488–503, 2
54. Shaahid SM, Elhadidy MA (2007) Technical and economic assessment of grid-independent hybrid photovoltaic–diesel–battery power systems for commercial loads in desert environments. Renew Sustain Energy Rev 11:1794–1810, 10
55. Andersen G (2014) Integration renewables. In: National conference of state legislatures, United States of America
56. Aalami HA, Moghaddam MP, Yousefi GR (2010) Modeling and prioritizing demand response programs in power markets. Electric Power Syst Res 80:426–435, 2010/04/01/
57. Mondal AH, Denich M (2010) Hybrid systems for decentralized power generation in Bangladesh. Energy Sustain Develop 14:48–55, 3
58. Tina G, Gagliano S, Raiti S (2006) Hybrid solar/wind power system probabilistic modelling for long-term performance assessment. Solar Energy 80:578–588, 5

Thermoelectric Air Conditioners for Tropical Countries

Balaji Bakthavatchalam and Khairul Habib

Abstract Fuel and electricity consumption for buildings is a crucial aspect of the tropical countries' policy for limiting national electricity use and thereby reducing atmospheric carbon emissions. Currently, the energy demand for space conditioning is significantly higher, dominated by vapor compression air conditioning systems that contribute 30% of the global electricity consumption in tropical zones. Application of thermoelectric cooling modules can be regarded as one of the favorable options to solve the issue of high electricity consumption and CO_2 emissions. This chapter reviews the growing importance of thermoelectric air conditioning systems in buildings. Moreover, thermoelectric coolers that work on the principle of Peltier effect possess certain advantages such as compact size, no moving parts, easy cooling rate control, long life span, and no liquids or gases. In this study, the present scenario of energy supply, energy demand and energy consumption in different sectors of tropical countries is discussed in more detail. Furthermore, the working principles of conventional air conditioning systems are presented and compared with thermoelectric air conditioning principle. This chapter also includes the recent investigations on the advancements and developments in thermoelectric air conditioning for buildings. Finally, this study tends to understand how efficiently thermoelectric cooling systems generate cooling, and its potential to replace traditional air conditioning systems.

1 Introduction

The innovations of energy supply and usage practices in the future are becoming extremely imperative despite the rising anxiety regarding the possible social and economic effects of climate change. Lighting, electrical appliances, and cooling

B. Bakthavatchalam · K. Habib (✉)
Department of Mechanical Engineering, Universiti Teknologi Petronas, 32610 Seri Iskandar, Perak, Malaysia
e-mail: khairul.habib@utp.edu.my

B. Bakthavatchalam
e-mail: balajibp1991@gmail.com

devices' energy consumption was less studied and regulated for summer cooling. Due to the high demand for energy in recent years and the rising thermal load in houses,advanced cooling system implementation is needed. For instance, in commercial industries and residential buildings, 59% of the electricity is used for cooling and heating systems. In terms of global energy consumption, 15% of power attributes to refrigeration and air conditioning technologies [1]. Figure 1 depicts the total energy consumed by different sectors from 1970 to 2040.

Since 2000, the cooling energy usage in buildings has increased and is the second rising end-use of tropical countries with higher weather and growing development. The surge in wealth and increasing population, particularly in tropical countries that possess hot climates, tend to use air conditioners more often. Indeed, about a fifth of the overall energy spent in buildings is consumed by air conditioners and cooling fans and attributes to 10% of total electricity usage in the world. The use of an air conditioner is expected to increase in the next three decades, becoming one of the world's leading energy drivers. To reduce the requirements of rapidly growing electricity demand for cooling, the energy performance of the air conditioners or the level of present technology must be improved. Without performance improvements, with the increased operation and usage of the air conditioner, space cooling energy consumption could be more than double from now until 2040. The energy efficiency for cooling in the Efficient World Scenario mitigates the impacts of climate, business, and structure to limit cooling energy growth to 19% by 2040. Heating ventilation and air conditioning (HVAC) applications for residential and industrial buildings are mainly dependent on vapor compression technology. Nevertheless, the refrigerants used in vapor compression system have harmful environmental consequences. The contribution of CO_2 emission observed in different sectors is presented in Fig. 2.

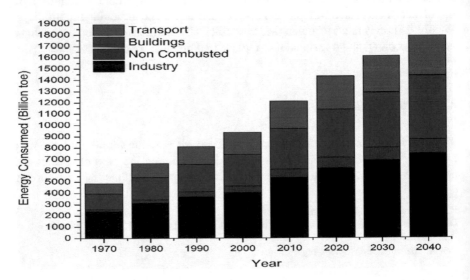

Fig. 1 Energy consumption by different sectors from 1970 [2]

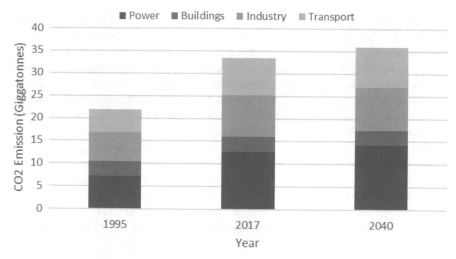

Fig. 2 Carbon dioxide emission by different sectors [2]

The International Energy Agency (IEA) predicted that the global electricity demand for buildings would rise from 31,983 to 51,253 TWh by 2050. Residential buildings (houses) consume more than half of the energy consumed by buildings (industries). The housing sector in China has a market value of 14 billion USD in 2012, which may increase to 84 billion USD by 2020. Around 188 billion square feet of new residential buildings will be built during this period. These points imply that energy consumption in residential buildings has to be analyzed seriously. The total energy consumed by different devices in a typical residential building is displayed in Fig. 3.

Fig. 3 Overall energy use of
buildings in tropical
countries [3]

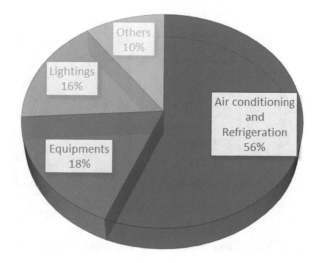

Based on cost and performance, conventional air conditioners may be beneficial, but it is not environmentally friendly and might be restricted in the future.
Some researchers have therefore suggested a novel thermoelectric air conditioner
that works on the Peltier principle together with the photovoltaic system. The law of
thermoelectric modules and the Peltier effect is described in Fig. 4. Thermoelectric
cooling, as a solid-state system, has advantages of high efficiency, no coolants, no
moving parts, and silent operation. Also, Freon coolants are a significant cause of
greenhouse emissions in traditional HVAC systems.

Thermoelectric space conditioning is regarded as one of the green cooling
processes that could substitute traditional HVAC systems for the successful reduction
of Freon. A schematic representation of a thermoelectric air conditioner installed in
a test room is shown in Fig. 5. Moreover, thermoelectric cooling systems can be
conveniently combined with sustainable systems directly utilizing direct current.
Based on the Peltier effect, the thermoelectric system transforms electrical energy

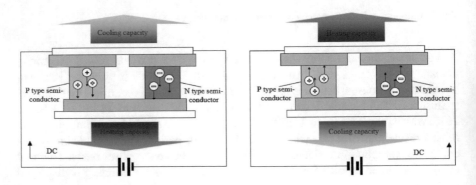

Fig. 4 Working principle of TEM and Peltier effect [4]

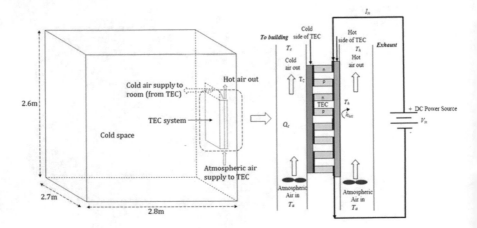

Fig. 5 Typical thermoelectric cooler installed in a building [5]

into a temperature gradient where the modes of cooling and heating can be changed easily by reversing the current input direction. The primary purpose of this research is to discuss the newly developed thermoelectric air conditioner in building sectors in a view to enhance the energy performance of buildings. Moreover, this study reflects on the advancement of thermoelectric cooling over the last decade, with particular emphasis on nanofluid and PCM-based thermoelectric air conditioners in tropical countries.

2 Thermodynamic Cycles of Air Conditioners

Regulation of temperature, humidity, and air quality in a constrained space such as commercial, residential, and industrial places is known as air conditioning, and it works on the principle of conservation of mass and energy. Air is passed through hot or cold coils or water spray to control the temperature and humidity. The essential processes of air condition are presented in Fig. 6. The primary purpose of this system

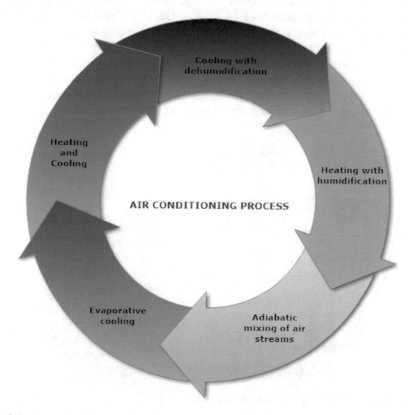

Fig. 6 Basic process of air conditioning

is to have a pleasant working atmosphere and to make the workers in workplaces, commercial buildings, and industrial plants more relaxed and competitive. A cooling unit, heat rejection device, air, and water distribution systems are the main parts of an air conditioning system. Air conditioning systems work either through the compression cycle or absorption cycle or evaporative cycle or through the thermoelectric cycle.

2.1   Vapor Compression Cycle

A halocarbon substance or ammonia as the coolant is used throughout this cycle. A conventional single-stage vapor compression refrigeration system consists of four components, namely condenser, metering device, compressor, and evaporator. As shown in Fig. 7, at low pressure and temperature, the heat from the atmosphere or object to be cooled is absorbed by the refrigerant through the process of evaporation. The absorbed heat is released to any available cooling medium like air or water, where the coolant tends to condense, which is made to pass through a condenser and to a metering device, where the liquid refrigerant's pressure is decreased to that of the evaporator leading to the cycle completion. In the cooling tower, the rate of circulation of water should be around 3GPM per ton of refrigeration along with − 12 °C temperature reduction to obtain proper cooling. The compressor, which works under a specific pressure head, is the leading power consumer for the cooling device, and this pressure is almost equal to the pressure of the liquid refrigerant in the condenser, which is also termed as "high head pressure."

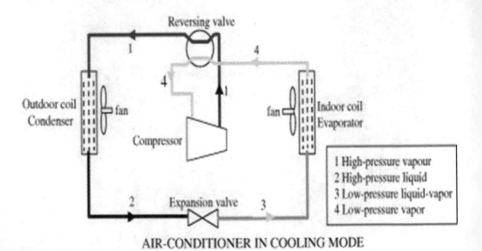

AIR-CONDITIONER IN COOLING MODE

Fig. 7   Schematic of the vapor compression cycle [6]

2.2 Absorption Cycle

In this cycle, low-pressure steam or hot water is used as a source of energy, while water and lithium bromide are used as refrigerant and absorbent, respectively. An absorption cycle uses external heating of the working fluid to produce a cooling effect. The system components are more complex and more expensive in an absorption system compared to a vapor compression system, so absorption systems are used mainly in large industrial and commercial building applications and in situations where electricity is not available, such as vacation homes, mobile homes, and trailers. There are two working fluids in an absorption cycle, an absorbent and a refrigerant. The absorption cycle uses the solubility of the refrigerant gas in the absorbent liquid to reduce the pumping energy required to compress the refrigerant. To absorb 1200 Btu/hr, an absorption cooling device needs a thermal input of 18,000 Btu/hr to power the absorption cycle, which implies that heat dissipation at the cooling tower is roughly 30,000 Btu/hr. In general, the thermal rejection of an absorption device requires drainage of around 4 GPM water (1 ton of air conditioning) with a decrease in temperature of 8 °C through the cooling tower where 3.7 gph per ton evaporation of water occurs. Besides the coolant pumps and solution, the absorption system contains no moving parts. While this is a cost-efficient design benefit, it is still essential to recognize the risks of supplying the required low-pressure steam or high-temperature water. Figure 8 shows the schematic layout of a vapor absorption cycle.

2.3 Evaporative Cycle

Evaporative cooling is used when low humidity air is available and can be cooled by the evaporation of water sprayed into the air stream. It is less expensive to install and maintain relative to a vapor compression system and has lower power consumption since no compressor is needed. The working fluid is water, not a halocarbon refrigerant. However, the temperature decrease of the air stream is smaller when compared to vapor compression cooling. The simplest and common form of air conditioning is evaporative air conditioning (EAC). Usually, this device uses a fan to pull hot air through the building using a transparent precipitation medium. The coolant water absorbs heat when it evaporates from the pore precipitation medium, leaving the EAC with a decreased temperature. The cooling performance is dependent on the efficiency of the precipitating source, the fan, the size, and the structure of the device. The use of water is a significant part of EAC. For instance, in small residential coolers, the range may vary from a few liters to hundred liters per day. The significant difference between the vapor compression systems and EAC is the fact that EAC needs more understanding of the airflow orientation, circulation, and extraction to render active cooling. Spot cooling EACs have been implemented in some places, but due to the lack of sufficient guidance, they have been used only in closed areas like vapor compression systems. The EAC systems have not worked

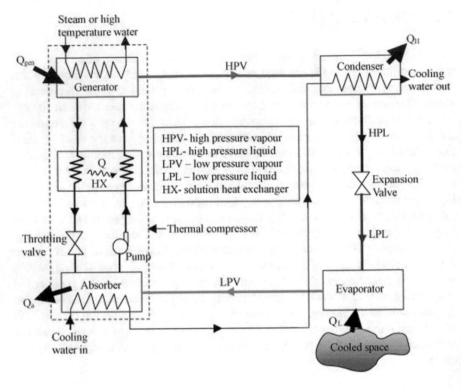

Fig. 8 Schematic layout of vapor absorption cycle [6]

well, creating consumer dissatisfaction and disappointment on the market because of the insufficient outside air movement. These points illustrated that EAC needs to be implemented with sufficient consumer knowledge and skilled employees. As shown in Fig. 9, a cubic box with vertical filters, a bottom sump, a fan operated by an electric motor, a water pump, and a water delivery device are connected to form an evaporative air conditioner. Water is circulated through a pump to reach the wet pads (precipitation medium) where the ambient air is forced to pass through it with the help of a fan, and the drained water from the wet pads reaches the sump by gravity. The obtained air is supplied to a specific room through ducts to remove the heat from the space.

2.4 Thermoelectric Cycle

In 1834, Jean Peltier, a French watchmaker, published an article on temperature anomalies observed in the vicinity of the boundary between two different conductors when a current was passing through them. The phenomenon was first observed by Peltier, and therefore given the name of the Peltier effect, consists of the generation

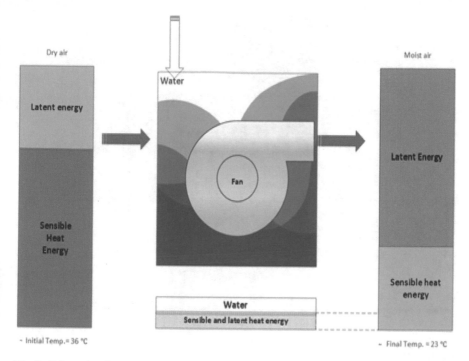

Fig. 9 Schematic of evaporative cyclic process

or absorption of heat at the junction between two different conductors when a current flows through them. This concept is the baseline of thermoelectricity and is currently employed in all the thermoelectric cooling modules. In simple, the Peltier effect of heat absorption at the junction of two conductors is used in thermoelectric cooling by the passage of a current. The primary condition governing the efficiency of a cooling device as well as of a generator is the heat flux across the thermophile. In a generator, this mainly depends on heat conduction. The amount of heat transferred to the hot junctions is only slightly higher than that removed from the cold terminals. The better the heat transfer conditions between the intersection and the surroundings and lower the temperature difference concerning the surroundings required for transferring the power, the lower is the temperature difference at the junctions of the thermoelement and the higher its coefficient of performance.

Thermoelectric generators are based on the use of the Seebeck effect. If the junctions in a circuit consisting of two different conductors are maintained at different temperatures, an electromotive force (emf) appears in the circuit. Thermoelectric cooling makes use of the reverse phenomena, namely the Peltier effect. When an external source of emf is connected to such a circuit, heat is generated at one junction and absorbed at the other terminal. There also exists a third thermoelectric phenomenon, the Thomson effect, the nature of which is as follows: when there is a temperature drop along a conductor through which an electrical current is

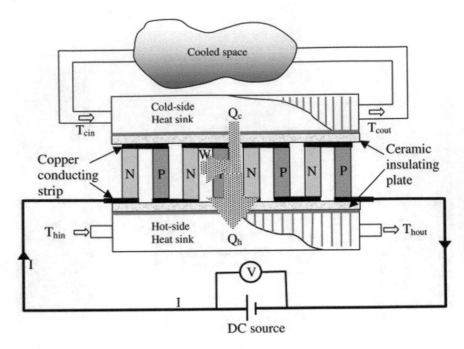

Fig. 10 Schematics of the thermoelectric cycle [6]

flowing, then, in addition to Joule heat, Thomson heat is generated or absorbed within the conductor. The schematic representation of a thermoelectric cooling system integrated into a building for space conditioning is depicted in Fig. 10.

3 Sustainable Energies in Tropical Countries

Historically, analysts have neglected the association between regional influences and economic growth. This section explores the significance and availability of sustainable energies that dominate the economy of tropical countries. In general, tropical countries are also considered as developing countries that have a low per capita volume of goods with a non-diversified economy. Economy and social growth will be the highest goal for tropical countries in the pursuit of sustainable prosperity and on more environmentally responsible roads. This suggests that it is crucial to look at the problem of climate change from the viewpoint of human progress. The efficient use of energy improves production, fosters economic developments, and boosts living standards. The outlook on future energy is very unclear. The availability of fossil fuels may not be completely depleted, but their extinction is more likely due to market pressures rather than resource limits. Due to the scarcity of these conventional energy resources, human culture is now fighting to find alternative sources of energy. From the recent investigations, it was found more economical to utilize available energy resources effectively to minimize waste instead of creating new energy sources to satisfy energy needs. Tropical countries are areas of the earth that exist in between the Tropic of Cancer and Tropic of Capricorn, near to the equator.

In fact, the tropical countries that are situated between the tropic of cancer and Capricorn have abundant resource of solar energy. This geographical environment has elevated temperatures throughout the year that attributes to the continuous fall of solar radiation which acts as an excellent resource for photovoltaic cells. These solar photovoltaic systems can supply direct current for thermoelectric cooling modules more efficiently and effectively. Therefore, energy saving is accomplished by utilizing the abundant solar energy available in tropical countries that prohibits global warming, radioactive wastes, and other problems increasingly influence energy policies in tropical nations. The heat transfer takes place from one side to another with the application of the low voltage DC source in the thermoelectric module resulting in cooling on the one side and heating on the other side. In addition to the manufacturing and transport industries, the building sector is one of the prominent energy users in tropical countries. Global primary energy demand is projected to grow by 1.6% per year or 45% over the next 21 years between now and 2030. Malaysia is expected to have a power demand of 18,947 MW by 2020 and 23,092 MW by 2030, which is almost 35% more than the previous year's 14,007 MW. This building sector accounts for approximately 32.2% of total energy sales in Malaysia, which was around 7750 GWh and is projected to rise in the future [7]. The recent increase in annual energy usage in building sectors is expected to continue to add substantially to the country's greenhouse gas emissions unless energy quality is increased. An increase leads to the rise in energy use in new construction, inefficient electricity usage, and the growing stock of electrical equipment in existing and new buildings. In general, the scope for improving the energy efficiency of buildings in Malaysia is very high. Photovoltaic-based thermoelectric air conditioners can help to use the available solar radiation regarding enhance the energy utilization efficiency of buildings.

4 Up-to-Date Investigations on TEAC in Buildings

Thermoelectric cooling modules are tiny systems made up of p-type and n-type semiconductor elements, placed between two ceramic plates, which are electrically coupled in series and thermally in parallel, set by soldering. Bismuth telluride and antimony telluride are commonly used semiconductors as their figure of merit is very high when compared with others. Due to the optimal cost and performance ratio, Al_2O_3 is chosen as ceramics. However, other elements, like aluminum nitride and beryllium oxide, which possess better thermal properties, are also used as ceramics, but they are costly. Thermoelectric anomalies were found about 150 years ago, but they have been used widely only over the recent years. Commercial thermoelectric modules are evolving for recent times following two significant paths of technological development, namely photonics and electronics. Due to the numerous advantages of thermoelectric coolers, they are mainly implemented in electronic devices, and photovoltaic panels, and only a small number of researchers were reported on the application of TEMs in space conditioning which is presented in Table 1.

5 Thermoelectric Cooling with Nanofluids and PCM

The most challenging task of the thermoelectric module is to remove the heat dissipated on the hot side of the thermoelectric module, which results in the COP enhancement of the system. Many researchers proved that the integration of heat sinks and energy storage materials such as phase change materials (PCM) and nanofluids attributed to the efficiency improvement of thermoelectric modules. The use of nanofluid as a coolant is an enticing way to improve the heat transfer between the cold side of thermoelectric modules and coolant. Nanofluid is a fluid that contains nanometer-sized particles and possesses fascinating thermophysical properties to be used for heat transfer systems. The advantage of nanofluid-assisted thermoelectric devices is illustrated in a recent review conducted by Bakthavatchalam et al. [21]. Soltani et al. [22] used air, water, SiO_2/water, and Fe_3O_4/water nanofluid coolants to examine the efficiency of a photovoltaic-based thermoelectric system. The authors found that nanofluids resulted in maximum efficiency enhancement of 3.35% (SiO_2/water) and 3.13% (Fe_3O_4/water) when compared with air and water coolants. The cooling performance of thermoelectric coolers with Al_2O_3/water, TiO_2/water, and SiO_2/water nanofluids was investigated by Cuce et al. [23]. Authors realized that thermoelectric coolers combined with nanofluids are much better than conventional water-cooled units as they achieved a 55% temperature reduction with Al_2O_3 nanofluid. Parsa et al. [24] analyzed the performance of a solar still using silver/water nanofluids-based thermoelectric module that resulted in 100.5% improvement in the total yield of the solar still. Nazari et al. [25] attempted to use copper oxide nanofluid in a thermoelectric cooling channel to improve the efficiency of single slope solar still where they achieved a maximum temperature difference of 3.9 °C. A typical schematic layout of the nanofluid cooled thermoelectric system for electronic cooling is illustrated in Fig. 11. Many studies are analyzing the efficiency of nanofluids in heat exchangers, solar stills, photovoltaic systems, and electronic devices. However, there is not any research about how to use thermoelectric air conditioners with nanofluids for space conditioning in buildings.

Another successful approach in the performance enhancement of thermoelectric coolers is to integrate phase change materials with the thermoelectric modules. In a recent study, Jiang et al. [27] examined the performance of a battery impregnated PCM-based thermoelectric cooler that led to the reduction of battery temperature and enhanced life span. The experimental setup of the PCM-based thermoelectric system is presented in Fig. 12. Manikandan et al. [28] integrated PCM materials on the hot side of the thermoelectric modules to improve the thermal performance of a thermoelectric cooler. They found that the hot side temperature of the TEM decreased from 52 to 30 °C and the cold side of the TEM lowered from 25 to 12 °C with the

Table 1 List of thermoelectric cooling investigations in buildings

Authors	Climate region	No. of modules and type	Operation current/voltage	Cooling capacity	COP	Area of application
Looi et al. [8]	Perak, Malaysia	9 (Ferrotec 9500/391/085B)	5 A	181 W	1.67	Room (3.6 m^3)
Irshad et al. [9]	Tronoh, Malaysia	24 (TEC1-12730)	6 A	498.6 W	0.679	Room (2.8 m × 2.7 m × 2.5 m)
Shen et al. [10]	Wuhan, China	888 (TEC1-12706)	1.2 A	5670 W	1.77	Virtual office space (10 m × 7 m × 3.5 m)
Ibañez-Puy et al. [11]	Pamplon, Spain	16 (RC12-8)	12 V	600 W	0.78	Residential Building (1.05 m × 1.895 m × 0.135 m)
Zhao et al. [12]	Los Angeles, USA	101 (ZT8-12)	–	3000 W	1.87	Residential building (223 m^2)
Zhao et al. [13]	USA	1 (LairdTech ZT8-12-F1-4040-TA-W8)	1.7 A	10.2 W	0.58	Human body
Martín-Gómez et al. [14]	Spain	84 (RC12-8)	1.4 A, 12 V	3000 W	0.6	Room (24 m^2)
Kim et al. [15]	France	60 (Ferrotec 9501/242/1603)	0.9 A	4600 W	1.983	Residential building (120 m^2)
Rincón-Casado et al. [16]	Spain	12 (RC12-6L)	120 V	480 W	0.77	Climatic chamber
Cai et al. [17]	China	8 (TEC1-12730)	2 A	265.96 W	2.57	Building
Manohar and Adeyanju [18]	West Indies	3 (TEC1-12730)	12 V	286 W	0.465	Plywood Enclosure (1.641 m^2)
Cheon et al. [4]	Seoul, South Korea	9 (HMN 6040)	1–3.2 A	300–504 W	0.73	Office building (3000 m^3)
Liu and Su [19]	Taipa, Macau	1 (TE-127-1.4-1.5)	0–10 V	–	0.25–0.31	Small enclosed space

(continued)

Table 1 (continued)

Authors	Climate region	No. of modules and type	Operation current/voltage	Cooling capacity	COP	Area of application
Lim et al. [20]	Seoul, Korea	7 (TEC DT12-4)	0.9 A	55 W	–	Aluminum box (0.6 m × 1.2 m × 0.3 m)

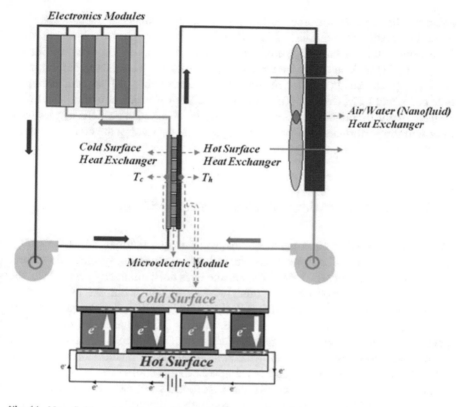

Fig. 11 Nanofluid-assisted thermoelectric cooling system [26]

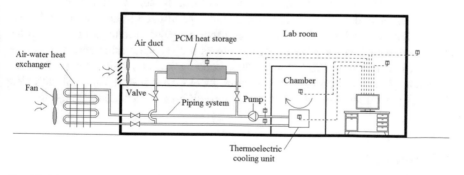

Fig. 12 The typical prototype of PCM-based thermoelectric cooling system [33]

use of PCM under a cooling load of 0.03 W equipped with two thermocouples. Zhao and Tan [29] investigated the effect of PCM in a thermoelectric system for space conditioning and pointed out that COP enhancement was significantly dependent on the energy storage material (PCM). They obtained a maximum COP of 1.22 and saved 35.3% electrical energy consumption with the use of PCM. Skovajsa et al. [30]

enhanced the thermal comfort of buildings using thermal panels equipped with PCM-based thermoelectric modules. Their findings proved that this technique improves the building's thermal capacity and can also be used to heat and cool actively. Souay-fane et al. [31] took an extensive review of phase change materials in buildings for cooling applications. In a different study, Farzanehnia et al. [32] prepared nanopar-ticle combined PCM (MWCNT/Paraffin) to analyze the efficiency of an electronic chip thermally. In comparison to PCM, the nano-PCM-based thermoelectric modules improved the performance of the thermal system by decreasing the cooling time up to 6%.

6 Current Applications of Thermoelectric Cooling

Thermoelectric coolers have many benefits, which may sometimes affect their judg-ment even for large-scale applications. For instance, this cooling system is tiny, silent operation, no moving parts, long life span, and does not contain any gases. Also, the cooling cycle can be modified very quickly by changing the current, as the reverse current path converts the cooling unit to the heater with a COP of more than 1. All the applications listed in this section use one or more of the unique benefits of thermo-electric cooling. Thermoelectric cooling modules application includes a wide variety of fields such as military, medical, telecommunications, and laboratories as shown in Fig. 13. The type of use ranges from simple beverage coolers to very complicated temperature control devices in missiles and spacecraft. Thermoelectric coolers can be preferred for applications requiring heat dissipation from milliwatts to several 1000 Watts. Many thermoelectric coolers in one stage module (including the high and low current modules) are capable of pumping 3–6 W of the surface area of module per centimeter. Several thermoelectric modules may be utilized to increase overall heat pump efficiency. In the past decade, many thermoelectric modules in the kilowatt range were equipped for advanced applications such as submarine and railroad car refrigeration. Furthermore, thermoelectric cooler allows an object temperature to be lowered below atmospheric temperature level than a simple heat sink and also stabilizes the temperature of objects subject to different environmental conditions.

7 Summary

Air conditioners are bound to have a significant effect due to large-scale applications in buildings and industries. At present, precise information on alternative technolo-gies on space conditioning is minimal. Therefore, this chapter aims to convey a specific idea of using alternate technology and energy sources to solve the issues of greenhouse gas emissions, and high energy consumption resulted from conven-tional air conditioning systems. In recent researches, thermoelectric air conditioners (TEAC) which are reliable, Freon free, and noise free proved to be a right solution

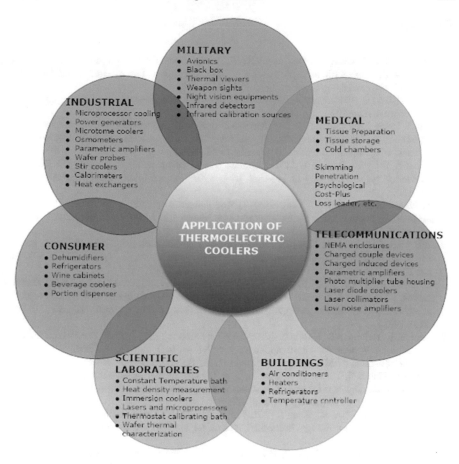

Fig. 13 Typical application of thermoelectric coolers

over conventional air conditioners are critically reviewed in this chapter. Despite the advantages, thermoelectric cooling modules could not be as useful as compressor units when the needed cooling capacity is high. Therefore, thermoelectric cooling cannot be considered as a substitute for traditional cooling solutions but as a complementary technique. Thermal comfort and thermal load must be studied in-depth for improvements relative to conventional cooling systems to support their positive potential. Finally, regulatory bodies must be formed to lay down rules and acceptable limits to ensure the implementation of air conditioners without creating environmental harm and depletion of energy sources.

Acknowledgements Thank you to the Institute of Sustainable Building (ISB) in Universiti Teknologi PETRONAS for their support on the research facilities provided to the authors.

References

1. Liu ZB, Zhang L, Gong G, Luo Y, Meng F (2015) Experimental study and performance analysis of a solar thermoelectric air conditioner with hot water supply. Energy Build 86:619–625. 01 Jan 2015. https://doi.org/10.1016/j.enbuild.2014.10.053
2. I. E. A. (IEA) Global Energy Review 2020 [Online] Available: https://www.iea.org/reports/glo bal-energy-review-2020
3. Cao X, Dai X, Liu J (2016) Building energy-consumption status worldwide and the state-of-the-art technologies for zero-energy buildings during the past decade. Energy Build 128:198–213. 15 Sept 2016, https://doi.org/10.1016/j.enbuild.2016.06.089
4. Cheon S-Y, Lim H, Jeong J-W (2019) Applicability of thermoelectric heat pump in a dedicated outdoor air system. Energy 173:244–262. 15 Apr 2019, https://doi.org/10.1016/j.energy.2019. 02.012
5. Manikandan S, Kaushik SC, Yang R (2017) Modified pulse operation of thermoelectric coolers for building cooling applications. Energy Conv Manag 140:145–156. 15 May 2017, https:// doi.org/10.1016/j.enconman.2017.03.003
6. Riffat SB, Qiu G (2004) Comparative investigation of thermoelectric air-conditioners versus vapour compression and absorption air-conditioners. Appl Therm Eng 24(14):1979–1993. 01 Oct 2004, https://doi.org/10.1016/j.applthermaleng.2004.02.010
7. U. N. D. P. (UNDP) (2019) Achieving Industrial Energy Efficiency in Malaysia. United Nation Development Programme, Malaysia. https://www.my.undp.org/content/malaysia/en/ home/library/environment_energy/EEPub_IndustrialEnergyEfficiency.html (accessed 28-06-2020, 2020)
8. Looi KK, Baheta AT, Habib K (2020) Investigation of photovoltaic thermoelectric air-conditioning system for room application under tropical climate. J Mech Sci Technol 34(5):2199–2205. 01 May 2020, https://doi.org/10.1007/s12206-020-0441-8
9. Irshad K, Habib K, Thirumalaiswamy N, Saha BB (2015) Performance analysis of a thermoelectric air duct system for energy-efficient buildings. Energy 91:1009–1017, 01 Nov 2015, https://doi.org/10.1016/j.energy.2015.08.102
10. Shen L, Xiao F, Chen H, Wang S (2013) Investigation of a novel thermoelectric radiant air-conditioning system. Energy Build 59:123–132. 01 Apr 2013, https://doi.org/10.1016/j.enb uild.2012.12.041
11. Ibañez-Puy M, Bermejo-Busto J, Martín-Gómez C, Vidaurre-Arbizu M, Sacristán-Fernández JA (2017) Thermoelectric cooling heating unit performance under real conditions. Appl Energy 200:303–314. 15 Aug 2017, https://doi.org/10.1016/j.apenergy.2017.05.020
12. Zhao D, Yin X, Xu J, Tan G, Yang R (2020) Radiative sky cooling-assisted thermoelectric cooling system for building applications. Energy 190:116322. 01 Jan 2020, https://doi.org/10. 1016/j.energy.2019.116322
13. Zhao D et al (2018) Personal thermal management using portable thermoelectrics for potential building energy saving. Appl Energy 218:282–291. 15 May 2018, https://doi.org/10.1016/j. apenergy.2018.02.158
14. Martín-Gómez C, Ibáñez-Puy M, Bermejo-Busto J, Sacristán Fernández JA, Ramos JC, Rivas A (2015) Thermoelectric cooling heating unit prototype. Build Serv Eng Res Technol 37(4):431–449. 01 Jul 2016, https://doi.org/10.1177/0143624415615533
15. Kim YW, Ramousse J, Fraisse G, Dalicieux P, Baranek P (2014) Optimal sizing of a thermoelectric heat pump (THP) for heating energy-efficient buildings. Energy Build 70:106–116. 02 Jan 2014, https://doi.org/10.1016/j.enbuild.2013.11.021
16. Rincón-Casado A, Martinez A, Araiz M, Pavón-Domínguez P, Astrain D (2018) An experimental and computational approach to thermoelectric-based conditioned mattresses. Appl Therm Eng 135:472–482. 05 May 2018, https://doi.org/10.1016/j.applthermaleng.2018.02.084
17. Cai Y, Wang W-W, Liu C-W, Ding W-T, Liu D, Zhao F-Y (2020) Performance evaluation of a thermoelectric ventilation system driven by the concentrated photovoltaic thermoelectric generators for green building operations. Renew Energy 147:1565–1583. 01 Mar 2020, https:// doi.org/10.1016/j.renene.2019.09.090

18. Manohar K, Adeyanju AA (2020) Design and analysis of a thermoelectric air-conditioning system. J Sci Res Reports 26(4):1–11. Art no. JSRR.55582, https://doi.org/10.9734/jsrr/2020/v26i430243
19. Liu Y, Su Y (2018) Experimental investigations on COPs of thermoelectric module frosting systems with various hot side cooling methods. Appl Therm Eng 144:747–756. 05 Nov 2018, https://doi.org/10.1016/j.applthermaleng.2018.08.056
20. Lim H, Kang Y-K, Jeong J-W (2018) Thermoelectric radiant cooling panel design: numerical simulation and experimental validation. Appl Therm Eng 144:248–261. 05 Nov 2018, https://doi.org/10.1016/j.applthermaleng.2018.08.065
21. Bakthavatchalam B, Habib K, Saidur R, Saha BB, Irshad K (2020) Comprehensive study on nanofluid and ionanofluid for heat transfer enhancement: A review on current and future perspective. J Mol Liq 305:112787. 01 May 2020, https://doi.org/10.1016/j.molliq.2020.112787
22. Soltani S, Kasaeian A, Sarrafha H, Wen D (2017) An experimental investigation of a hybrid photovoltaic/thermoelectric system with nanofluid application. Solar Energy 155:1033–1043. 01 Oct 2017, https://doi.org/10.1016/j.solener.2017.06.069
23. Cuce E, Guclu T, Cuce PM (2020) Improving thermal performance of thermoelectric coolers (TECs) through a nanofluid driven water to air heat exchanger design: an experimental research. Energy Conv Manag, 214:112893. 15 June 2020, https://doi.org/10.1016/j.enconman.2020.112893
24. Parsa SM, Rahbar A, Koleini MH, Aberoumand S, Afrand M, Amidpour M (2020) A renewable energy-driven thermoelectric-utilized solar still with external condenser loaded by silver/nanofluid for simultaneously water disinfection and desalination. Desalination 480:114354, 15 Apr 2020, https://doi.org/10.1016/j.desal.2020.114354
25. Nazari S, Safarzadeh H, Bahiraei M (2019) Performance improvement of a single slope solar still by employing thermoelectric cooling channel and copper oxide nanofluid: an experimental study. J Cleaner Prod 208:1041–1052. 20 Jan 2019, https://doi.org/10.1016/j.jclepro.2018.10.194
26. Mohammadian SK, Zhang Y (2014) Analysis of nanofluid effects on thermoelectric cooling by micro-pin-fin heat exchangers. Appl Therm Eng 70(1):282–290. 05 Sept 2014, https://doi.org/10.1016/j.applthermaleng.2014.05.010
27. Jiang L, Zhang H, Li J, Xia P (2019) Thermal performance of a cylindrical battery module impregnated with PCM composite based on thermoelectric cooling. Energy 188:116048, 2019/12/01/ 2019, doi:https://doi.org/10.1016/j.energy.2019.116048
28. Manikandan S et al (2020) A novel technique to enhance thermal performance of a thermo-electric cooler using phase-change materials. J. Therm. Anal. Calorim. 140(3):1003–1014. 01 May 2020, https://doi.org/10.1007/s10973-019-08353-y
29. Zhao D, Tan G (2014) Experimental evaluation of a prototype thermoelectric system integrated with PCM (phase change material) for space cooling. Energy 68:658–666. 15 Apr 2014, https://doi.org/10.1016/j.energy.2014.01.090
30. Skovajsa J, Koláček M, Zálešák M (2017) Phase change material based accumulation panels in combination with renewable energy sources and thermoelectric cooling. Energies 10(2). https://doi.org/10.3390/en10020152
31. Souayfane F, Fardoun F, Biwole P-H (2016) Phase change materials (PCM) for cooling applications in buildings: a review. Energy Build 129:396–431. 01 Oct 2016, https://doi.org/10.1016/j.enbuild.2016.04.006
32. Farzanehnia A, Khatibi M, Sardarabadi M, Passandideh-Fard M (2019) Experimental investigation of multiwall carbon nanotube/paraffin based heat sink for electronic device thermal management. Energy Conv Manag 179:314–325, 01 Jan 2019, https://doi.org/10.1016/j.enconman.2018.10.037
33. Tan G, Zhao D (2015) Study of a thermoelectric space cooling system integrated with phase change material. Appl Therm Eng 86:187–198. 05 July 2015, https://doi.org/10.1016/j.applthermaleng.2015.04.054

Innovations to Reduce Air-Conditioning Energy Consumption

Shaharin A. Sulaiman, Abdul Hakim Bin Abdullah,
Adam Bin Muhammad Yusof, Ummu Nadiah Binti Suhaimi,
Rangjiv Dharshana, Amirul Aizad Bin Nor Hamedi, and Kim Leong Liaw

Abstract Home air-conditioning for space cooling has become a necessity in today's world due to its affordability and the need for comfort, especially in tropical countries. The hot and humid climate makes it difficult for many people to concentrate with their indoor activities. Nevertheless, the air-conditioners consume large amount of energy. As a result, home owners who operate air-conditioners will usually receive high electricity bills. From the society's point of view, the high energy consumed for the air-conditioners would result in high amount of emission by the electric power

S. A. Sulaiman (✉)
Block 17, Department of Mechanical Engineering, Universiti Teknologi Petronas, 32610 Seri Iskandar, Perak, Malaysia
e-mail: shaharin@utp.edu.my

A. H. B. Abdullah
Kayangan Height, 19, Jalan Merah Mawar U9/6, Seksyen U9, 40150 Shah Alam, Selangor, Malaysia
e-mail: abdulhakimabdullah97@gmail.com

A. B. Muhammad Yusof
Kampung Warisan Condominium, Block Rumbia 5-H, Jalan Jelatek 2, Au 1, 54200 Kuala Lumpur, Malaysia
e-mail: adam.yusof@gmail.com

U. N. B. Suhaimi
IPG Kampus Bahasa Melayu, 3632, Kuarters Pensyarah, Lembah Pantai, 59990 Kuala Lumpur, Malaysia
e-mail: ummunadiahsuhaimi@gmail.com

R. Dharshana
Taman Tuanku Jaafar, Sungai Gadut, No. 1215 Jalan TTJ 2/18, 71450 Negeri Sembilan, Seremban, Malaysia
e-mail: rangjiv18@gmail.com

A. A. B. Nor Hamedi
Taman Saujana Impian, No. 4 Jalan Impian Makmur 1/4, 43000 Kajang, Selangor, Malaysia
e-mail: amirulaizad71@gmail.com

K. L. Liaw
No. 3, Jalan Muhibbah, 98850 Lawas, Sarawak, Malaysia
e-mail: reban_leong@hotmail.com

© The Author(s), under exclusive license to Springer Nature Singapore Pte Ltd. 2021
S. A. Sulaiman (ed.), *Clean Energy Opportunities in Tropical Countries*,
Green Energy and Technology, https://doi.org/10.1007/978-981-15-9140-2_15

plants. Furthermore, although generally affordable to many people, certain group of people can afford to own and operate air-conditioners due to their low income level, which deters them from buying and maintaining the operation of air-conditioners. Most home air-conditioners are the split unit type. An issue with these conventional air-conditioners is that they cool large rooms, which is occupied by one or two persons, even when the occupants are just sleeping or sitting down for a long period of time. In actual fact, only a very small space, surrounding the occupants would need to be air-conditioned. Thus, the energy consumed for home air-conditioning can be reduced significantly if there is a personalized air-conditioners that cool air within a small volume of space which surrounds the individuals especially when they are idling due to sleeping or watching television. This chapter delves into the idea of personalized air-conditioners, including their potentials and challenges

1 Introduction

An air-conditioning system cools a building by removing the heat from the indoor and transfers it to outdoors. The climate in tropical countries is hot and humid all the year round. In Malaysia, for example, during hot season the average high air temperature usually reaches up to 33 °C [1]. The high heat poses serious consequences to human health and life, which makes the air cooling to be very important. Home air-conditioners usually come in small sizes in the range of 1–2 hp. They are normally in operation after working hours, when most people are at home trying to rejuvenate from long and exhausting hours in work place or schools. Based on the results of national census showed that the total number of households with air-conditioning in Malaysia increased dramatically from 13,000 in 1970 (0.8%) to 229,000 in 1990 (6.5%) and 775,000 in 2000 (16.2%) [2]. The increase can probably be associated with a few aspects such as increase in buying power due to economic growth, and also due improved in technology, which causes air-conditioners to become more affordable. Most of the countries that are experiencing rapid urbanization and population growth today are the developing countries. The energy consumptions in these countries have been increasing tremendously in the last few decades.

The high number of household air-conditioners results in high demand for electricity. For the consumers, their electricity bills can be high due to the large energy consumed by the air-conditioners. At the same time, the high consumption of energy can be harmful to the environment through emission of greenhouse gases and nitrogen oxides at power plants. In addition, the leaked refrigerants, which contain chlorofluorocarbons (CFCs) and hydrochlorofluorocarbons (HFCs) [1], during installation or uninstallation of air-conditioning unit is also hazardous to the environment. They give a negative impact on the environment as they are part of the greenhouse gases that trap heat, and they also lead to thinning or depletion of the Earth's ozone layer. In view of these issues, it would be ideal to reduce the rate of growth of consumption of energy from air-conditioners, especially those that are using refrigerants containing CFCs and HFCs. This can probably be attained by optimizing the air-conditioners by just cooling the required space or through the use of low-energy thermoelectric cooling (TEC).

The air-conditioner market in ASEAN-6 countries (comprising Indonesia, Malaysia, the Philippines, Singapore, Thailand, and Vietnam) is currently in the growth stage and is expected to expand at a strong compound annual growth rate, CAGR of 7.0% between 2018 and 2023. Increased urbanization and rising awareness about energy-efficient air-conditioners is increasing the demand. Poor outdoor air quality and hot, humid climatic conditions further support this demand growth. While new air-conditioners contribute a large percentage of the demand, replacement air-conditioners are also seeing fast growth owing to concerns over energy efficiency. The market is being led by Japanese manufacturers who are well challenged by other Korean, American, and Chinese competitors [3].

Companies populating the competitive landscape of the ASEAN-6 air-conditioner market include Daikin, Panasonic, Mitsubishi Electric, Johnson Controls, LG, and Samsung among others. Singapore is the most mature market for air-conditioners in the Southeast Asian region, Indonesia, and Vietnam are the biggest markets in terms of revenue as well as forecast growth rates [3].

The market for split air-conditioners is growing the fastest, representing 88 per cent of worldwide room air-conditioner sales. The market for window air-conditioners is declining or disappearing in some countries. Driven by increasing temperatures and higher incomes, the Southeast Asian region is set to see a skyrocketing of air-conditioner sales. The overall number of air-conditioner units in 2040 could rise from 40 million units in 2017 to 300 million units in 2040 [4].

2 The Requirement for Comfort

According to ISO7730 Standard [5], thermal comfort is defined as the condition of mind which expresses satisfaction with the thermal environment. The factors that affect the thermal comfort include temperature, humidity, air motion, and radiation source. There are also other comfort factors, though not thermal in nature, such as noise, vibration, odor, and dust. The physical conditions of human body can also be considered as factors in achieving thermal comfort, which include gender, age, size, and activity. Thermal comfort can be analyzed and interpreted by using predicted mean vote (PMV), predicted percentage of dissatisfied (PPD) as well as local comfort criteria which differ according to the environmental conditions [6].

Hypothalamus is part of the brain that controls and regulates the temperature of human body, which is normal in the range from 36.1 to 37.2 °C [7]. A human body has heat sensors and cold sensors. When the body is warmer than 37 °C, the hot sensors will send out hot signal, and when the skin temperature is lower than 34 °C, the cold sensors will send out cold signals [8]. The signals are sent to the hypothalamus to inform the brain about the thermal condition of the body. When the hot signal is higher than the cold signal, the brain will determine the condition as hot. Inversely, the brain will interpret condition as cold when the cold signal is higher than the hot signal. When it is hot, cooling mechanism will be triggered. Skin blood flow increases to encourage heat loss from blood to the environment. Sweating is also a way to enhance heat loss from body. The sweat glands secrete water to the surface, then the water will absorb heat during the phase change resulted by evaporation of the sweat. When it is cold, warming mechanism is engaged. The skin blood flow is reduced to keep the heat from easily loss to the environment. Shivering also helps to keep the body warm. Muscles activity for shivering requires adenosine triphosphate (ATP) to move the muscle. When the body produces ATP, there are other products that will be produced simultaneously through the same process. One of the products is heat. That is the reason why body shivers when cold. The heat balance for human body can be written as [8]:

$$S = M - W - E - (R + C) \tag{1}$$

where heat storage rate, S, is zero when the heat production is the same as the heat loss. Metabolism rate, M, is the rate of heat generated chemically from human body. The metabolism rate depends on the muscle activity and the surrounding environment. High muscle activity increases the mechanical work, W. The evaporative heat loss rate, E, is the loss that occurs by mean of sweating and heat release through respiratory vapor. The dry heat exchange through radiation, R, and convection, C, are another forms of heat loss from the body.

In assessing thermal comfort, the predicted mean vote (PMV), which is an index that predicts the mean value of the votes of a large group of persons, is used. The PMV index has a seven-point thermal sensation scale of Hot (+3), Warm (+2), Slightly Warm (+1), Neutral (0), Slightly Cool (−1), Cool (−2), and Cold (−3) [5]. ASHRAE Standard 55-2017, which requires that at least 80% of the occupants feel satisfied with the thermal conditions, also uses the PMV model in determining the requirements for indoor thermal comfort [5]. Nevertheless, the evaluation of comfort depends on individuals due to many factors. Therefore, another index known as the predicted percentage of dissatisfied (PPD) is used to predict the number of people who are dissatisfied in a given environment. The value of PPD depends on the value of PMV. When the PMV value is equal to 0, the PPD is 5%, indicating that there are 5% of people who are predicted as dissatisfied with the given environment.

Home air-conditioners emit noise especially from the compressor. The compressor is mounted in the outdoor unit so that the indoor can be felt relatively quiet by the occupants. Nevertheless, it is still a noise pollution for anyone who is present outside the building. As for the indoor environment, actually it is still a little noisy since the sound from the compressor is very loud. As the equipment ages, the noise becomes more prominent. Certain design considerations can be made during installation to minimize the noise. One approach is to install the outdoor unit farthest away from the room, but within the allowable limit of distance to avoid loss in the fluid energy. Wall insulation using noise absorbing materials is also another possible option to minimize the noise.

The concern on noise should not be underestimated. It was estimated by the WHO [7] that in European countries of 340 million people, at least 1 million healthy live years have been lost annually due to hearing loss caused by uncontrollable noises. The most common downside associated with noise is hearing loss, which can be caused just by one exposure to an intense impulse sound or by steady-state long-term exposure to sounds from 75 to 85 dB [9]. Besides that, being exposed to sounds of 85 dB or higher for more than 8 h could also contribute to hearing loss, where 85 dB is equivalent to the noise of heavy truck traffic on a busy road. Musicians and those working at entertainment places usually experience hearing loss. The impulse noise exposure should never exceed 140 dB in adults and 120 dB in children [10]. Hearing loss causes inability to understand speech thus contributing to social effects. It can even cause accidents and fall; 10 to 20% of mortality within the last 20 years happened due to hearing loss [9].

There is a term called noise annoyance which is a phenomenon that results from interference of noise with daily activities, feelings, thoughts, sleep, or rest. There would be changes in everyday behavior, social behavior, and daily mood [11]. Such effects include anger, displeasure, exhaustion, and stress-related symptoms as well as affecting wellbeing and health [8]. A person's memory could also be slowed down due to noises, and even moderate-intensity noise could affect verbal memory tasks [12].

3 Conventional Home Air-Conditioners

Most home air-conditioners are designed to cool individual rooms. The system is flexible that homeowners can decide which rooms to provide air-conditioner or even if the entire house is to be air-conditioned would be possible too. Conventional air-conditioners use vapor compression refrigeration cycle, which is illustrated in Fig. 1. The cycle comprises a closed-loop piping which has four main components: compressor, evaporator, condenser, and expansion valve. Details on how the vapor compression refrigeration cycle works can be found in thermodynamics textbook and will not be elaborated here. For home air-conditioners of air-cooled split unit type, the system comprises of an indoor unit, an outdoor unit and a set of closed-loop piping which connects the two. The indoor unit has the evaporator, which takes out

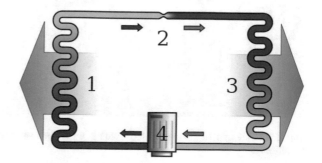

Fig. 1 Schematic of the vapor compression refrigeration cycle: (1) condensing coil, (2) expansion valve, (3) evaporator coil, (4) compressor

heat from the room. It also has a blower to spread cool air into the room. The outdoor unit, which is usually mounted just outside the air-conditioned room, has in it the condenser and compressor. A typical photograph of the outdoor units is shown in Fig. 2. A reason why the condenser is built in the outdoor unit is due to its reasonably loud noise. The red color in Fig. 1 denotes hot section due to the heat taken from the indoor. This heat is dissipated to the outdoor with the help of a fan in the outdoor unit. The blue color denotes the cool section as a result of the cool refrigerant. The blower in the indoor unit blows through the cold cooling coil (evaporator) to provide cooling in the air-conditioned room.

Large air-conditioners for offices take ambient air from the outdoor environment, filter it from unhealthy fine particles, and then cool it to a desired temperature by passing it through cooling coils. For home air-conditioners, the air from the room is used instead of taking air from the outdoor. Heat in the room is transferred to the cool air by convection due to the air blown by the blower so that the desired room air temperature can be attained. In tropical countries where the outdoor air is humid, condensation will most of the time occur at the cooling coil. Consequently, the resulting cool air usually has low humidity. While cooling is happening in the room, there is also increase in the evaporation rate of the water on the surface of the occupants' skin. The enhanced evaporation rate can be regarded as another mechanism to achieve thermal comfort in the room. Most home air-conditioners recycle the air in the room so that the energy consumption can be optimized. This is usually harmful for the occupants if the unit is operating for long hours. For homes, this is

Fig. 2 Photograph of floor-mounted outdoor units that are part of air-cooled split unit air-conditioners

overcome by allowing a period of natural ventilation by opening the windows when the occupants are usually out for work, school, or recreation.

Another type of air-conditioner is the evaporative cooler. Also known as swamp cooler, it is not commonly used as compared to the conventional vapor compression air-conditioners. Its utilization is probably better for places with dry climates since evaporative coolers increase the air humidity in order to obtain reduction in air temperature. It pulls the outdoor air to pass through moist pads where the air gets cooled via evaporation. The cold air then gets circulated throughout the room. This system can save up to 75% of cooling costs during hot seasons since the fan is the only mechanical component that uses electricity. It also costs about half to purchase as compared to a central air-conditioner.

An issue with home air-conditioning is that the level of comfort of occupants differs. This may have with two factors, which are large space to be cooled and high number of occupants. Large space may result in long transient time to reach steady conditions. High number of occupants may result in thermal discomfort for some group of people who have different feeling about the thermal condition of the room at a given time. At the same time, high energy consumption rate of the air-conditioner is also an issue as it affects the electricity bills. Critical factors that affect the electricity consumption of a home air-conditioner are outdoor temperature, set-point temperature, thermal insulation of the room, and air infiltration through openings like windows. The condition of air-filters is also important as dirty filter tend to result in more work done by the blower in order to maintain the room's comfortable temperature and consequently the higher electricity bills.

Another issue on home air-conditioner is the requirement for high amount of energy to unnecessarily cool down the whole room, as depicted in Fig. 3. The cooling

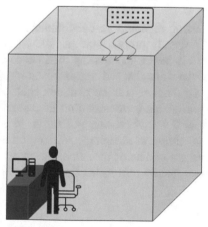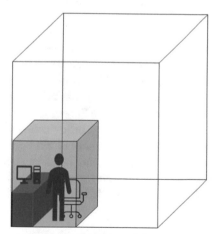

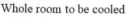
Whole room to be cooled   Space need to be cooled

Fig. 3 Comparison of size of the space that actually need to be cooled with the space that is normally cooled; a ratio of about 1:4

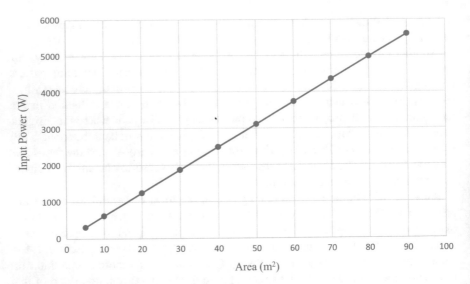

Fig. 4 Relationship between air-conditioned area and input power

energy can be saved if the air-conditioner is set to cool down only specific occupied area in the room where cooling is needed (e.g., the office desks that are occupied only). In fact, the air-conditioner is intended to provide thermal comfort for the occupants only; it will not be necessary to condition the air at space where occupants are not present such as the space near the ceiling. The space that the occupants occupy inside the room would be smaller than the whole room. The unoccupied space that is not used by the occupants does not require cooling, which is otherwise a form of energy waste.

Based on rule of thumb method [13] for sizing of home air-conditioners, the relationship between input power and area is directly proportional, as shown in Fig. 4. For example, when the size of the room is larger, the cooling load needed will be larger also. This is because the area of the walls that will transfer heat into the space is smaller and the volume in the space that needs to be maintain cool is smaller. In relation to Fig. 4, it is therefore reasonable to assume that the actual cooling energy can be brought down to 25% or lower if cooling is focused to a space where the occupant could be idling (e.g., sitting or sleeping). Although the outdoor environment and the indoor heat sources are kept being the same, the energy consumption for removing the heat load is lesser just by decreasing the size of the space need to be cooled.

4 Public Perception on Personalized Air-Conditioners

It has been explained earlier that energy can be saved by reducing the cooling load of air-conditioners by downsizing the space to be cooled. This can be attained by cooling the space only to surrounding of the occupant rather than the whole volume of the room being occupied. Such an idea may be awkward, and furthermore, it is not easy for people to change something that they have been used to. In order to understand the expectations and perception of users, a survey was conducted among a selected local community in Malaysia.

The survey involved 223 respondents in the town of Seri Iskandar. It was conducted in around January and February 2020. The questionnaire set used in the survey is shown in Fig. 5. The survey set comprised 12 questions, which basically covered the demographics of the respondents as well as their preferences of an ideal air-conditioned compartment. Questions 1, 2, and 3 are about the demographics of the respondents. Question 4 is about the respondents' opinion on whether they would need a small personalized air-conditioners. The respondents' usage of home air-conditioners is asked in Questions 5 and 6. These would provide an insight on pattern and purpose of their usage of home air-conditioners. Questions 7 to 9 are about the respondents' perception on home air-conditioners' market. Finally, the respondents provide their expectations on the type of air-conditioners that they would need through Questions 10 to 12.

The demography of the respondents is represented by the pie charts shown in Fig. 6, which considers their gender, age, and occupation. As shown in the figure, the

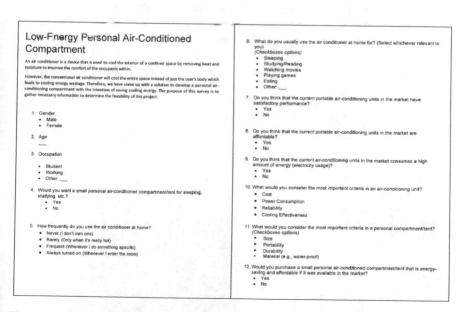

Fig. 5 Questionnaire set that was used in the survey

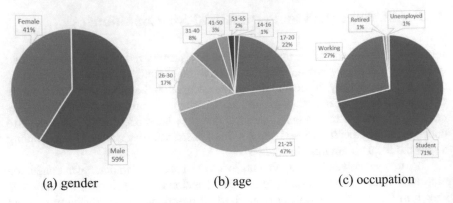

(a) gender   (b) age   (c) occupation

Fig. 6 Gender, age, and occupation of respondents

respondents comprise a nearly balance gender with 59% male respondents and 41% female respondents. The respondents were mainly dominated by those at the age of between 21 and 25 years old (47%). The next largest groups of respondents were those of 17–20 years old (22%) and 26–30 years old (17%). The high percentage of respondents who were 30 years old or younger was mainly caused by the fact that the survey was conducted in a university town. This is supported by the third pie chart, which indicates that 71% of the respondents were students.

The respondents were asked on whether a small personal air-conditioned compartment is acceptable for personal use such as sleeping or sitting. The result is shown in Fig. 7. It is indeed encouraging to note that 86% of respondents would accept the idea of small personalized compartment. This could probably be due to the promising benefits from such concept such as affordability, portability, and space-saving. On the frequency of usage of home air-conditioners, the result is shown in Fig. 8. The result shows that the majority of respondents (41.2%) rarely use their home air-conditioners unless when it gets really hot. This is followed 32.6% of respondents who responded frequent usage when they are doing something. Quite a small portion of the respondents (17.6%) admitted that they would definitely turn on their home air-conditioner whenever they occupy the room. The smallest group of respondents

Fig. 7 Response on whether a small personal air-conditioned compartment is acceptable for sleeping or studying purpose

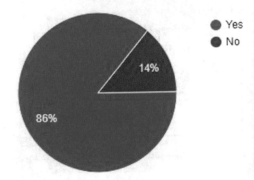

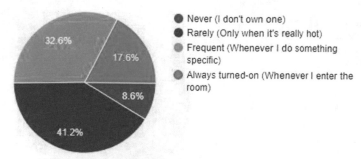

● Never (I don't own one)
● Rarely (Only when it's really hot)
● Frequent (Whenever I do something specific)
● Always turned-on (Whenever I enter the room)

Fig. 8 Response on frequency of usage of home air-conditioner

(8.6%) replied that they did not own any air-conditioner unit at home. Overall, it can be deduced from Fig. 8 that most of the respondents rarely use their home air-conditioners unless the weather gets too hot. This probably has to do with the high electricity consumption resulted by the use of air-conditioners.

The respondents were asked on their reasons for turning on their home air-conditioners so that their priorities could be understood before any new concept could be proposed. Figure 9 shows the results for the questionnaire. It is shown that the sleeping had the highest responses (81.9%), which implied as the utmost reason for the respondents to turn on their home air-conditioners.

Apart from sleeping, 58.8% responded studying as the reason for turning in the air-conditioners, 51.1% for watching movies, 31.2% for playing games, and 12.2% for eating. As shown in the figure, there are other reasons too though the percentage is very small. In general, the results in Fig. 9 imply that people turn on home air-conditioners to rest or relax, which is a norm after a long-day working in office or spending time in campus.

The next three questions emphasize on the respondents' opinions on the air-conditioning units in the present market. Majority of the respondents (63.7%), as depicted in Fig. 10, agreed that home air-conditioners in the present market deliver satisfactory performance. At the same time, in Fig. 11 it is shown that the majority

Fig. 9 Reasons for using air-conditioner at home

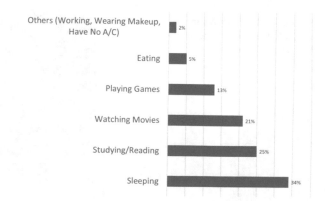

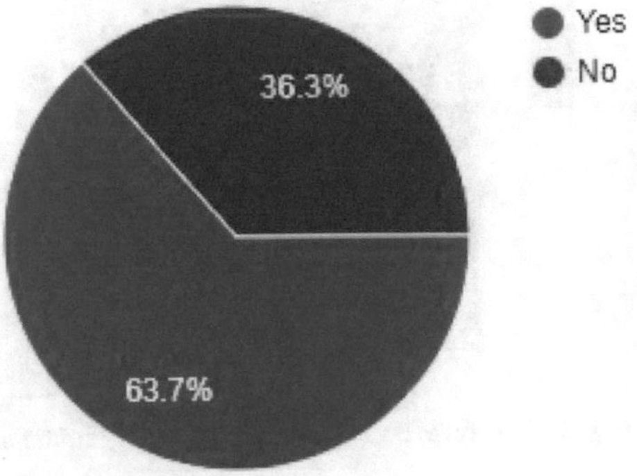

Fig. 10 Response on
performance of portable
air-conditioning units in the
present market

Yes
No

36.3%

63.7%

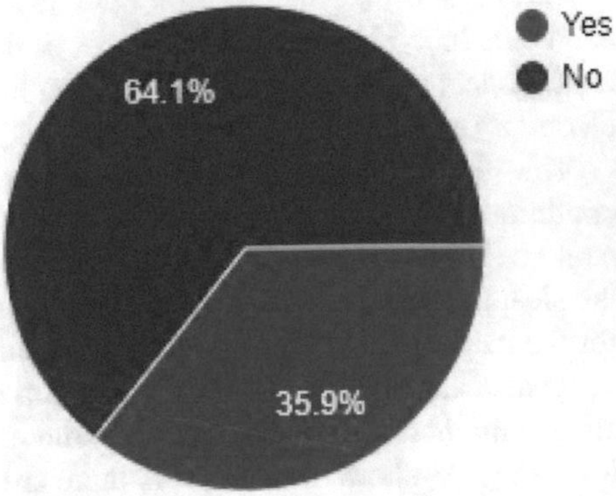

Fig. 11 Response on
affordability to buy portable
air-conditioning units in the
present market

Yes
No

64.1%

35.9%

of the respondents (64.1%) replied that they could not afford to buy or bear the
operating cost of home air-conditioners. This outcome is somewhat in agreement
with the belief by outstandingly high number of respondents (90.6%), as implied
in Fig. 12, that home air-conditioners in the market consumed high electricity, and
would therefore result in high electricity bills.

Fig. 12 Response on
whether air-conditioners in
the present market consume
high amount of energy

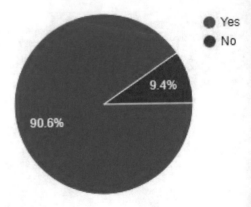

Yes
No

9.4%

90.6%

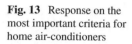
Fig. 13 Response on the most important criteria for home air-conditioners

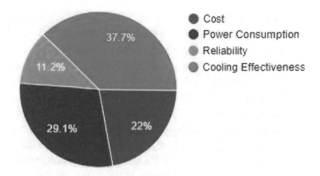

In understanding the users' requirements pertaining to air-conditioners, the users were asked about their priorities in the survey. They were given only four options for their answers, which were (1) cost (2) power consumption (3) reliability, and (4) cooling effectiveness. Figure 13 shows the result for this questionnaire. Interestingly, 37.7% of the respondents believed that the cooling effectiveness of the air-conditioner would be the most important criteria. The next important priority is shown the figure to be 29.1%, which is about the power consumption. The cost is shown in the figure as the third priority, with response by 22% of the respondents. The last priority, despite its importance, is reliability (11.2%). The result in Fig. 13 deduces that the utmost concern of the respondents in considering a prospective air-conditioner is its effectiveness in delivering the cool condition of the occupied area.

The last aspect of the survey was pertaining to the proposed idea of personal compartmented home air-conditioners. Of course, this chapter is not intended to elaborate about any design or detailed feature of personal compartmented home air-conditioners. Suffice to say, the equipment would be small to fit and serve one or two persons, and would cool the small space that envelopes a person so that the energy consumed would be small. But at the same time, the user need to accept that as soon as he needs to move around, the cooling effect may be temporarily lost, unless it is a type of clothing. The air-conditioner may be most suitable for activities such as sitting or sleeping. Figure 14 shows the result pertaining to the important

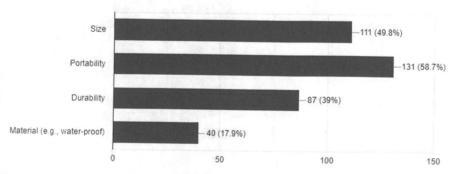

Fig. 14 Most important criteria in a personal compartment air-conditioner

criteria expected by the respondents on this type of air-conditioner. It is shown in the figure that portability is the most important feature. This is probably because the present society has been very used to mobility through innovations such as mobile phone, television, laptop computer, and e-hailing; thus being able to carry the air-conditioner unit to various places would be valued. The next important feature is size, perhaps considering the relatively smaller space designed for in new accommodation properties nowadays. The other concerns are durability and material, which has to do with the sustainability of the equipment. Today's societies have gained awareness on the importance of environment conservation, and therefore, have started to look into after-life factors when they buy a new equipment or appliances.

The final question in this survey was on whether the respondents would be willing to purchase a personal compartmented home air-conditioner described earlier. The result is shown in Fig. 15, where it is interesting to note that the majority of the respondents (88.8%) would consider buying such equipment if it is available. The high percentage denotes that people are looking forward to improve their thermal comfort especially if the cost incurred is low. This also gives hope to inventors who may want to explore into the idea of compartmented personal air-conditioners or the like.

All in all, it can be concluded that majority of the respondents are interested with the idea of compartmented personal air-conditioners. This is most likely due to the fact that the equipment can be applied for personal space instead of one whole room. Such feature can save energy and more importantly can help to reduce carbon footprint and save electricity bills. The small-sized compartment may somehow help to reduce space consumption, and it can be sold at a far lower price than the conventional air-conditioners sold in the market presently. The majority of them also would consider to purchase such air-conditioners if available in the market. Therefore, it can be seen that the idea of personal compartmented air-conditioner can be promising if properly developed by taking into consideration the consumers' expectations.

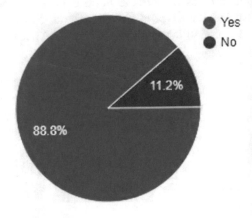

Fig. 15 Response on probability of purchasing a personalized compartmented home air-conditioner

5 Efforts in Reducing Energy Consumption of Small Air-Conditioners

In addition to advancing low-global warming potential refrigerants, the air-conditioning industry has steadily improved the energy efficiency of air-conditioning systems through a combination of technological innovation and market transformation strategies. Today, manufacturers have incorporated these component technologies into their product line-up, such as ductless mini-split air-conditioners using inverter-driven variable-speed compressors and fans. Multi-stage and variable-speed drives are one of the main components which help in reducing the consumption of energy in air-conditioners. Electronic motor controls have enabled substantial energy efficiency improvements on compressor and fan motors. By modulating motor speed on compressors, the air-conditioning system can more closely match the part-load cooling demand and improve seasonal efficiency by reduce cycling losses that are common during the majority of the cooling season when the system's full capacity is not required. Similarly, variable-speed controls operate fans motors at their most efficient setting to meet the airflow needs of the system.

Air-conditioning compressor efficiency and performance has steadily improved as manufacturers incrementally improved current designs (e.g., high efficiency reciprocating) and introduced entirely new compressor technologies (e.g., scroll, rotary). In addition, larger systems use multiple compressors to stage capacity and improve part-load performance. Moreover, manufacturers have increased the size of heat exchangers to improve system efficiency, especially during part-load operation for transferring heat between the refrigerant and air. Advanced heat exchanger designs, such as microchannel heat exchangers and other small diameter designs, have further improved system efficiency, while also reducing refrigerant charge, fan energy consumption, and physical size.

High efficiency fans also help in reducing the energy consumption of air-conditioning. Fan energy consumption has decreased over time as many systems have incorporated more aerodynamic component designs (e.g., fan blades, condensing unit housing), high efficiency motors, and variable-speed controls. Manufacturers have applied these innovations to axial and centrifugal fans for both heat exchange and distribution of conditioned air throughout buildings. Furthermore, electric motors are core components for air-conditioning compressors and fans, and improved motor designs have a significant impact on overall air-conditioning system efficiency. Electrically commutated motors (ECM) have higher efficiencies than permanent split capacitor (PSC) motors for air-conditioning fans and operate at a wider range of conditions using electronic controls.

The air inside a building receives heat from a number of sources during the cooling season. If the temperature and humidity of the air are to be maintained at a comfortable level, this heat must be removed. The amount of heat to be removed is the cooling load. One opportunity to reduce cooling load and increase comfort is by installing wall insulation. Insulation helps increase thickness and R-value for walls and attics that reduce conduction heat gains. Moreover, installing multi-pane glass windows with inert gas fill and low-emissivity coating helps enhance the window's resistance to heat. Dynamic glazing technology may further reduce solar heat gains by filtering out infrared radiation while continuing to provide natural lighting. Furthermore, thermal bridges help eliminate gaps in building insulation that allow increased conduction through poorly resistive materials. Installing light colored roofing helps increase albedo and reduces solar heat gain. Surface orientation must also be looked into by designing building layouts to avoid large sunlit exposures not shaded by vegetation or building outcroppings. In addition, installing light-emitting diode (LED) lighting with occupancy sensors also helps reduce waste heat, which will otherwise contribute to high cooling energy. Lastly, infiltration technique also helps minimize air exchange between conditioned space and outdoor environment while still providing adequate fresh air. These measures can help to reduce the cooling load and at the same time reduce energy consumption.

With the rapid development in socio-economy and the greatly improved living standards of people, power users' requirements of power supply reliability have also improved. Workers who work under distribution network is in charge of the task of distributing power to various users and is the most direct and critical link to ensure the quality of power supply. Distribution network working with power uninterrupted is an effective means to improve power supply reliability and ensure people's normal life, which has been widely carried out across various countries. However, the working environment is extremely poor in hot weather for the distribution network working with power uninterrupted, and the temperatures are often as high as 40 °C, and at the same time, operators have to wear closed insulated clothes when working in heights. In this case, it is of high risk for operators to appear hyperthermia, trance, heat cramps, heat stroke, and other diseases [14], which can cause personal and power grid security incidents. In this case, air-conditioning suit for distribution network working with power uninterrupted has been looked into and have improved the working environment for the operators effectively, which is of great significance to improve the working condition with power uninterrupted, reduce the probability of accidents and enhance the reliability of power supply. This system of air-conditioned suit for distribution network working with power uninterrupted mainly includes insulated garments, vortex tube coolers, gas pipe, high-pressure gas source, insulated bucket arm car.

Insulated bucket arm car can meet the needs of different height of overhead work in distribution network, and at the same time, high-pressure gas source mounted on the insulated bucket arm car instead of requiring the operators to carry, which significantly improve the scope of work and reduce labor intensity. The vortex tube cooler is fixed to the waist of the insulation suit, and there is a cooling pipe lay into the insulation suit. The high-pressure gas source, vortex tube coolers, and the cooling pipe laid in the insulation suit were connected by gas piping and quick connectors, which has a compact structure, and installed easily.

The cooling pipelines are arranged inside the insulated suit, which includes a trunk pipe, an arm pipe and an armpit pipe, and the trunk pipe includes a horizontal pipe and a vertical pipe. The horizontal pipe is connected with the vortex pipe, and the vertical pipe is connected to the horizontal pipe through the joint, and reasonably arranges these pipes according to the degree of human sensitivity to the temperature to improve the cooling effect in the sweaty parts like armpits and others. The trunk pipe is connected with the vortex pipe, and one end of the arm pipe is connected with the trunk pipe, while the other end closed. Both ends of the armpit pipe are connected with the arm pipe, and each pipe is provided with a vent hole, the hole was staggered layout in the pipeline on both sides of arm pipe, vertical pipe and armpit pipe, the vent holes on the horizontal pipe are opened upward, and the density of the vent holes increases with the distance between the pipe and the vortex pipe, so that air can reach every parts of the body equally, and improve the cooling effect.

The widely use of the air-conditioning suit for distribution network working with power uninterrupted can significantly improve the work environment, which is of great significance to improve the operation level of distribution network working with power uninterrupted all over the world, reducing the probability of accidents occurring and improving the reliability of power supply.

6 Summary

The necessity of air-conditioning devices is becoming more and more apparent due to the need of comfort especially in hotter climates and their increasing affordability over the years. With the industry steadily improving the energy efficiency of air-conditioning systems through a combination of technological innovations and market transformation strategies, the primary concerns regarding traditional air-conditioning systems such as power consumption and environmental effects are now less alarming. However, an issue with these conventional air-conditioners is that they would cool large rooms or spaces when in reality, only a very small space surrounding the occupants would require the said effect. To achieve this, an air-conditioning system with some form of personal compartment with minimum space applicable would be required while considering the basic requirements of comfort. Since the idea would be different from conventional means, a survey was necessary in order to understand the expectations and perception of users and from it, a majority of the respondents within the target market within the local community in Malaysia are interested in such. Therefore, this idea might prove to be able to change the current method of air-conditioning for the better in terms of reducing the operating cooling energy required.

References

1. Weather Atlas (2020) Monthly weather forecast and climate Kuala Lumpur, Malaysia. https://www.weather-my.com/en/malaysia/kuala-lumpur-climate (accessed 18 Jul 2020)
2. Kubota T, Jeong S, Hooi Chyee Toe D, Ossen DR (2010) Energy consumption and air-conditioning usage in residential buildings of Malaysia. In: 11th international conference on

sustainable environmental architecture SENVAR 2010, Skudai, Johor, Malaysia
3. IEA (2019) The future of cooling in Southeast Asia, IEA, Paris https://www.iea.org/reports/the-future-of-cooling-in-southeast-asia
4. Frost, Sullivan (2019) Air conditioner market in ASEAN-6, forecast to 2023 increasing urbanization and rising demand for energy-efficient air conditioners driving market growth
5. "Ergonomics of the thermal environment. Analytical determination and interpretation of thermal comfort using calculation of the PMV and PPD indices and local thermal comfort criteria," The British Standards Institution 2020. https://doi.org/10.3403/30046382
6. Fabbri K (2015) The indices of feeling—predicted mean vote PMV and percentage people dissatisfied PPD. In: Fabbri K (ed) Indoor thermal comfort perception: a questionnaire approach focusing on children. Springer International Publishing, Cham, pp 75–125
7. "Body temperature norms: MedlinePlus Medical Encyclopedia." https://medlineplus.gov/ency/article/001982.htm (accessed 16 Jul 2020)
8. "4.6 Thermal Comfort (I)—The Hong Kong University of Science and Technology," Coursera. https://www.coursera.org/learn/intro-indoor-air-quality/lecture/JsQnw/4-6-thermal-comfort-i (accessed 16 Jul 2020)
9. Basner M, Babisch W, Davis A, Brink M, Clark C, Janssen S, Stanfeld S (2014) Auditory and non-auditory effects of noise on health. Lancet 383(9925):1325–1332
10. Goines L, Hagler L (2007) Noise pollution: a modern plague. South Med J 100:287–294
11. World Health Organization (2011) Burden of disease from environmental noise. WHO Regional Office for Europe
12. Stansfeld S, Haines M, Brown B (2000) Noise and health in the urban environment. Rev Environ Health 15:43–82
13. https://www.mrfixitbali.com, Air conditioner size calculation. https://www.mrfixitbali.com/air-conditioning/air-conditioner-calculator-12.html (accessed 09 Apr 2020)
14. Muhammad FMA, Norashikin Y, Irraivan E (2020) Automatic Arrhythmia detection algorithm using statistical and autoregressive model features, platform. J Eng 4:41–49. Available at: http://myjms.moe.gov.my/index.php/paje/article/view/5898. Accessed: 28 July 2020

Printed in the United States
by Baker & Taylor Publisher Services